SOCIAL FABRICS OF THE MIND

Social Fabrics of the Mind

edited by

Michael R. A. Chance
(assisted by Donald R. Omark)

*Social Systems Institute,
Birmingham, U.K*

LAWRENCE ERLBAUM ASSOCIATES, PUBLISHERS
Hove and London (UK) Hillsdale (USA)

Lawrence Erlbaum Associates Ltd., Publishers
27 Palmeira Mansions
Church Road
Hove
East Sussex, BN3 2FA
UK

British Library Cataloguing in Publication Data

Social fabrics of the mind.
1. Social psychology
I. Chance, Michael R.A. II. Omark, Donald R.
302

ISBN 0-86377-097-5

Typeset by Acorn Bookwork, Salisbury
Printed and bound by A. Wheaton & Co. Ltd., Exeter

DEDICATION

Most of us have a sense of "self", which we think identifies us as a person, but this is far from true. Aldous Huxley wrote: "The number of completely unified personalities is small. Most of us go through life incompletely unified—part person, the rest a mere collection of discontinuous psychological elements."

Bill and Clare Russell in their book *Human Behaviour: A new approach* set this awareness in a biological (Ethological) frame of thought and, at the same time, hinted at the unifying role of intelligence within the personality. This unifying principle was finally specified by Bill and Lucille Gray when they drew attention to the existence of a systems-forming faculty within the brain—so pointing to an organic system, as defined by Hall and Fagen, capable of transforming the personality in the way envisaged by Aurelio Peccei, the founder of the Club of Rome, in his book *The Human Quality*. He has exhorted us to cease taking our nature for granted and to actively improve it.

This book is dedicated to all these Pioneers.

Contents

Preface

In the Autumn of 1983 Donald Omark and I sat on a balcony overlooking Newport Bay in California discussing the need to put together a volume outlining the evidence that would substantiate and, in some measure, define the evident uniqueness of human beings as the dominant species on which all life on earth now depends for its existence—and to do so in a way which recognises the evolutionary origin of our species. It was not our intention to speculate further on how this came about. What seemed to us much more in need of understanding was just in what way we differ from our nearest simian relatives, who like us are social. In this feature, which we share with them, we were convinced would be found the answer to why our nature so often impedes our social endeavours. The editor is profoundly grateful to Donald for this early discussion, which initiated and gave birth to this volume, and latterly to the contributors whose work has brought it into being.

I am deeply indebted to Anthony de Reuck whose intellectual companionship has all along provided valuable support and criticism in depth, and also for the expert clerical assistance and informed understanding of the material by Diana Palmer during the work to assemble the manuscript.

All along I have been encouraged and helped by the comments and advice of my wife Mariella, and latterly by Brenda Tyler, Martin Edwards and Felix Wedgwood-Oppenheim who commented on the text of the Introduction—to them my thanks for easing the work of integration. In the last stages a final boost to the enterprise came from Paul Gilbert, to whom I am very grateful for a great advance in understanding. Repeated discussions with John Price also helped me to revise the Introduction up to the last moment.

We acknowledge with thanks a donation from the Rowntree Charitable Trust towards expenses and the publishers Brunner/Mazel of New York for giving permission to reprint Chapter 1 from the *American Journal of Social Psychiatry*. We also thank the Editors of the *Quarterly Review of Biology* for their permission to reproduce the contents of Chapter 5 of this volume. Quotations from *The Social Contract* by J. J. Rousseau in Chapter 6 are reprinted by permission of A. D. Peters and Co.

Michael R. A. Chance

List of Contributors

Carol Barner-Barry
5972 Grand Banks Road, Columbia, Maryland 21044, U.S.A.

Michael R. A. Chance
12 Innage Road, Birmingham B31 2DX, U.K.

Frans B. M. De Waal
Wisconsin Primate Research Centre, University of Wisconsin, 1223 Capital Court, Madison, 53715-1299, U.S.A.

Gary R. Emory
6050/15 Henderson Drive, La Mesa, CA 92041, U.S.A.

Russell Gardner, Jr.
1,200 Graves Building (D29), University of Texas Medical Branch, Galveston, TX 77550, U.S.A.

D. Godard
INSERM/U.70, 388 Rue du Mas Prunet, 34070 Montpellier, France

Junichiro Itani
Centre for African Area Studies, Kyoto University, Yoshida Sakyo, Kyoto 606, Japan

Theodore D. Kemper
Department of Sociology, St John's University, Jamaica, New York, NY 11439, U.S.A.

D. Laurent
Faculté des Sciences, Université de Franche-Comté, Route de Gray, Besançon 25030, France

Roger D. Masters
Department of Government, Dartmouth College, Hanover NH 03755, U.S.A.

H. Montagner
INSERTM/U. 70, 388 Rue du Mas Prunet, 34070 Montpellier, France

D. R. Omark
7035 Galewood, San Diego, CA 92120, U.S.A.

Thomas K. Pitcairn
Department of Psychology, University of Edinburgh, 7 George Square, Edinburgh EH8 9JZ, U.K.

Margaret Power
1969 Kings Avenue, W. Vancouver V7V 2B6, Canada

John Price
Department of Psychiatry, Milton Keynes Hospital, Standing Way, MK6 5LD, U.K.

A. Restoin
Faculté des Sciences, Université de Franche-Comté, Route de Gray, Besançon 25030, France

D. Rodriguez
Faculté des Sciences, Université de Franche-Comté, Route de Gray, Besançon 25030, France

Peter Scott Lewis
Broadmoor Hospital, R.G.11. 7EG, Berkshire, U.K.

V. Ullmann
Faculté des Sciences, Université de Franche-Comté, Route de Gray, Besançon 25030, France

M. Viala
Faculté des Sciences, Université de Franche-Comté, Route de Gray, Besançon 25030, France

Felix Wedgwood-Oppenheim
Institute of Local Government Studies, University of Birmingham, Birmingham B15 2TT, U.K.

Introduction

Michael R. A. Chance
Social Systems Institute, Birmingham, U.K

This book provides much new information; new not only because each contribution is original, but because the complex of knowledge so displayed is new. From this novel perspective, which my introduction outlines, individuals or groups may learn to tackle problems in a fresh way—a way that subverts the all too familiar clash of cultures and ideologies by suggesting that to understand the fabric of the mind is to understand what we all have in common.

Humanity early on made sense of the movements of the planets, stars, sun and moon, that part of nature furthest removed from us. It has taken the whole span of civilisation for our focus of interest to be turned in upon our own nature, at least from a scientific rather than a religious, philosophical, or artistic point of view. The lens that makes possible this scientific examination of personality is our affinity with other apes and simians. These relatives are social animals and like us spend the greater part of their lives in social pursuits. This book sets out, therefore, to distinguish what we share with these zoological relatives and what distinguishes us from them at least in degree; what drags us back and what potentially sets us free by setting free our intelligence.

Our intelligence has evolved in large part to cope with the demands of social existence. In this, two disparate processes have been at work; for, as this book demonstrates, our present minds are formed, and in turn help to form, two antithetical types of social system with markedly different functional thrusts. By studying the behaviour of the higher primates, and our evolutionary history, it has become clear that we tend to function in one of two mental modes. As adults we may become stuck in one or other mode, unconsciously moving back and forth between them, or we may have our personality constructed by movement between them during our upbringing.

In one, *the agonic mode*, we are primarily concerned with self-security, and our attention is much taken up with being part of a group and with what others think of us so as to assure acceptance by the group. We become concerned with rank hierarchy, convention, and maintaining good order, as an expression of this inbred security mechanism. In this mode our concerns are predominantly self-protective and engage information-processing systems that are specifically designed to attend, recognise and respond to potential threats to our physical self, status, and social presentation.

In the other, *the hedonic mode*, we are more free to form a network of personal relationships that typically offer mutual support. Then we can also give free rein to our intelligence, our creativity, and the creation of systems of order in our thoughts and in our social relations. This is because attention, when released from self-protective needs, can be used to explore and integrate many new domains. Our mind may thus process information in two quite different ways.

This is a powerful observation for it premises that social life and a corresponding mentality are constructed on a relatively simple underlying framework upon which we weave our affairs. An understanding of this duality is important to those who wish to see the world's peoples living in greater harmony with each other and with the environment, and who desire to be more effective in influencing events to this end. So this book is particularly addressed to workers in social philosophy, psychology, education, ethics, social planning and the social sciences, and to any organisation, especially political, that recognises change as the order of the day. The fifteen contributors, each aware of this social and mental duality, approach a possible unification of disciplines from their individual viewpoints. Theodore Kemper (see Chapter 13) independently discovered this duality of mind in the macrostructure of modern society, so providing a link with the social relations in small groups from which most of the remainder of the evidence for it has come. Other contributors by implication point to its existence.

One purpose of this introduction is to help the reader select the point at which they may most easily begin their acquaintance with the main theme. To this end, a synthesis of the separate subjects and hence the links between them is first outlined. When the whole has been absorbed, best achieved by suspending judgement until the theme has emerged clearly, it should prove possible for the reader to test the theory in practice. An overall perspective is sketched here; more exact language is used in each chapter.

The broad perspective is derived from a number of recent discoveries, of which three will be summarised now. The first is that each of the two modes already referred to is, at one and the same time, a property of our minds and of the corresponding way in which we relate to those about us. The second is that each mode predisposes us to deploy our attention in ways quite different from the other, so that we are thereby either prevented from or enabled to develop our intelligence. This process may bring into being, to a greater or lesser extent, a systems-forming faculty, our possession of which is the third, recent discovery. This faculty is a uniquely human characteristic that forms our personality as it develops from childhood to adulthood within the limits set by our experience of the two modes. We shall now look at the distinction between the two modes—agonic and hedonic.

THE TWO MODES

It is important to realise that the two modes are two separate systems. As will be discussed later, a switch from the state where the agonic controls the personality to one of hedonic well-being has been uncovered by juxtaposing the work of several of the contributors to this book.

The Agonic Mode

To understand both modes and the potential switch between them requires a complete perception of the structure of hedonic chimpanzee society, and this is made possible in Chapter 4 by Margaret Power. The marked contrast between that society and hierarchical agonic-type structures, however, can only be appreciated if the reader is now given more detail about such hierarchies. The characteristic of ranking in the agonic mode is seen clearly in non-human primate societies in which the individuals are arranged in a series of levels, one above the other; individuals being of the same or different rank. This rank differentiation between individuals is manifested in the structure of attention between them, as each individual accords and receives attention as a function of their rank, Omark et al. (1980). Higher-ranking individuals accord less and receive more attention than those lower in the social scale. In this way, channels of attention develop, binding those who accord more attention to those of higher rank so that lower-ranking individuals have most of their attention directed to those above them.

This has several consequences. The first is that dominant members of the society are able thereby to exert control over those lower in rank, simply, but not solely, through the proximity of the lower-ranking to the centrally dominant figure, and because the channel of communication is always open to the subordinate from the dominant. Emory (see also Chapter 2), has assessed the attention structure in similar groups of monkeys, caged in identical conditions in the San Diego Zoo. He found that the amount of attention paid to the dominant male was correlated with the nearness of each individual to him—this is centrally orientated (centric) attention. That the degree of centric attention and of group cohesion was due to an in-built tendency to return towards the dominant male (influenced to some degree by that individual's attention to others of the group) was made evident through the difference between the two species.

This work has verified the existence of what Chance and Jolly (1970) termed "reverted escape"—the return of the individual towards the dominant male after withdrawing from a threat by him. This can be seen in any group of Indian Macaque monkeys or African Savannah baboons. Mason (1965) has shown how the mechanism underlying the construction of this response is related to arousal. A high level of threat arousal produces escape, but intermediate levels produce momentary withdrawal followed by *reverted* escape. In this way the spatial arrangement is generated around a central dominant individual in hierarchical agonic-type societies. This type of cohesion is, however, reinforced by the dominant individuals who, when they threaten subordinates in such societies, reactivate their arousal and so their reverted escape and their attentiveness. However, social attention also mediates a higher level of communication as the dominant male demands a specific indication from subordinates, who may move away on their own initiative, that they recognise where he is. Thus, there are mechanisms to maintain cohesion in the face of conflict, but we shall see that reconciliation, the prevention of conflict, is an alternative.

Having discussed the structure of social cohesion, let us now look at how an individual fits into it, orients itself and uses it. One of the recurrent themes in the literature on this topic is unprovoked aggression. Although my own studies on a colony of long-tailed macaques confirm that it is essentially unprovoked, the occurrence of this type of aggression early in the day, when taken together with the fact that its direction is predictable from any one individual, suggests that it is used as a means of testing the stability of the system. For if the reaction of the recipient of threat has changed in any way this may alert the threat-giving individual to some instability in his own social rank. This was observed before, and for some period after, a rebellion in the Uffculme Colony of long-tailed macaques (Chance, Emory & Payne, 1977.)

Pitcairn (Chapter 3) describes attention structure in a colony of long-tailed macaques at the Basel Zoo studied from 1969 to 1971. He found that each female had positive and negative attentional referents among the other monkeys. The negative referent was constantly avoided (withdrawn from), whereas a female would frequently approach her positive referent with whom she could relax, sitting in close proximity to them. The positive referent may also be her "backer" or ally in a conflict; from them she may sally forth to explore or threaten another individual. So a positive referent acts as a safe base where the individual's escape motivation is consummated and where she feels secure.

This is an example of what Kummer (1987) has called "protected threat". A female will position herself in front of a high-ranking male from which vantage point she will threaten a subordinate. This is a frequent strategy employed by females coming into a sexually receptive state, who then set up a consort relationship with a male by this means.

Essentially the same strategy is employed by the maturing individual to upgrade their rank at appropriate transitions from infancy to youngsterhood, then to juvenile status and finally to sub-adulthood. The infant who explores away from mother, (both monkey-infant and young human child) instantly runs back to her on being startled. As the infant matures it will set up a positive referent relationship with another more adult monkey of the same sex. This is often done by mildly threatening a prospective positive referent to gain its attention, then rapidly showing ambivalent postures alternating with submission gestures, so as to tease the more adult monkey into recognising the friendly nature of the approach. Skilful youngsters may then, by this means, secure positions of influence more rapidly than others. Here the negative (agonistic = conflictual) elements of behaviour are used to alert the prospective positive referent, and then are skilfully integrated with approach elements to establish an hedonic bond (see also de Waal, Chapter 5). (The distinction between "agonistic" and "agonic" is clarified by the subsequent text.)

Protected threat is also used to downgrade an opponent; the threat may then escalate to actual chases and, in extreme instances, capture and biting. When this happens, knowing the role of a backer, an observer has been able to control the aggressor by removing the positive referent (backer) for a day or even just a few hours. In some instances this has curtailed the aggression of the aggressor toward the still-present aggressee for several months. A likely explanation is the need for the aggressor to realign its position in the group, and to make propitiatory approaches to another prospective positive referent. Hence the aggressor's attention is deflected away from aggression to the aggressee and turned towards affiliation with another monkey, thus replacing agonistic motivation by an hedonic activity.

So, in a typically cohesive, hierarchical social system, in which reverted escape brings back the low-ranking individuals towards the source of threat and hence back into the society, escape itself is consummated by proximity to a supportive referent. If the positive referent is of the same rank the individual can relax close to them or, by gaining an affiliation with an individual of higher rank, may use that referent as an ally to maintain or achieve higher rank vis-à-vis another individual through the deployment of protected threat. This strategy may be used when females are competing for a male consort or during the maturation of the individual in the society. The maturing male, in the wild, is displaced to the periphery of the group, and may leave it altogether (to which the term exit may be applied) to join bachelor bands, become a lone male, or re-enter another group later on. When escape is no longer reverted, a break in this fundamental social vector occurs and exit takes place. This may occur as a result of persistent persecution by a more dominant monkey using protected threat, or simply through the high propensity for escape in the low-ranking individuals, as their sensitivity to threat increases.

When exit occurs the individual comes under the influence of a number of in-built escape mechanisms. These constitute a stereotyped escape reaction involving a number of escape defensive elements strung together in a sequence (Chance, 1957; 1963). This stereotyped escape reaction in mammals is but one manifestation of a wider category of protean (unpredictably varied) escape behaviours that are also recognisable, in birds and reptiles, as anti-predator defensive strategies, evolved long before the arrival of the social primates. Larger mammals evolved from small mammal-like reptiles, so the existence, in contemporary small mammals, of convulsive escape mechanisms, which involve not only escape itself but catalepsy and automatic counter-aggression, makes the presence of such mechanisms in our own inheritance very likely. The manifold consequences for primates of having evolved a reverted-escape system are outlined in Chapter 1.

The existence of the agonic and hedonic modes was discovered by comparing and classifying the structures of similar societies of non-human primates. A subsequent study by Emory (see Chapter 2) has revealed that the classification can be extended to form a continuum related to the degree of cohesion in the societies of different species. Pitcairn (Chapter 3) has shown that the precise type of attention paid to another individual is typical of each mode.

Thus, in the *agonic mode*, which is illustrated by the way of life of the Savannah baboon and the Rhesus Macaque, individuals are always together in a group yet spread out, separate from one another, keeping their distance from the more dominant ones to whom they are constantly attentive. They are ready, at an instant, to avoid punishment by reacting to those threats that are dealt out from time to time down the rank order. This they do with various submissive and/or appeasing gestures, and by spatial equilibration (see Chapter 2) which, arising from withdrawal followed by the reversion of escape, serve to prevent escalation of threat into agonistic conflict, yet with tension and arousal remaining at a high level.

The continuous high tension, without the accompanying agonistic behaviour, is the unique characteristic of this mode, for which the term agonic is reserved—*as arousal must be balanced by inhibition to preserve this state* (Chance, 1984).

The evidence for a neurophysiological mechanism capable of sustaining the arousal characteristic of the agonic mode has been reviewed by Paul Gilbert (1984). He supports others in referring to this neurophysiological state as one of "braking", because it ". . . implies an unabated state of arousal which does not provide any effective behaviour as long as the powerful brakes (controlled by the hippocampus) are applied" (pp. 109–111). Presumably this mechanism is nearer to the centre of the central nervous system than the one that maintains a braced musculature, discovered by Whatmore and Kholi (1974). The "braking" state, accompanied by the fixing of attention towards more dominant individuals from whom threats are anticipated, distinguishes the agonic situation in which those individuals find themselves from that of their counterparts in hedonic societies.

The Hedonic Mode

The chimpanzee *hedonic-type society* is quite different. It is very flexible: Margaret Power (Chapter 4) and Frans de Waal (Chapter 5) show how the assuaging qualities of appeasement become transformed into reassuring and reconciliatory gestures between individuals who are mutually dependent. This is seen as the group splits up into twos or threes to go foraging, when the less confident individuals seek and are offered reassurance by contact gestures usually from older, more confident, leaders.

After the chimpanzees have been foraging in small groups they will come together in response to calls, when "carnivals" of *competitive display* will focus attention upon the most demonstrative individual. Display actions include jumping up and down, or throwing things. These congregations are one of many different occasions when contact greeting takes place between group members. The most significant occasion for maintaining socialisation follows threat from a more dominant individual although, if Power is right, this is not a frequent occurrence in the wild. Then either party may initiate reconciliation through touch. As de Waal explains, touching and especially kissing, bring about reconciliation among chimpanzees; the contact then reduces the tension and relaxation occurs. This means that except during moments of excitement *the arousal level of the individual is low—this is the hedonic condition* (Chance, 1980), and is responsible for the flexibility of the hedonic mode. This flexibility is a manifestation not only of absence of the fear of punishment in the relationship between individuals, but also of a freeing of an individual's attention from being the medium or channel of the social bond between them and the rest of the society. Because it is no longer active as a bonding element, attention is freed for detailed investigations and manipulations of objects in the physical environment, thus facilitating the development and expansion of intelligence (Chance, 1984).

So we can now see the nature of the hedonic mode in humans—the healthy human individual has a flexibility of arousal and attention that allows time for integration of reality, inter-personal relations, and private feelings and thoughts, providing prerequisites for the operation of a systems-forming faculty, as I will describe later. Conversely the condition required for preventing the operation of intelligence is the agonic mental mode, in which attention is fixed on preoccupations of security that bring about rank awareness, and when distancing and reverted escape dominate the behaviour, these perpetuate estrangement.

The extensive ethological study of children in playgroups made by Hubert Montagner (Montagner et al., 1970, and Chapter 10) is of particular relevance. There are (hedonic) "leader-type" children who, like the wild chimpanzees, do not escalate threat into aggression and who, by actively appeasing their followers, are the initiators of play and co-operative pursuits. In contrast, there are some four categories of agonistic-type children, in each of which behaviour is dominated by flight (social withdrawal tendencies). Such children are agonistic, not agonic, because these actual tendencies shape their behaviour and are not inhibited as in agonic adults. The agonistic children also differ from the *hedonic leaders* in having a high urinary output of adrenal-gland (adreno-corticosteroid) hormone. This hormonal change reflects a physiological aspect of the agonic response to stress. Of great signifiance, in relation to this finding, is the work of McGuire and Raleigh (1985) and of von Holst, Kevern and McGuire (Note 1), who have shown that different blood chemistry and hormonal levels correlate with the social status of monkeys and that, in particular, they can distinguish between submissive individuals who are truly subordinate, exhibiting reverted escape and remaining in the presence of the dominant, and those who escape and try to hide. These physiological aspects of social rank persist for weeks after removal of the individual from the social group but, as the Uffculme rebellion mentioned earlier reveals, are overruled by changes in social structure.

In the light of Montagner's findings and the ubiquity of agonistic tendencies in the social behaviour of birds and fishes (Morris, 1970), human mental illnesses, as listed by Vaillant (1977), can be reassessed. It can be seen that they show evidence of flight (social withdrawal) tendencies, suggesting that existing stylised methods of describing mental disorder could be replaced by a more accurate description of what are essentially security-programmed behaviours (Chance, 1984). This approach is expanded by Gardner's original observations in Chapter 8 of the identification of social arousal modems or *psalics*, and by Price (Chapter 7) in his definition of catathetic yielding behaviour, as it relates to depression.

GROUP STRUCTURES

All the contributors to this book have been chosen because they are looking at the societal edifice of which we are all a part from separate points of view in an attempt to discern the outline of its structure. The structure is a social one, not only for humans, but also for the higher non-human primates. So the primate way of life helps us to understand how we have been constructed to live in small groups.

This book presents the harvest of the first five years of the 1980s. Its planned context was the hypothesis that within the propensities evolved by the primates as a class, in order to keep group members together while they roamed the plains and forests, would lie not only the infrastructure of communication, but also the very essence of mental structure itself. This has proved to be so in far greater depth than was conjectured when the project was first envisaged. For now it is clear that the position in which individuals place themselves in relation to the other group members is itself a crucial statement of how they see themselves. The foundation of character, the limits of capability, the capacity to expand mental horizons with age, to occupy roles or have expectations fulfilled—all of these reside in the individual's way of relating to group structure, including the option of leaving the group. This option, so perilous a path in the non-human primate environments, may be turned into a sanction in human society. What was once, earlier in evolution, a desperate act now becomes one of many choices that may represent the operation of intelligence by individuals free to act in defence of their interests and in order to influence the behaviour of others. That these diverse phenomena are of the same essence, moulded out of the basic structural features of social relations, is here laid bare by the individual initiatives of writers who have, in most instances, not known of each other's work. They have toiled in widely separate disciplines that have kept them apart. The purpose of this volume is to juxtapose their work so that the reader can sense an emerging synthesis in the belief that bringing this new synthesis into being is what the rapidly growing demands of knowledge, and self-knowledge in particular, require. It is what the present stage of human development needs.

A differentiated hierarchy arises among primates partly because individuals at all stages of maturation will be present together at any one time. Typically these are arranged concentrically around one or two dominant males. Close to these are consort females with whom they are mating, their young, and pregnant females. Outside these are out-of-season females with young and, finally, on the periphery, sub-dominant adult males offering protection to the group. This spatial hierarchy is the most compact way of keeping together a set of individuals whose age and sex predisposes them to adopt various functional roles at different times of life. The question addressed here is how far, and in what way, are these fundamental facts relevant to the establishment of an infrastructure of communication.

Emory (Chapter 2) shows the importance of the social structure of attention that exists within a group. Tom Pitcairn (Chapter 3) and also Russell Gardner (Chapter 8), variously indicate that the developmental states reached at various ages by each individual—sometimes reflected in roles, but also in and of themselves—stamp on their behaviour the unmistakable characteristics of the positions they occupy and the roles they may be carrying out in the social structure. These signal corresponding messages to other members of the group.

Interest in the structure of groups, an intrinsic part of social anthropology, arose afresh in 1964 with the publication of *T Group Theory and Laboratory Method*, a series of essays edited by Bradford, Gibb and Benne (1964). This interest spread and led to the formation of the group-psychotherapy movement. It also spread into education, social work and industry, which took the new findings as an aid to their training methods.

T Group Theory set out to display what was then known about the way in which members learn within a group where didactic methods have been replaced by attempts at co-operative learning. Without doubt this was a response to the technological innovations of the post-war years and to the opportunities that an expanding economy gave for a more liberal climate within the whole community. The emphasis then was on change, and change of a kind that could be comprehended by teachers who were seeking new ways. From this new awareness of the social context of learning, emphasis was placed on changing a person's ideas; focusing on what were the best conditions within a group to bring this about. It is not surprising that these groups were seen as learning laboratories. As so often happens in a new venture no one at the time thought it worthwhile to look outside their own particular concerns to the wider fields of knowledge that might be relevant. Hence the concept of a natural group evaded their notice, although it must be admitted that the relevant knowledge was then scanty.

Bion's experience in therapy groups had recently been published (Bion, 1959) and this drew attention to the social relations that impeded understanding. Undoubtedly Freud (1959), in *Group Psychology and the Analysis of the Ego*, had provided a touchstone to start this work, but the theoretical background was heavily wrapped in Freudian theory. Despite this, however, Warren Bennis, in his writing on patterns and vicissitudes in T Group development (in Bradford et al., 1964, p. 251), was able to formulate a conclusion that goes some way to describing the features of groups, developed and further investigated here, namely:

> . . . that the core of the theory of group development is that the principal problems or issues the group must solve are to be found in the orientations towards authority and intimacy which members bring to the group. Rebelliousness, submissiveness, or withdrawal, as the characteristic responses to authority figures; destructive competitiveness, emotional exploitiveness, or withdrawal as the characteristic reponse to peers, prevent consensual validation of experience.

Bion (1959) contrasts a group in which these features predominate with a task-orientated group. Although he confined himself to observing, recognising, and reporting human irrationality at work, particularly in small groups, he nevertheless revealed that there is a strong tendency in any given group to the development, at any given time, of a powerful over-dependency on the nominal leader who is in some inchoate way, felt to be in possession of all knowledge and problem solutions if only they could be pressurised, propitiated, sacrificed to, entreated, etc. to vouchsafe them. Such magical and god-like powers (projected wishes and attributes of the self) were, in large measure unconsciously, being attributed to the leader. This was accompanied by a corresponding, unconscious, irrational, profoundly powerful devaluation of the abilities and potential capabilities of the rest of the membership. Despite Bion's great achievements as a psychiatrist, his findings did cause him to pause and consider whether, and in what way, the regressive or progressive ways of group structuring could reflect basic human predispositions, derived from features that originally evolved to fit us for the primitive groups of hunter–gatherers. This phase of our existence is supremely relevant, as it constitutes more than 90% of the time that *Homo sapiens* has been around, to say nothing of the influence inherited from even further back in our primate ancestry.

James Woodburn (1982) has identified two forms of the hunter–gatherer phase of human existence. The simplest (type 1), based on immediate consumption, is without the ranked hierarchies of the later civilizations. The second type of hunter–gatherer society accumulates some surplus and shows the beginnings of social rank in relation to who controls and distributes the surplus. Woodburn deduces this state of affairs from an extended comparison of many hunter–gatherer societies. If, as Margaret Power (Chapter 4) so cogently argues, wild chimpanzees live in a non-hierarchical form of society then they have a way of life in many respects similar to Woodburn's human type-1 societies. So non-hierarchical as well as hierarchical groups must have existed well before the earliest phases of *Homo sapiens*.

In the hedonic societies of the chimpanzee and gorilla, close-contact social relations are frequently sought after and indulged in for long periods, and therefore are pleasurable. By implication they are also voluntary. Reassurance and reward go hand-in-hand as parts of the reciprocity that is at the core of this mental mode. This is fully illustrated and explained by Margaret Power, and discussed in the context of equality by Itani (Chapter 6).

The mutuality of social relatedness forms the hedonic/emotional infrastructure of the Old World apes. Superimposed upon this, and also drawing affective strength from it, is the specifically human form of social relatedness—co-operation in a task. The execution of a task by more than one individual requires the integration of the participants as well as of the separate activities involved. This is well achieved in Woodburn's type-1 societies, in which social relations are very flexible and roles are adopted in relation to immediate requirements, often placing control in the hands of those most competent in the skills then required. It is implicit that the skills of those co-operating should match one another in competence, and that tasks are voluntarily undertaken. Any act voluntarily accorded to another, and acts of courtesy and politeness between people, imply voluntary accord of status to the other, as would acts that recognise standards held by another. Therefore, formal cultural recognition of an individual's skills also constitutes the basis upon which status is accorded to individuals in present-day Western democracies. This is Theodore Kemper's message in Chapter 13, and status-accord provides an understanding of how the social dimensions, found essential for learning by Bion and his associates, are provided within the framework of institutions in a modern society. Kemper contrasts *status* ranking with that based on *power*, which in essence directs and restrains the individuals in modern societies—and those in Woodburn's type-2 hunter–gatherer societies.

FORMS OF SOCIAL STABILITY

What additional evidence is there for the nature of the two mental modes (agonic and hedonic), and what evidence is there about the factors that either stabilise each mode or make change from one mode to the other possible? The first question is answered in two ways: each mode is a property of *both* the social relations and of the mental structure that relate the individual to the rest of the group (or society) by direct interaction, so each source will provide relevant information.

Many of the authors here address the question of the hierarchical nature of society and provide evidence for the rigidity or flexibility of social relations when ranking is present. An exception is Margaret Power (Chapter 4), whose most original contribution not only introduces a new and more precise understanding of the social relations of chimpanzees living in the wild, but also relates this structure to Woodburn's non-wealth accumulating hunter–gatherers (type 1) as described earlier. She also provides a solution to much of the confusion that has arisen from the picture of a distorted chimpanzee society in the Gombe Stream Reserve after 1965—hitherto regarded as one of the most important sources of behavioural observations. This illustrates how a knowledge of the two modes alerted her to the possibility that interference by the observers there had unwittingly influenced the nature of the information then obtained (see Power, 1986). Her work will now be further discussed.

First let us be clear about the meaning of the term "inbuilt". In different species, some of the rigid ranked relationships (together with any species-specific characteristics), may be established by two distinct but interwoven mechanisms. Either genetic programmes establish rank-related behaviour, allowing of little modification from early in life or, after the relaxation of their constraints, learning can reinforce a weaker hereditary disposition and achieve the same result. Both of these mechanisms are covered by the term "inbuilt".

Learning can also establish new patterns of behaviour. As suggested by Margaret Power's contribution, the structure of the social relations is brought into being wholly by learning when hedonic relations prevail throughout the society. Indications of what is meant by these hedonic features are provided here by Power, de Waal, Itani, Montagner, Pitcairn and Price. They were initially pointed to in early field work on chimpanzees by Vernon Reynolds, Kortlandt, Itani and other Japanese workers, as well as by Jane Goodall before 1965, (as now discussed).

Margaret Power argues that the social relationship between wild chimpanzees is essentially one of awareness *of* rather than of reaction *to* each other. She is the only authority to provide an explanation of the breakdown of the social order among chimpanzees at the Gombe Stream Reserve in Tanzania after they had been brought into intense competition for bananas that were locked away for part of the day—which started in 1965. After this practice had been in existence for some time, one group started hounding and killing members of another group. The hedonic relations, which previously existed, broke down. In the hedonic state, each individual monitors its relationships with others, so that they are able to act together under the guidance of "charismatic" individuals, who can thereby control a bully. The startling breakdown in social group relations, although made known to me only from several T.V. programmes produced by Jane Goodall, demands interpretation, which an integration of information supplied by several of the contributors to this book makes possible. Margaret Power's chapter is not primarily about those events, but the interpretation of them is crucial to her rejection of the evidence about chimpanzee society that came out of Gombe after 1965.

Her main concern is to demonstrate just what is unique about chimpanzee society in its natural wild state. She argues that their natural state (the hedonic society) represents a qualitatively different and stable state and she defines what it is very clearly. The stability appears to rest on a learnt reciprocity that is the consequence of the negotiated acceptance of one individual by another. Such negotiations are initiated and repeatedly reinforced by reassuring and affiliative gestures that form a network of social attention.

We know that proximity between submissive and dominant originally came about by reverted escape. Afterwards, as de Waal has shown (Chapter 5), proximity can be further maintained by repeated appeasement or greeting gestures that reaffirm the individuals' status *after* which, sharing and other relaxed forms of affiliative relations can develop. Therefore greetings are essential for relaxed relations, which means paying attention to another individual until the greeting is reciprocated. Prolonged and frustrated competition reduces the opportunities for greetings as: (1), attention is diverted from social companions to the object of that competition; (2), prolonged, close proximity to the object competed for does not provide the space required for fluctuating approaches to allow repeated greetings between neighbours (see Chapter 4); and so (3), arousal begins to be sustained at a raised level. Thereby the oft-rewarded greetings are eliminated so that the learnt reinforcement of the network of social relations is broken.

These chimpanzee societies, therefore, rest alone on some form of rewarded and hence learnt stability, which explains why, when the opportunity to practise mutual reassurance was prevented at Gombe through a critical period of prolonged competitive provocation, the network of social attention underlying the social relations broke down. This learnt stability is not dependent upon inbuilt rigid support in the way of an agonic structure that will reassert itself after the social links are temporarily broken.

In effect, Margaret Power has shown that Jane Goodall uncovered a general, learnt infrastructure of social relations that is based on the greeting behaviour of chimpanzees, and which is the essential prerequisite for subsequent, learnt social relations. In fact the stability of the society rests entirely on the continuity from one generation to the next of reassurance practices that mediate the diversified forms of attention within the social structure. Before we can understand how vulnerable chimpanzee society is to the destructive influence of reinforced competition we have to consider again the role of inhibition in forming the agonic mode.

In this mode, as we have seen, all the agonistic components that lead to withdrawal, such as violent aggression and escape, are continually activated but not expressed because at the same time they are fully inhibited. In this way they are kept under control to be released instantly when required. By contrast, in the hedonic mode, the relaxed state drops the level both of arousal and inhibition. As has been explained, fluctuating arousal, related to the activity in hand, promotes investigation and incorporates learnt elements into behaviour. When hedonic learning alone is relied upon to maintain the social structure from one generation to the next, then the situation identified by Margaret Power can arise. The infrastructure of hedonic social relations is constantly maintained by appeasement, reassurance and contact, resulting in arousal reduction, which enables the relations between one generation and the next to be smoothly integrated. This integration was broken by continuous competitive provocation for bananas at Gombe, and the chimpanzees, having lost much of their power of neural inhibition because it was not built in by the hedonic learning provided during their ontogeny, during evolution, were wide open to permanent social destruction.

Perhaps something similar happens when a human culture, which is also learnt, collapses not just through physical destruction by war, but when the mental strain of prolonged, total war proves too much. If so, we may gain deeper understanding of how recent events in Vietnam and Cambodia destroyed cultures and caused reversion to self-killing dictatorship—a dark portent for the future of humanity if we come to rely on the technology of force to control terrorism and other forms of retaliatory aggression without recognising the fragility of the infrastructure on which freedom and democracy depend. Both Itani and Power here step across the bridge linking us with our ape relative and pose the question: Should we not look long upon the apes to understand what it is in us that we most need to preserve?" It would seem that not only the continued incorporation into our culture of the value of ritual greetings is important, but also that a way of life providing ample opportunities for repeated greetings is something of extreme value.

THE BIMODAL SWITCH

Of focal importance for understanding the relationship between the agonic and hedonic modes, and the mechanism that may underlie the way in which the personality comes to be controlled by one or other, is the work of John Price (see Chapter 7). He bases a far-reaching ethological interpretation of depression on the flight component of agonism, which as was pointed out earlier, is exaggerated in mental disorder. He suggests that "depression has evolved out of the yielding component of ritual agonistic behaviours seen in birds and fishes"—the individuals who yield and become subordinate may retain their position by a reduction of affect and desire as an alternative to astute maintenance of status. As he says: "Ritual agonistic behaviour is a form of signalling between two individuals whose function is to create, readjust and reinforce complementarity of behaviours; converting a symmetrical behavioural relationship, which has the potential to escalate into severe conflict, into a stable complementary asymmetrical relationship with an 'agreed upon' winner and loser."

A factor of crucial significance is his identification of a second simultaneous message, a paracommunication, from the subordinate to the dominant, arising out of the comportment of the submissive. The existence of the two provides the primordium for a switch mechanism between the modes, as I will now describe.

The term paramessage is applied to two or more messages expressed simultaneously in different representational systems—or (more usually) *in different output "channels"*. I emphasise this distinction (which is etymologically correct) between the two variants of the term because, when the dominant animal can see from the comportment of the subordinate that it poses no threat, this is a message of awareness and is in a different "channel" from that through which it receives a signal of submission. Indeed we can now see that a submissive signal is required if, and only *if*, the subordinate individual remains close to the dominant and gives out a complementary signal, making it clear that its maintained proximity is no threat. Withdrawal also indicates that the subordinate, by this very act, poses no threat. De Waal also approaches this concept when he told me that an individual does not have to fear its subordinates, except when the communication of submissiveness is becoming less self-evident.

Let us look at this distinction more closely. All social encounters take place with varying amounts of space separating the participants. Submission by definition involves two things: the maintenance of proximity to the dominant by the subordinate and, at the same time, a signal given by the subordinate that switches off the aggressive arousal of the dominant so that attack does not ensue—this is the submissive signal. If, alternatively, the subordinate is prepared to put distance between it and the dominant *and the dominant perceives this act for what it is*—that thereby the subordinate no longer constitutes so much of a threat—then the aggressive arousal automatically drops away. Sometimes, however, a subordinate who moves away or runs away is chased, presumably because the earlier, close proximity was above a threshold at which withdrawal could no longer be perceived, thereby *ipso facto* removing the threat which the earlier proximity of the subordinate presented to it. The very act of withdrawing, which I earlier described as a root component of the agonistic complex, will reduce arousal in both participants and thus can introduce a mutual awareness that would enable them later to come together under the influence of any other, non-agonistic, approach drive. This is why, in the chimpanzee, sexual components play such a prominent part in behaviour that provides approach capabilities at a relatively high level of arousal. In combination with reassurance gestures, these lead to the development of a fully developed affilitative behaviour.

It is de Waal's argument that the hierarchy, which permits reconciliation, has served as a backbone for social primates. If so, then hierarchy came about through the conversion of withdrawal into submission, so enabling a flight-motivated individual to remain close to a dominant. The original rank order was thus converted to the hierarchy that permits reconciliation only when flight or escape are reverted, so reactivating all the social propensities. The great contribution by de Waal is to specify that positive attraction as well as the agonistic components of flight and aggression, is always reactivated by this reversion, and thereby makes hedonic relations possible.

Equipped with reverted escape, social exploration, and sexuality, the chimpanzees are able to rely on hedonic, social approach motivations provided they can keep down aggressive and escape tendencies, which they do by reassurance and reconciliation. The subtlety of de Waal's thesis is in the way he shows how this amounts to converting rank (agonic) into status (hedonic) recognition by the very act of reaffirming submission; thereby making reconciliation and reassurance possible.

Hans Kummer (1987) has also studied how similar conflicts are resolved in encounters between male baboon strangers caged together so that they become companions. Their reconciliations proceed by the appearance of the socially positive elements of sexual behaviour (presenting and mounting), followed later by care behaviour (e.g., grooming). As Kummer points out, this is the sequence of reproductive behaviour in lower vertebrates at the breeding season, now at the service of a higher faculty: simple sociality.

Kummer deduces from such observation, and from the appearance of other forms of non-aggressive approach by the dominant, that the dominant baboon is actively using these gestures to promote resolution of the conflict between himself and a subordinate who cannot at first approach, despite the dominant's frequent sexual advances (presentation). The prominent initiative taken by the dominant may be because the subordinate, under the enforced proximity, is unable to look at him and interpret his positive mood until actively prompted into awareness of his affiliative intentions.

Moreover, Kummer, notes that the dominant would be able to assess the strength of the subordinate's conflicting motivation from the subordinate's signs. So, on the one hand, the standard resolution is automatic through signals and responses, and on the other, some conflicts are creatively resolved by the dominant creating ways to achieve it, especially when the subordinate appears to be stranded in the early phases. Both features of the interaction—the automatic responses to signals, and the contrasting created resolutions—together with the existence of a time at which subordinates can become stranded, suggest that there is a switch point at which approach becomes possible, and that a qualitative switch occurs when signal/response dependency is replaced by a state of awareness.

John Price (Chapter 7) independently addresses this point by suggesting that in all conflicts of this kind two types of information are always available: signals that trigger set responses; and awareness through which the reality of the situation can be assessed. An evidently non-threatening behaviour in itself will cause less arousal than the submission signal, and indeed can lead to enhanced awareness. Signs of reduced tension (arousal) in the dominant will also have a calming effect on the submissive; and this in turn will enhance the subordinate's ability to perceive the reciprocated lack of tension and hence the lack of incipient threat. Thus a process is initiated that could lead eventually to the state of "mutual dependence" discovered by Margaret Power, in which reciprocal (interdependent) roles are adopted by the "charismatic leaders" and "less confident members" of exploratory, foraging groups of chimpanzees.

Is this the actual outcome of the "switch mechanism" suggested by Price's work? It would appear to be so; for the asymmetry of rank order first reduces the impact of the agonistic signal, and then the reassurance gestures promote re-establishment of hedonic relations. The other important implication of Price's insight is that the dominant can draw the same inference from seeing a lower-ranking monkey withdrawing from his presence. If, as has been argued, awareness is the substance of the hedonic faculty, then the reciprocal is also true—where awareness can be inferred, an hedonic influence is operating on the mood of the dominant. Put another way, a lower-ranking animal who remains close but also signals submission is likely to produce more arousal than one who moves away. Therefore withdrawal while still remaining in the group, which hitherto was regarded as an integral part of the agonistic signalling system, is in reality an arousal-reducing component built into the agonistic rank relations. Thus withdrawal is the component that promotes the "mutual dependency", as identified by Power, in truly wild chimpanzees.

De Waal, Itani, and Price show us that throughout the vertebrate phylogenetic tree, rank order does indeed appear to act as an imposed backbone upon which differentiation of social roles can and does occur. According to de Waal, this order will reassert its role when slight overcrowding introduces an element of social stress into the life of chimpanzees. Three elements can now be seen to operate: the freedom to withdraw; contact gestures; and reassurance gestures—the last two sometimes merging into a single act. A group of humans undertaking a task need to remain together in order to integrate their separate efforts. I have said that Montagner's study of leader children (see Chapter 10) shows us how the agonistic withdrawal tendencies of the children who co-operate with them in their pursuits are overcome by reassurance and constituency behaviour, (as defined by Barner-Barry, Chapter 11). Now, the origins of this behaviour can be seen in chimpanzee social life—but this is only the infrastructure of co-operative behaviour. Supplementing this togetherness is the systems-forming faculty, which is directed towards the conception and completion of a task.

METHOD AND ITS CONSEQUENCES

Gregory Bateson (the biological anthropologist) has discussed how we may begin to understand what is going on between the members of a group about which we know nothing. In his studies of porpoise groups (Bateson, 1974), he insisted that the first essential was to draw a baseline, and he suggested that the sleeping pattern would serve this purpose because it was a recurrent pattern of social relations. Without this baseline, no idea of the order between individuals could be obtained and without that the subtleties and complexities of group behaviour would remain unseen.

How knowledge of social behaviour is acquired is shown here by some of the contributors, especially Emory (Chapter 2), Montagner (Chapter 10) and Masters, (Chapter 12), who illustrate observation techniques in operation. To follow these authors through their studies will do much to assist the reader who may want to devise methods for observation or investigation in other fields. It will also be found an intellectually rewarding exercise in its own right.

Itani's contribution (Chapter 6) does much to highlight the inappropriateness of the sociobiological approach (which I discuss later) in studies of primate social behaviour, where the structure of ranking has been insufficiently explored *before* introducing cost/benefit analysis. In the only study so far in which the nature of individual advantage arising from social strategies has been assessed, it was found that as much as 50% of aggression was deployed to ensure the individual's membership of the group (Kaplan, 1978). Itani points out how the structure of social cohesion found by organised observation reveals consequences other than those detectable by reductionist analysis. The reader may then see that the material of this book is set out in a way that will acquaint him or her with a new dimension in both biology and human sociology.

What is clear is that members of a non-human primate group are programmed to be part of, and operate in, their system. By contrast humans possess a systems-forming capability (outlined in the next section), evident not only through the creation of scientific hypotheses, by which understanding and awareness are extended, but also through the creation of diverse social and industrial organisations. As a reflection of this capacity, here are brought together authors who accept the need to recognise our evolutionary origin as a way of understanding in what ways we are unique. I maintain that because organic systems possess unique features, the correct procedure of investigation for a human survival strategy is to give as much weight to the emergent aspect of knowledge, our systems-forming capability, as to the survival criteria demanded by hereditary mechanisms that gave rise to the capability itself. Clearly, the operation of intelligence is revealed in group goal-facilitation behaviour, which is Bion's category of work-oriented group behaviour, as already mentioned. This is the hedonic dimension in which the group's individuals show deep concern about the acceptance of all the members into the group by being agreeable to one another. In the agonic dimension, the group splits into leaders who coerce and excessively control followers who are over-dependent.

In democratic societies power is recognised only as an aspect of the civil authority, so power is taken to be synonymous with authority, and it is natural for us to think power means authority of office, when really it is the other way round. Authority is power given to a person by virtue of office acquired through culturally prescribed methods; the individual will wield this power either in an agonic/agonistic or a hedonic fashion dependent on their personality. This is how the prescriptions of the civil authority will have different consequences—the way these are carried out will be heavily influenced by the agonic and hedonic personalities and by social structures. The agonistic/hedonic bimodality has also been uncovered in the differences between the behaviour of children in playschool groups by the two ethological psychologists, Hubert Montagner (see Chapter 10) and Barbara Hold (1976), and re-emerges in the class room, as shown by Barner-Barry (Chapter 11). This bimodality is thus deep-seated in our nature and, on the balance between the two modes in the adult, will depend his or her capacity to initiate and sustain a task in adult life.

THE HUMAN SYSTEMS-FORMING FACULTY

The distinctive feature of being human is the possession of a systems-forming faculty. The distinctive nature of this faculty is best understood through the realisation that evolution has created very many types of systems, which are the species of plants and animals of the world. We are one of these systems, but we also possess the ability to formulate systems. Our faculty to construct systems is evident in the game of chess, for example: not only is the game and its rules an invention—a uniquely human construction—but also the planning ahead within these rules illustrates our ability to form systems, by constructing, altering and remodelling new strategies as the game progresses. Analyse what is going on in the game and you will see that each player is trying to match the patterns of the opponent's chess pieces on the board with another one, which invites certain responses. If the first player has made a move in one of several possible ways this may produce checkmate. At an opposite pole of activity we are not just able to be simply aggressive to each other but can chart a course of aggression and execute it with all malice aforethought—we set our aggression within a planned system. *This ability to form systems arises, therefore, in our formulating faculty*.

That faculty itself arises from the recursive properties of the mind (see Chapter 1), which enables it to match up one pattern with another. If we can now become aware that there is a logic in the formation of systems, that needs understanding, this will itself constitute a recursive process of a higher order. Through our ability to pick up patterns we can pick up the logic underlying the construction of a system from a pool of diverse elements. When patterns are picked up through awareness from the external world they can be combined to make an operational system, the nature of which is independent of any elements of one's own nature. This gives an autonomy of operation to one's thoughts that is recognised in the objectivity of science.

The formulating faculty becomes a system-forming faculty because the internalised recursive processes can bring together, for comparison, elements or patterns observable in the outside world with elements or patterns already established internally in the ideational system. The combination can then be retained, altered or rejected. In this way, the correspondence between the two elements/patterns can be established and a hypothesis worked out, which can then be tested. Or elements drawn from observation or internal construction (phantasy) can be combined to produce a working model of an invention—or a work of art.

Organic systems are complex, open and self-organising, made up of self-developing processes, e.g., systems deriving a supply of energy from the sun. Self-organising systems are now a recognised phenomenon in an active field of research, sometimes referred to as autopoetic systems (Jantsch, 1987). The human systems-forming faculty is a unique example of these. It is important to outline as far as possible what is already known about its scope, especially how it can effect the personality. This amounts to asking how the systems-forming faculty is embedded in the human brain.

The central nervous system, throughout its length, is divided into a motor, executive, lower half consisting of a hierarchy of co-ordinated actions, and a sensory, flexibly integrating, upper half. Paul MacLean (1982) emphasises that the neocortex, which is the largest and most highly developed part of the sensory nervous system in simians, evolves primarily in relation to sense organs that receive and process information from the external world. It is also known that the motor systems are represented in the neocortex in two ways, which both provide us with information about our bodies. The first is from the proprioceptors (pressure bulbs) in the muscles, which inform us at any one time of the state of tension in the muscles, and from which, for example, we can tell the position of a limb without using our eyes. The second is derived from executive motor centres, which send instructions to the muscles to bring about movement. These simultaneously send "output copies" (Mittelstaedt, 1970) of the instructions to the neocortex to inform the formulating faculty of what is about to be undertaken. So it is clear that there is a neurophysiological basis for the systems-forming faculty to operate with information from within ourselves, as well as from without. This mechanism of internal awareness is likely to provide the neurological basis of relaxation techniques, embodied in Zen and Yoga practices, whereby the mind is cleared and the body relaxed (Crook, 1980). These allow a reorganisation of the personality, free of agonically activated defensive elements, (see Gardner, Chapter 8; and Price, Chapter 7) and, in conformity with a rehabilitated awareness of self-reality, may transform the organisation of the personality from the agonic to the hedonic mode.

The Person in Society

When we allow ourselves to become the agents of aggression (e.g. when going to war), we allow the social system of which we are a part to determine our actions. This illustrates that we live as part of a particular social system, which often influences subtly how we value different aspects of our lives. How we each give attention to others and how we react to and perhaps perceive the attention that they give us, powerfully affects the development of our character, our capabilities, and how far we find fulfilment. The social structure of attention amongst a group of people is both influenced by the individual attentions and influences how individuals will give their attention (Chance, 1976) Thus their whole behaviour is affected and, by changing behaviour, they affect their social environment. Each person, as a system, is a subsystem of the social system, and by our individual decisions we can come to affect the larger whole. There is no such thing as freedom of the will to act except that which is made effective because it is based on an understanding of the social system of which we are a part. Otherwise, we have no choice but to be influenced in ways we do not recognise and so are compelled to conform to the system in which we live.

Therefore, if we are to be free to think or to act, we can do so only as part of a social system that nevertheless can be changed in ways we desire, if we act in accordance with the rules of self-regulatory systems. These are only just beginning to be understood. Self-regulatory systems in part involve feed-back loops, but this is only a simple form of a much more diverse set of what are known under the collective title of recursive processes, and which have already been mentioned.

The Mind in Biology

A fully developed science of the mind and personality must be based on biological knowledge because we are the product of evolution but, as biology developed later than the physical sciences, it is still largely a distorted system of knowledge. Research in this field has so far been dominated by attempts to explain events within organisms rather than by attempts to become aware of what is unique about each organic system. Hence we are preoccupied with reductionist explanations and evolutionary mechanisms rather than coming to grips with the nature of biological systems. We have found it very difficult to look at and see ourselves for what we are because inevitably we approach such study from a position dominated by scientific preconceptions and are also influenced by convention. Ethology[1], however, offers a new starting point in biology, which requires that we describe the features we possess before starting to analyse them. When we do this we find, as I have argued, that the most remarkable feature we possess is the ability to create systems—that the human mind represents the evolutionary emergence of an organ capable of formulating systems; this is unique to our species. So now finally we must look at the nature of organic systems.

The Nature of Organic Systems

Let us be clear how organic systems are described. For this we turn to Hall and Fagen (1956).

> *A System consists of objects with properties which cohere.* The relationships that are brought about by this coherence not only tie the system together, but create the conditions from which the properties arise. The importance of the concept of a system is that within an environment a set of objects can be seen to cohere and interact in such a way that their attributes define the nature of the system, and may *create properties which the system alone manifests*. The coming together to form a system is called "systematization" and any *tendency* of the objects to fall apart is called "*segregation*".
>
> *In biological systems these two fundamental aspects are in balance with one another.*

Our systems-forming capability is itself based on the exaggerated development in the brain of one process typical of all organic systems, namely the ability of parts once brought together to fall apart again. Many structures are apparently stable but not actually so: For example, bone is

[1]Ethology is the science based on description of social behaviours by *observation without interference* either in natural or semi-feral environments.

not simply made of fixed deposits of calcium phosphate on an organic matrix, but is a dynamic structure resulting from the simultaneous deposition and removal of calcium phosphate by living bone cells. At the innumerable nerve endings in the human brain, structures are built up, some of which fall apart again as the individual matures. Like clouds these structures can take on any form but, because of the propensity at one time rather than another for certain processes to predominate, they will tend to take on typical forms; such as cumulus, stratus etc. so it is with humans, who may develop any one of a myriad of personality forms. What we are concerned with is to identify the processes that tend to influence in a major way, the formative properties of the systems-forming capability.

This capability is clearly a flexible aspect of brain function and is certainly greatly enhanced by, if not located largely in, the neocortex—the most recently evolved part of the cerebral cortex. What is certain is that it is an integral part of the organism so that it will interact in some yet unspecified way with deeper layers of the brain and with the properties resident therein, such as the various inbuilt social propensities (psalics; see Chapter 8) attention mechanisms; arousal packages; and information-processing modules, etc. These will be accessible to the systems-forming faculty to the extent that preformed linkages can be broken by segregation. We must, therefore, look for evidence of such segregation and reformation of the personality. This is a very different concept from that of the "onboard computer", (see next section), which would be no more than an additional exploratory organ, as I contest in Chapter 1.

In fact, the process of segregation can be used by the reader to help them take in the full meaning of this collection of essays. This is done by allowing their existing pattern of knowledge to fall apart and to be reconstructed from the patterns arising from out of the collection.

A SYSTEMS-BASED SOCIAL SCIENCE

This book can be seen as an attempt to address the problem of establishing a single behavioural science by understanding the nature of the biological systems. It reaches out especially to those who would surmount inter-disciplinary barriers. For them it should be possible to understand what is meant by our formulating faculty, and how this offers a way of reconstructing the foundations of our thought by systems-thinking rather than causal-thinking.

Methodology, as I earlier outlined, is not just about differences in technique, but more fundamentally involves adopting alternative assumptions, which may be suggested by new evidence or found to change the emphasis so that new correlations are revealed. Many biologists are not systems analysts, though the formative aspect of evolution, which is dealt with by systems thought, is clear for all to see (Whyte, 1965) and needs to be brought more fully into the theory of evolution before that theory can be regarded as complete. Biologists have, therefore, begun to look at social phenomena without realising that a radically new methodology is required. This is because social relations are always part of a social system, and because we need to prepare ourselves in order to be able to construct systems of which we are a part, rather than to be content with single inventions, such as providing simple pumps for the Third World or discovering a new drug. These constructions rather than natural selection will become the agent of systems formation. So far the systems we can form through our thoughts are mostly inanimate but already we can insert new genes into organisms, and it will not be long before we may manipulate the membranes forming the various envelopes that surround the genes and then we will have invented a new single-celled organism (already the nuclei of two species of amoebae have been successfully exchanged)— to say nothing of patenting new species by cross-breeding or genetic transfer. Clearly, our mental capabilities are the present, most powerful and indeed omnipotent, formative agent on this earth. Already we behave in practice as if this were and yet biology does not place mental capacities amongst its first theoretical concerns. How has such an incongruous situation come about?

Briefly, it is because many biologists consider the human brain and mind too complex for investigation (Dawkins, 1987, Note 2, appears to believe this), but they also ignore the research proceeding in other disciplines because:

1. They are preoccupied with the theory of evolution as an all embracing framework of biological thought to the exclusion of other possibilities. If they looked around them they would see that there is much new information from other fields of knowledge, many closely associated with medicine, which has made great advances in understanding not only the brain, but also the mind.
2. Biologists have a poor opinion of the social sciences and so ignore what is currently being discovered. (To rectify this is the purpose of this book.)

Moreover, since the middle of the last century biologists under the impetus of Darwin's Theory have moved further and further away from the belief that "Man is made in the image of God" and towards understanding our place amongst other living creatures. Also over the last half century ethologists, (with the notable exception of John Crook, 1980), who observe behaviour, have rapidly eroded the barrier separating the human mind from the "beast" and, in the process, have ignored what distinction nevertheless remains.

Richard Dawkins (Note 2; my transcription from television of his exact words) is unswerving in his attempt to focus our attention "away from the individual as agent towards the immortal gene—the replicator". That is a legitimate process of thought, but he then proceeds to move away from that position back towards one where humans are agents, when he asserts that: "individual humans are there, like any other kind of life, to propagate their genes". This stance, in the past, has brought him under attack from those who brand his views reactionary, deterministic, even authoritarian. To defend himself Dawkins also admitted (in the same T.V. programme) that humans possess a great deal of flexibility in their repertoire and thought. In fact he went so far as to define how this could have come about through the evolution of an "onboard computer", which because it is essentially a flexible instrument, can be turned to other uses than those for which it originally evolved. This is an evolutionary process long ago recognised by Julian Huxley as pre-adaptation, and presented in his *Evolution*: *The modern synthesis* (1944). So, in self-defence, Dawkins has arrived at the same position as is being put forward here, namely that we possess an organ in our brains capable, through combinatorial competence, of being used for a variety of purposes, one of which is inventing and formulating systems—indeed a systems-forming organ.

There is, however, a crucial difference between the way Dawkins defines the flexible properties of the mind and the way I place it as part of an organic system in which it has reflexive relations with the rest of the organism. In the hedonic state, when segregation is possible, the systems-forming faculty would be able to bring about a restructuring of the personality. This possibility is already being explored by those therapeutic counsellors, psychotherapists and thinkers who solicit the aid of insights from Eastern philosophers, as John Hurrell Crook (1980) is doing, so making possible the conscious enhancement of the power of a person's personality as envisaged by Aurelio Peccei (1977). I hold that this possibility is of primary concern for the correct development of biology as it relates to survival, not only of our own species but also of the world ecological system. It is equal to if not greater in importance than research into hereditary mechanisms. This approach would involve study of how the system-forming ability can be potentiated, through knowledge of how the systems-formulating capacity is related to other brain propensities. In doing this we would be shifting the emphasis of investigation away from concern for the individual solely as agent and onto the relationship between the individual and the environment, which is expressed through the individual's behaviour. For example, through systems-based awareness, environmental selection may be redefined as a recursive process whereby the particular re-pairing of haploid sets of chromosomes from sperm and egg within the new individual (zygote) is determined externally by the selective aspects of the environmental niche.

That our intelligence incorporates a systems-forming capability within the brain implies that the attributes of the individual operate at more than one level. Investigation of the influence of the genes on behaviour, which is done from below, so to speak, by sociobiology, must be supplemented by investigation of the systems-formulating faculty, because this incorporates a link to the environment, part of which is the influence of reality awareness. How incongruous the sociobiologist's use of the term "altruistic" seems, when referring only to assistance given to genetic relatives rather than to describe the faculty that enables people to recognise another's plight and go to their assistance without reference to their heredity when another's plight can be seen for what it is. This is part of their social relations and, as has been pointed out, the study of social behaviour helps us to become aware of the social contexts which either help or hinder us in the use of our intelligence.

Within the brain there are many packages of instruction for behavioural traits that predispose us to occupy social roles, such as being a child; being a mother; undertaking daring and dangerous projects in adolescence and early adulthood. All of these and more may be either integrated into a smooth sequence or become deployed as deception devices under agonic control. The perfectly healthy propensity toward daring exploits may become manifest as vandalism when the environment provides no opportunity to exercise this propensity. Vandalism is the work of street gangs; similarly, Scott Lewis's story (Chapter 9) of a drug users' group secreted within an established institution of an industrial society is but another example of the group-forming propensity. There it went undetected until someone came along who could recognise that the problem of drug addiction was linked to the existence of a social structure embedded within the institution.

Russell Gardner in his Chapter 8, however, goes further. He reveals that without a conceptual scheme based in a knowledge of social structure we may remain blind to the fact that, for example, genes associated with manic-depression also contribute to the characteristics of the alpha personalities who take up leadership-roles in our society. Biologists, by excessively influencing psychiatrists by their own genetic preoccupations, are contributing to a distortion of the truth. Instead ethologists need to reassess the theoretical worth of the psychoanalytic approach, which points to the existence of innate propensities. This is done well by Paul Gilbert (1984), who points out that the failure to determine what constitutes evidence in psychoanalytic procedures is largely due to the lack of an appropriate biological investigative procedure. This is rectified by the methods of observation developed by ethology. Gardner's chapter specifically tackles this problem.

Living persistently in the agonic mode of inhibited arousal predisposes many to mental disorder, and some to tyrannical rule as will be understood from Russell Gardner's chapter. Liberated in the hedonic mode the intelligence blossoms and eventually leads to the expansion of the personality under the guidance of the systems forming faculty which promotes health, happiness and competence in life. Our unique formulating faculty thus gives hope for eventual and complete self-determination. How can we reconcile this creative aspect of our mental powers with the nature that we have inherited from our evolutionary past, and why is such a reconciliation essential for our future?

Application of the formulating faculty to productive processes has given the Western democracies a remarkably generous lifestyle during the last century. The emergent industrial civilisation has given more and more individuals a freedom of association of speech and of action that had no parallel in the social life of any previous community. In the modern, impersonal societies, however, the society's business has become separated from the reciprocal personal relations, whereas these were once dovetailed together in the hunter–gatherer societies. Without an understanding of this division it could be that we and future generations will be engulfed in a resurgence of the hierarchical forms of the agonic mode, which lie just below the surface of our civilisation. The resurgence of authoritarian forms of religious fundamentalism, for example, or the aggressive self-defence of small, embattled states vie for supremacy with freer forms of association on the global stage. Religions which were the first forms of institutionalised authority (but supportive of the individual) take many forms, and often twist the political machine out of the hands of forward-looking radicals.

But it appears that the progressives are almost totally ignorant of the social systems of personal interdependence which are the foundations of the freedoms that they enjoy. Liberal-minded radicals in much of the Western World stand amazed and aghast at the widespread suppression of human rights and of access to information, which, in so-called democracies, amounts to an insidious renewal of authoritarian government. What is more serious is that intellectuals (with the notable exception of the Columbus Centre) have not produced any new political methods to prevent the spread of authoritarian practices, and they have few insights as to how this might be done.

The failure of such progressives to develop a scientific philosophy that incorporates any understanding of how to promote reciprocity between peoples has left the stage wide open for a rampant individualism that runs away with the wealth of society, without an obligation towards the needs of society. If we are to re-establish a philosophy that enables us all to enjoy the beauty and wealth of this world but without making excessive demands upon it, we must rediscover the sources of contentment that lie within us. We must recognise the nature of our creativity that will enable us to achieve this.

At present we are without this understanding because knowledge is at a dangerous hiatus. The physical sciences have unlocked many of the principles governing the nature of physical entities but biology has done little to illuminate the nature of our humanity. Science is knowledge of the universe as an evolving system and so we shall have to let ourselves be seen as part of this system before we may wisely influence the course of events.

It is often supposed that humankind have been humbled—first by astronomy, which removed the Earth from the centre of the Universe; and then by Darwinism, which placed us in nature amongst the Apes—but in truth this apparent reduction has done nothing to enlighten us about the stuff our minds are made of, or made us aware of our ignorance on this matter. Worst of all it has failed to teach us how to act in our self-ascribed role of the arbiter of life on earth. Philosophers still discuss airily the choice between democracy and autocracy, or between freedom and dictatorship, without even a side-long glance at how this came to be the choice that we have been presented with down the ages. What is needed to fill this gap in our knowledge is a fully developed scientific understanding of the mind in society, which must start, as ethology does, by revealing the structure of the mind in its social setting, and hence uncovering the mental templates that operate in our social life.

HUMAN IMPLICATIONS

In their conscious planning for everyday contingencies, groups of people enter into what is for our purposes, too loosely referred to as co-operative behaviour. Early in this introduction I pointed out that the cussedness of human beings, who often fall foul of themselves in pursuit of a worthwhile endeavour, is largely due to inbuilt parts of our nature, which we share with other animals, especially the Primates. Success in conscious collaboration always means a differentiation of roles so that the various required activities are done by different group members in ways which interlock. Joint consultation can set up a plan, which must then be broken down into several technically distinct tasks. Frequently this means putting someone in charge of the overall plan in order to co-ordinate the activities and intiatives of the others who act, under this direction, their accepted roles. In essence this sets up an hierarchical, ranked form of complementary social relations, structured by the nature of the task, and this structure shares its feature of ranking and even its hierarchical form with the type of social relations set up by agonic systems. Hence whenever attention to a collaborative task wanes, there is always the danger that the task differentiation of roles will be reinterpreted as control by an agonic-type, authoritarian command.

In industry, for example, changing technologies or diversification cause a waning of tasks. When this happens, greater attention should be paid to the promotion of social activities—to sociality for its own sake—as a means of making individuals consciously aware of their membership of the workforce as a social entity. In this way, hedonic conditions can be maintained, not as before through good management and consultation, but by invoking the natural capacity of a well-organised group to maintain its cohesion on the basis of mutual friendship. The group is thus made aware of itself as a complete entity, and reassured by these fundamental constituency mechanisms before being faced with the need for reconstitution.

What I am implying is that with these two identical systems, present in both monkey and human, it is possible that both we and they have an unconscious and a conscious state of awareness, existing in a para-relation to each other. By examining various insights along the way, provided by de Waal, Price, Itani, and Power, we have arrived at the same conclusion argued by Pitcairn—that as well as reacting to each other, individuals in a group of primates can be aware of the role behaviours adopted by other members of the group and through this, group awareness can be established.

Russell Gardner's propensity states (psalics; Chapter 8), which programme humans to accept emotionally various social roles, evidently instigate the appropriate social signals that will be perceived or reacted to accordingly. For example, mechanisms exist that enable subordinate individuals to adopt a "nurturant recipient" role, often as a ploy to remove them from participation in the adult scene. Gardner is saying that we possess neuronally organised packages, which are energised by being aroused and which place us in socially expectant states of mind. These are normal to our development and programme us to respond to others of the social group that our age and sex prescribe for us.

What then, as deduced from Gardner's work, is the progress of events to which normal development from infancy to adulthood predisposes us? It consists of a series of stages, each one demanding of us an appropriate focus of attention, activity, and emotional endowment (see also Whatmore and Kholi, 1974) The phases can be designated as follows, based on what Russell Gardner writes about in his chapter. Mental development takes the infant from the nurturant–recipient phase to a place in the adult rank structure, and then to a sexual phase—if female, then potentially to a nurturant state of motherhood. To complete the picture I think we need to introduce the notion of a pre-adolescent "chumship" stage (see Heard & Lake, 1986; Pearce & Newton, 1969) in which ranking propensities as well as co-operative faculties are being laid down. Depending on the experience of this phase, which ends in adolescence, the individual may reach adulthood fully integrated in the hedonic mode or arrested by some agonic arousal.

We must now return to the other aspect of the two modes—the social structure. The nature of hunter–gatherer societies makes it clear that the equality principle with fluid social relationships and role integration promotes skill application in teamwork. This state of affairs prevailed for much of the existence of *Homo sapiens* on earth, and elements of this type of relationship can be seen in the societies of other great apes—the chimpanzee, gorilla, and orang-utan. So, in a very real sense, it appears that humanity has passed through and left behind a "Golden Age", and that much of what we perceive as "alternative culture" is an attempt to recreate the forms of association that the earlier societies enjoyed.

Nevertheless, the future of existing societies will depend not so much upon attempts to recreate alternative older forms of living as on the smoother running and stabilising of large, impersonal societies, made possible by a stable world population, and on our systems-forming ability to integrate our nature into the variety of the groups that comprise the larger societies.

Organisations like Friends of the Earth, Green Peace, and the various Green Parties act as pressure groups, motivated by an awareness of these types of relationship. However, in their attempts to change the immensely complex and powerfully engrained forms of the impersonal society, they are likely to fall prey to the same hierarchical pitfalls that all other societies have experienced, unless they are aware of the social forces, both agonic and hedonic, working within themselves. The split in 1987 within the Social Democratic Party (S.D.P.) of Great Britain involved those wishing to merge with the Liberals to form a new party and those wishing to stay with the one-time S.D.P. leader, Dr. David Owen. This is of great interest because, apart from the political issues, this split is likely to express an unconscious desire of part of the membership to gain personal reassurance either from joining an enlarged party or alternatively from a strong leadership. Two messages which would provide personal reassurance combined with a policy based on reality awareness might be a progressive solution. This will only happen when the leaders can reassure and affiliate party members by fostering community consciousness in its own right, while at the same time, and as a separate activity, educating the membership to formulate and carry through a flexible policy.

We come back to what was identified by Bion (see the beginning of the Introduction) as the two ways in which his group functioned. Either it was as a work group in the hedonic mode able to handle the material it was concerned with as a task, which also constituted its rationale for existence, or it "regressed", as he put it, to a rank dependent agonic state. In this the members see the leader as an all powerful figure onto whom they can project their concerns for their safety, and who is expected to provide solutions and shoulder all responsibility.

Hence, the task for an opposition with any chance of success is to solve these internal structural features so that instead of aiming for an ideal they can become able to make a feasible move forward in a stepwise process leading the nation away from the present competitive (ranked) individualism toward a more collective responsibility.

In important respects this leadership identity issue is one for all politicians who put themselves forward as people capable of achieving certain aims and who themselves *ipso facto* place themselves in leadership roles, and who gets votes not only by what they advocate but also by allowing the voter to identify with them.

Leaders themselves also have a problem. They tend to exceed their control functions when the primitive control tendency of the agonic mode becomes triggered under pressure of events and forces them further than prudence would advise. This is very common in heredity bosses and ambitious managers who lack education in the devolution of responsibility.

So, in general, there is a need to identify the primordial, inbuilt forms buried beneath the conscious efforts of organisations in pursuit of their reforming goals, as and when they occur. Barner-Barry (Chapter 11) and Montagner (Chapter 10) help us to see the politically relevant forms which these primordia take on in children's playgroups and in the classroom. There it is clear how constituency behaviour helps to bind followers into the group. Specifically, I will repeat that such behaviour is the various forms of reassurance extended from higher- to lower-ranking, potentially agonistic, individuals, who are thereby prevented from escaping from the group. Constituency behaviours are not taught to children, they appear spontaneously, and they are an extension of just those features of the hedonic mode that Power, Itani and de Waal have shown us lie at the base of what we have hitherto regarded as specifically human behaviour. So it appears that features long regarded as part of the civilising influence of western culture exist in that culture only through its encouragement of those particular aspects of our inbuilt social potential that can assist us, while at the same time deactivating agonic tendencies.

Significantly, appeasement behaviour in the electioneering stance of successful presidents of the United States serves a similar constituency function, as revealed by Roger Masters (Chapter 12), and this should be fully appreciated by forward-looking political parties if they are to succeed. Co-operative behaviour also emerges from the hedonic mode in the "leader" children observed in Montagner's study. This shows that hedonic types, who lead or occupy leader roles, are reacted to as dominants by other, predominantly flight-motivated, and, in part, aggressive children. Until these agonistic types are reassured by the leader's behaviour they cannot participate in games and tasks. Here are demonstrated the seeds of co-operation in later life, without which, as Bion also observed, "rebelliousness, submissiveness or withdrawal" may disrupt training groups, committees, trade unions and political parties, etc.

We shall see that Scott Lewis (Chapter 9) tells a remarkable story, and I have already outlined that, through an understanding awakened by knowledge of monkey hierarchies, he was able to explain how a drug-addicted group could live a separate, inmate existence while embedded in the fixed administrative hierarchy of a hospital institution. Relevant here is that he goes on to explain how he managed to act as a separate focus of their attention and so was able to undermine the credibility of the "maintainers" of the inmate culture that made life as comfortable as possible for the inmates. By this means he also was able to initiate therapy, and so integrate them back into the larger society.

The chimp jumping up and down waving the branches of a tree attracts the attention of those around him or, in more assertive mood, develops a charging display, in which he rushes through the group hitting out right and left indiscriminately, drawing the attention of the whole group onto himself. The same attention-capturing features of display emerge in the election campaigns of the impersonal democratic, societies of the industrial and electronic age. Roger Masters (Chapter 12) explores the development of this feature in establishing political legitimacy and character presentation of U.S. presidential candidates. If this centric structure of attention emerges in presidential election campaigns, it would also appear likely that, in the day-to-day conduct of government, primordial, face-to-face, hierarchical groups have been forming in the cabinets and decision-making bodies of, say, the White House, Whitehall, and Peking to mention but three contemporary, illustrative examples. In the White House at least, on two recent occasions, policy decisions seem to have been made more from criteria of social role-playing within the group and the requirements of image presentation to the public rather than from factual considerations—this would appear to be so if the books by John Dean (1976) and David Stockman (1981) are to be believed. Perhaps by now it will be easier for readers of this book to appreciate how political rulers may become victims of the group system they thought they controlled.

The origins of the way so many of us behave lie deep within; nevertheless the goal of improving collaboration and eliminating much discord is foreseeable provided only that we take into account the dual nature of the inbuilt tendencies that we have acquired as we grow up. These will tend either to assist us in this endeavour or frustrate us. They are operative now—in the present. One way we can be helped to identify and hence eliminate the frustrating elements, and also to take hold of our under-exploited assets, is to get a friend to help us to "see ourselves as others see us".

Our propensity to operate in one or other of the two modes is clearly demonstrated in the way organisations are constructed in large impersonal societies (see Wedgwood-Oppenheim, Chapter 14). Formal organisations either place our social relations into controlling hierarchical structures or create conditions that liberate the individuals so that their attention can be deployed for the requirements of a task. Note that the task-orientated culture appears out of the hedonic mode, as do the characteristics of the "leader children" of Montagner's study, who direct their energies to the invention and pursuit of games and other tasks. This link goes some way towards suggesting a solution of the problems encountered by Bion in directing a group towards the successful completion of a task.

In an age that is likely to see the destruction by man of as many species as died out with the dinosaurs, through our inability to live in an ecologically sound way, the aim and subject of this book is as realistic an endeavour as any that can be envisaged—lest we become the next dominant species to become extinct, and this time, by our own hand.

Aurelio Peccei, the founder of the Club of Rome, wrote in his book *The Human Quality* (1977) that if we are to survive we need to cease taking our nature for granted and improve it. This book is a window into a little-known area of knowledge that holds out hope that this may indeed be a possibility.

REFERENCE NOTES

1. von Holst, D., Kevern, E. B., & McGuire, M. (1986). *Physiology of leadership and subordination*. Report to the 11th International Primate Congress, Göttingen, July 1986.
2. Dawkins, R. (1987, March 1st). *Thinking aloud*. BBC 2 Television, London.

REFERENCES

Bateson, G. (1974). Observations of a Cetacean community. In J. MacIntyre (Ed.), *Mind in the waters*. New York: John Scrivenor.

Bion, W. R. (1959). *Experience in groups*. New York: Basic Books.

Bradford, L. P., Gibb, J. R., & Benne, K. D. (1964). *T-Group theory and laboratory method: Innovation in re-education*. New York: John Wiley & Sons Ltd.

Chance, M. R. A. (1957). The role of convulsions in behaviour. *Behavioural Science, 2*, 53–71.

Chance, M. R. A. (1963). A biological perspective on convulsions. *Colloques Internationale Centre de National Recherche Scientifique 12*. Paris: Seuil.

Chance, M. R. A (1980). An ethological assessment of emotion. In R. Plutnick., H. Kellerman (Eds), *Emotion: (vol. 1) Theories of emotion* (Pp. 81–111). New York: Academic Press Inc.

Chance, M. R. A. (1984). Biological systems synthesis of mentality and the nature of the two modes of mental operation: Hedonic and agonic. *Man–environment Systems, 14*, 143–157.

Chance, M. R. A. & Jolly, C. (1970). *Social groups of monkeys, apes and men*. New York, London: Jonathan Cape/E. P. Dutton.

Chance, M. R. A. (1976). Social attention: Society and mentality. In M. R. A. Chance & R. R. Larsen (Eds.), *The social structure of attention*. New York: John Wiley & Sons Ltd.

Chance, M. R. A., Emory, G., & Payne, R. (1977). Status referents in long-tailed macaques (*Macaca fascicularis*): Precursors and effects of a female rebellion. *Primates*, *18*, 611–632.
Crook, J. H. (1980). *The evolution of consciousness*. Oxford: Clarendon Press.
Dean, J. (1976). *Blind ambition; The White House years*. New York: Simon & Schuster.
Freud, S. (1959). *Group psychology and the analysis of the ego*. London: Hogarth Press.
Gilbert, P. (1984). *Depression; from psychology to brain state* London: Lawrence Erlbaum Associates Ltd.
Hall, A. D. & Fagen, R. E. (1957). Definition of systems. In L. von Bertalangly & A. A. Rapport (Eds.), *A general systems year book I*, 18–28 (Society for the Advancement of General Systems Theory).
Heard, D. H. & Lake, B. (1986). The attachment dynamics in adult life. *British Journal of Psychiatry*, *149*, 430–438.
Hold, B. C. L. (1976). Attention and rank specific behaviour in preschool children. In M. R. A. Chance & Larsen, R. R. (Eds.), *The social structure of attention*. New York: John Wiley & Sons Ltd.
Huxley, J. (1944). *Evolution: The modern synthesis*. London: George Allen & Unwin.
Jantsch, E. (1987). *The self-organizing universe*. London: Pergamon Press.
Kaplan, J. R. (1978). Fight interference and altruism in Rhesus monkeys *American Journal of Physical Anthropology*, *49*, 241–250.
Kemper, T. D. (1978). *A social international theory of emotions*. New York: John Wiley & Sons Ltd.
Kummer, H. (1987). In W. A. Mason & S. P. Mendoza (Eds.), *Primate Social Conflict*. New York: Alan R. Liss Inc.
MacLean, P. D. (1982). On the origin and progressive evolution of the triune brain. In E. Armstrong & D. Falk (Eds.), *Primate brain evolution*: Methods and concepts. London: Plenum Press.
McGuire, M. T. & Raleigh, M. J. (1985). Serotonin-behaviour interactions in Vervet monkeys. *Psychopharmacology Bulletin*, *21*, 458–463.
Mason, W. A. (1965). Sociability and social organisation in monkeys and apes. In L. Berkowitz (Ed.), *Advances in experimental social psychology, vol. 1*. (Pp. 278–302). London: Academic Press Inc.
Mittelstaedt, H. (1970). Reafferenzprinzip—Apolgie und Kritik. In W. D. Keidel & K. H. Plattig (Eds.), *Vorträge der Erlanger Physiologentagung*. (Pp. 162–171). Berlin: Springer Verlag.
Montagner, H., Henry, J. C., Lombardot, M., Restoin, A., Bolzoni, D., Darrand, M., Hubert, Y., & Moyse, A. (1970). Behavioural profile and corticosteroid excretion rhythms in young children, 1: Non-verbal communication and setting up of behavioural profiles in children from 1 to 6 years. In V. Reynolds & N. C. Blurton-Jones (Eds.), *Human behaviour and adaptation*. (Pp. 207–228). London: Taylor & Francis.
Morris, D. (1970). *Reproductive behaviour*. London: Jonathan Cape.
Omark, D. R., Stayer, F. F. & Freedman, D. D. (Eds.), (1980). *Dominance relations. An ethological view of human conflict and social interaction*. New York: Garland STPM Press.
Pearce, J. & Newton, S. (1969). *The conditions of human growth*. New York: Citadel Press.
Peccei, A. (1977). *The human quality*. New York: Pergamon Press.
Power, M. (1986). The foraging adaptation of chimpanzees and the recent behaviour of the provisioned apes in Gombe and Mahale National Parks, Tanzania. *Human Evolution, 1*, 251–266.
Stockman, D. A. (1981). *The triumph of politics: The crisis in American Government and how it effects the world*. New York: Harper & Row.
Vaillant, G. (1977). *Adaptation to life*. Boston: Little, Brown & Co.
Whatmore, G. & Kholi, D. R. (1974). *The physiopathology and treatment of functional disorders*. New York: Grune & Stratton.
Whyte, L. L. (1965). *Internal factors in evolution*. London: Tavistock Publications.
Woodburn, J. (1982). Egalitarian societies. *Man*, *17*, 431–451.

1 A Systems Synthesis of Mentality

Michael R. A. Chance
Social Systems Institute, Birmingham, U.K.

The idea that our intelligence can be regarded as an "onboard computer" (see Introduction) is gaining currency. By this analogy it may, therefore, be seen as just another organ but one possessing novel and broad adaptive significance. In essence it is then no more than another multi-purpose limb; one that may develop into forms of external intelligence.

This chapter sets out to show how erroneous this concept is because it neglects the "systems-based" properties of the organism of which intelligence is just a part. That organism, a human being, incorporates this new faculty, which therefore enters into a recursive (reciprocal) relationship with it. As this faculty amounts to a systems-forming organ, which I described in the Introduction, its power to segregate the components of the organism with which it interacts enables it to inaugurate a new and qualitatively different phase of evolution in the organism as a whole—that of the transformation of the personality itself. It follows that human nature need not be so bedevilled by the dichotomy of the two modes, agonic and hedonic, which are outlined here and considered by several of the contributors to this book.

Levine (Note 1) has drawn our attention to the operation of deep-seated, pervasive, intransigent, irrational forces in our mentality, which are not confined to the pathological but distort all aspects of our lives. After the publication of *Adaptation to Life* (Vaillant, 1977), it is clearly no longer possible to understand psychopathology without a clear picture of the healthy mind. This should not, however, be merely to establish the norm for mental health, as Coan (1974) attempted to do, but also to tell us how our expanding awareness will effect the evolution of mentality into the future. Gray (1981a; 1981b) has, in my view, pioneered how this can be done, pointing the way with two innovative heuristic ideas, which direct us toward an understanding of systems thought. These are the concept of Systems Precursors and the assertion that we possess a Systems-forming Organ in our brain, as one of a number of separate but interacting brain systems. I shall first discuss the location of the systems-forming organ in our mental structure before suggesting how systems precursors originate.

In my paper "Biological Systems Synthesis of Mentality" (Chance, 1984) I describe how the operation of intelligence and the emergence of a systems-forming capability in the recent evolutionary history of man have taken place within a framework of two essentially separate, but interacting, modalities of mental function. The *agonic* is self-defensive; it inhibits the operation of intelligence by a form of inhibited arousal, and also confines attention to social preoccupations. The *hedonic* on the other hand, by lowering arousal, deactivates self-security systems and so liberates attention. This permits the operation of intelligence and the emergence of the systems-forming capability.

This agonic/hedonic bimodality has independently been discovered through differences between the behaviour of children in playschool groups by ethological psychologists Montagner et al. (1970) and Hold (1976), and by the sociologist Kemper (1978). The psychiatrists Pearce and Newton (1969) also identified and described this bimodality in a review of their work with mental patients. They have come to see the self-system, as they call the adult personality in Western society, as divided between the "integral personality" (hedonic mode), capable of coping with and incorporating new experiences, and the personality that is bound into the "security apparatus". This binding is through the agonic component of the self-system, which is bent upon protecting the person from becoming involved in what experience has taught will bring it into irresolvable conflicts with custom and authority. The important feature of their understanding is that the two modes are two different systems, either of which may come to control the operation of the personality as a whole in different individuals (beautifully illustrated by Montagner's children), or may alternate as the operative system at different times in life (with movement into and out of mental disturbance), or may in yet other types of personality be more subtly integrated.

The nub of the argument for understanding the origins of mental illness lies in the statement made by Vaillant (1977) that Hans Selye was wrong: "It is not stress which harms us, but it is the effective response to stress that permits us to live." (To be convinced that Vaillant is right one only has to remember that Sky Divers deliberately put themselves through great stress in order to get an experience and feelings they otherwise would not have). The "effective response" to a stressful experience consists of the operation of intelligence and the re-synthesis of a person's way of life on the basis of reality awareness, so enabling them to remain healthy and enjoy life. This happens in the hedonic mode, but if the person is in the agonic state, self-defensive strategies come into play that always circumscribe their way of life, and *in extremis* the resultant isolation can lead to mental disorder. So, clearly the type of self-system which develops after any combination of bereavement, loss of status or of social support will depend on which mode the person is in. In this context the two modes, agonic and hedonic, act as preconditions for the synthesis of the personality—but we can be more specific.

Let us start with the definition of a system given by Hall and Fagen (1956): "A system consists of objects with properties which cohere." This is so self-evident that we usually overlook the implications to which Gray's concept of a "systems precursor" draws our attention, namely, that there is an emergent quality in the organic formation of a system. This is given in the second part of Hall and Fagen's definition: "The relationship between the parts, which are brought together by the coherence, not only ties the parts of the system together, but also creates thereby a structure from which arise properties *unique* to the system."; (for example, the properties of water arising from the combination of the gases hydrogen and oxygen).

Biological structures are often the result of a balance between processes of construction and disaggregation or segregation (for example, the normal deposition and removal of calcium phosphate in bone). Essentially the same processes go on at nerve endings, making and breaking synapses in the more labile parts of the central nervous system (C.N.S.). This leads to systems formation on the one hand and to segregation on the other, when the tendency to fall apart predominates. This well-known systematisation behaviour of neurones is supported by migrations of glial cells, the amoeboid neurone-support cells, into active areas of the C.N.S. that are powering any structures created by persistent mental activity (Rosenweig, Bennett & Diamond, 1972). Intelligence brings about new structures in the form of new hypotheses, skills, and personality growth. In the first two of these, at least, it does so by recombining existing concepts or motor acts. This is combinatorial competence.

My own studies have been on the nature and evolution of social structures in particular, because these have played a crucial role in the evolution of the higher primates. Such studies are important for understanding their enhanced mental powers of combinatorial competence, which are so evident in the problem-solving ability of the chimpanzees, as first studied by Kohler (1927), and now a recently revived topic. Fishes, birds, and reptiles, if threatened by a predator, all flee back to a refuge for safety, but the higher primates (notably baboons and macacques, and the apes, gorillas, and chimpanzees) clump together and gain safety from closeness to the dominant males of the group. Safety is thus gained, not so much in numbers, but in remaining a member of a group and maintaining group coherence.

The most important discovery I made in the 1950s was that this coherence came about by what Anna Harper (personal communication, 1985) has accurately described as "reverted escape". Whenever such animals are threatened by a higher-ranking individual they may move away temporarily only to turn back toward that individual; especially toward the dominant male, who periodically may reawaken this response by threats and so actively maintain group coherence. Reverted escape is a recursive movement of unique significance because it creates the precondition for primate social life, in which the state of remaining permanently together in a group has manifold consequences.

Careful comparisons of the social behaviour of rats and mice have shown that the essential property that enables rats to form rank orders (which mice do not) is their possession of submission postures, which enable them to remain next to an antagonist. It appears that the fundamental approach motivations (i.e., positive socialisations) remain the same, probably for all mammals, namely *aggression*, *mating*, and *social investigation*.

So, reverted escape creates a continuous society for monkeys and apes by, paradoxically, providing a fourth approach motive out of escape and, in so doing, reactivating the other three approach drives. Their society then becomes one in which the individuals are continually confronted by signals and arousals that may demand a choice between one or other of their motivations; based on internal desire or external opportunity. Such constant problem resolution must have been powerful in generating a selection pressure for the growth of intelligence in our ancestors, and it continues to exercise our own minds.

It might seem that the next logical thing to do would have been to examine whether the social situation of the higher primates generates other recursive processes. But this has had to await three things. Two together were the development of much greater understanding of the primate social situation, and my reading Hofstadter (1980), *Gödel*, *Escher*, *Bach*, in which he points out the recursive processes inherent in music, language, design, and now in computer science. The third and final factor was meeting with William and Lucille Gray who at once gave me the insight that these recursive processes are at least one of the processes involved in the creation of new systems; that in fact they are, *by their very nature, systems precursors*. Let us see how this is so. First of all, recursive processes increase coherence by bringing together parts of the already established system that had not previously been juxtaposed, and thus interactions within the system are increased. Secondly, depending upon the structural features of the system, its complexity may be increased.

I will now spell out those recursive processes in the social structure that, in the hedonic mode, lead to the growth of combinatorial competence. First, however, let us recall that reverted escape is a response to threat and, as such, binds the individual into a group and at the same time binds their attention to more dominant individuals, especially the central dominant male. Thus, the individual's attention is bonded as part of the centric attention of all individuals and so is not free to be deployed in exploration and the growth of intelligence. Only in the hedonic mode is this deployment possible. So the recursive processes I shall now mention are all manifestations of the hedonic mode of mental operation.

By reversion of escape, a complex social system becomes established within which the individual animal is forced to operate a system of choices appropriate to circumstance. This results in the evolution of a complex system of problem resolution. A further consequence is that the individual is faced perpetually with the need to adjust to other members of the group for long periods of time. What evidence there is suggests that half the time of an individual is devoted to socialising, forming alliances and avoiding conflict with others, apart from the dominant male (Kaplan, 1978). This illustrates what I think is a universal property, namely, that recursive processes bring the individual into contact with many more parts of the system and so are the crucial elements initiating precursors of further systems development.

If we compare a typical encounter of the baboon or macacque with a similar encounter between rats we see that the primates' encounters are much longer. The length of the encounter depends on the success of the strategies of the participants in resolving a particular dispute. In the rat, once the participants are familiar with each other, resolution is immediate. In old parlance, the rat habituates and the macacque does not. This concept is, however, inadequate because it neglects the formative consequences of the encounter.

The work of Bandler and Grinder (1975) on the relationship between person's language and the way interpersonal therapy can bring about change is a good example of this way of thinking about social relations. They point out that each individual has a perceptual model that structures their thoughts, and that the way they use language to express those thoughts sometimes does not convey the essence of their model. So, the other person gets a distorted view of what they are trying to say. This is because of the distortion of the communication process that originates from intervention by the security apparatus of their personality so as to prevent certain meanings from being transmitted. By focusing on the correspondence between the model of thought and the form of the communication they are showing their concern for the formative consequences of a person's speech.

A similar defect in comprehension is evident from the way the mother–infant relationship is regarded as a bond. Seen from the new standpoint of identifying recursive processes which go towards the establishment, in the adult, of a fully developed mind, this mother–infant bond is a systems precursor of the adult mind. For we now know that a very complex set of interactions begins between the newborn and its mother from the moment of birth. A mother who is free to act in private at the time of delivery immediately picks up the baby and, after a few minutes of looking at it face-to-face, touches its head, arms, and body with her hands. Later she will begin to whisper and coo to it. So, visual, tactile, and vocal channels of communication are set up, at first by the mother; later these processes, which are recursive because they go back and forth, can be initiated by either child or mother. The essentials of giving and receiving information and arousal inputs are thus established in all the sensory modalities. The ability to reciprocate is thus formed early in the child, provided the caretaker of later years also has this ability. Education continues the process in social interaction with other children and in the acquisition of language.

Grinder and Bandler (1976) have pointed out, however, that a feature of the language of the mentally disturbed provides evidence that they may be conceptualising and constructing their understanding in terms only of one sensory modality (visual, auditory, or kinaesthetic). Hence, in some incompleteness of the set of recursive processes between infant and mother may lie a source of mental incapacity and of eventual illness through aggravated misunderstanding.

The reciprocating process between mother and infant can later be seen between child and caretaker. This behavioural element occurs in the more sophisticated mental processes of the human primate as an activity we think of as mental rehearsal. The mind is able to repeat instances of experience to take us back to the source of threat, or pleasure, or embarrassment. This is a symbolic form of reverted escape that can be made to work in favour of a child's development by providing them with the opportunity to overcome the effects of isolation or the incomplete awareness of an experience.

Anna Harper (personal communication, 1985) tells me that as a child she had the habit of hiding at the top of the house if trouble occurred. A sharp word, a smack, and sometimes even heated exchanges between grown-up members of the family would send her hurrying upstairs—away from the source of threat. In her own words she continues: "My grandmother, who lived in our house, would always come to call me down—ask me what was the matter. She would then make me repeat what had occurred (even if she had been present at the incident). She would make me return mentally to the source of threat. She would make the right noises, either to comfort or admonish as I expressed my feelings of bitterness, anger, or even envy (I had a sister 18 months older than myself). Calling me back and comforting me gave me the opportunity to review my feelings about the episode."

I am sure there are still families where this sort of thing takes place, but sadly in our generation, with its speed and overactivity, not enough. (We all know only too well the need for a trusted friend in the early development of children.) My work with children has now ended, but from experience I have seen the desire to hide away acted out, and in almost all cases it has been coped with by restraint, or worse still, totally ignored. (No matriarchal figure here to teach return, albeit mentally, to the source of threat.)

Having conquered the fear in the mind of human beings, the physical return is so much easier. In mentally disturbed children—and I use the term disturbed advisedly for I realise there are many complicating factors in actual mental sickness—this concept of escape is an almost total living experience. I think these children are escape-motivated whatever the immediate environmental conditions—even the desire to communicate goes. (It then takes therapy to get them back.) In systemic family therapy courses, the evolutionary background to this behaviour is very useful to therapist and parents alike. I believe we are not just giving an ethological name to a behavioural activity, but have shown that returning to the source of threat in the familiar group situation is character building and socially important. (This, of course, cannot be practised where parents are physically cruel to children. There is a return under such conditions, but the outcome is devastating for the child.)

Psychiatrists and psychotherapists may say that they are talking people back all the time, that Freud started it many years ago. I agree, but if we use the concept of systems precursor we can see how overcoming the tendency to escape in the higher primates has enabled them, and us, to live in groups. We need each other: children must be led to recognise this, and to appreciate the benefits of recognising it.

Language leads to another recursive process when writing becomes established in the evolving civilisation. In judging the significance of this development in human history, emphasis has been placed on the way transfer of information from one generation to another was lifted from the oral to the written tradition, but from our point of view we can see that a new recursive process has been introduced, namely, the opportunity for writers to reflect upon what they have written. Every good tutor knows that as soon as students can be prevailed upon to write down their conclusions, and then are obliged to put their work away for a while before re-examining it for revision, they will suddenly see errors in their formulation which were unnoticed before and, in so doing, gain new insights into their powers of composition and expression. This is a form of reflection that can also bring insight into the nature of what was written about. This is a lesson in self-awareness that is encountered by serious writers, but has not yet been incorporated into our methods of education. It is indeed a very high level of self-reflective process. Let me emphasise the importance of revision by proposing that children should not be given marks for composition, but that they should be taught to reflect on what they have written from an early age, thereby also reducing the competitive element in education. As Anna Harper (personal communication, 1985) has pointed out to me, conversational correction has emotive as well as cognitive content, whereas essay revision, being without emotive content confines the attention to the adequacy of expression and by extension to the nature of the content. This is the final recursive process requisite for the emergence of scientific thought.

To conclude, let us see if a systems synthesis can be achieved out of what has been presented here. If the understanding of model making, as set out by Bandler and Grinder (1975), is placed in the context of the two modes, it is clear that thinking or scientific combinatorial competence comes about by a recursive process involving the projection of a model onto perceived reality to see if the patterns from reality fit those of the model. But this is only one consequence of the recursive process. Systematic encouragement of the development of such processes throughout an individual's life, with continuing recognition of their individual's particular circumstances within the social structure, can do much to promote their competence and progress towards greater stature of personality. Provided, of course, that the body politic, or local community, ensures that all its citizens or members participate in this growth process.

ACKNOWLEDGEMENTS

This chapter is expanded from a paper given at the Annual Professional Meeting of the American Association for Social Psychiatry, Boston, September, 1986, which was subsequently published in the *American Journal of Social Psychiatry* (1986), 7, 199–203. It is reproduced by kind permission of the publishers, Brunner/Mazel of New York.

REFERENCE NOTES

1. Levine, H. O. (1985, September). *Ate's meadow: Why can't we be reasonable about social issues*. Paper presented at the Annual Professional Meeting, American Association for Social Psychiatry, Boston, Mass.

REFERENCES

Bandler, R. & Grinder, J. (1975). *The structure of magic: A book about language and therapy*. Palo Alto, Calif.: Science & Behavior.

Chance, M. R. A. (1984). Biological systems synthesis of mentality and the nature of the two modes of mental operation: Hedonic and agonic. *Man–Environment Systems, 14*(2) 143–157.

Coan, R. W. (1974). *The optimal personality: An empirical and theoretical analysis*. London: Routledge & Kegan Paul.

Gray, W. (1981a). System forming aspects of general system theory, group forming and group functioning. In J. E. Durkin (Ed.), *Living groups: Group psychotherapy and general system theory*. New York: Brunner/Mazel.

Gray, W. (1981b). The evolution of emotional-cognitive and system precursor theory. In J. E. Durkin (Ed.), *Living groups: Group psychotherapy and general system theory*. New York: Brunner/Mazel.

Grinder, J. & Bandler, R. (1976). *The structure of magic II: Communication and change*. Palo Alto, Calif.: Science & Behavior.

Hall, A. D. & Fagen, R. E. (1956). Definition of system. In L. von Bertalanffy & A. Rapoport (Eds.), *General systems: Yearbook of the Society for the Advancement of General Systems Theory I*. (Pp. 18–28). University of Louisville Press.

Hofstadter, D. R. (1980). *Gödel, Escher, Bach: An eternal golden braid*. New York: Vintage.

Hold, B. C. L. (1976). Attention structure and rank specific behaviour in pre-school children. In M. R. A. Chance & R. R. Larsen (Eds.), *The social structure of attention*. New York: John Wiley & Sons Ltd.

Kaplan, J. R. (1978). Fight interference and altruism in rhesus monkeys. *American Journal of Physical Anthropology, 49*, 241–250.

Kemper, T. D. (1978). *A social interactional theory of emotions*. New York: John Wiley & Sons Ltd.

Kohler, W. (1927). *Mentality of apes*. London: Methuen.

Montagner, H., Henry, J., Lombardst, M., Restoin, A., Benedini, M., Godard, F., Boillot, F., Pretot, M., Bolzoni, D., Burnod, J., & Nicolas, R. (1970). Behavioural profile and corticosteroid excretion rhythms in young children. Part I. Non-verbal communication and setting up of behavioural profiles in children from 1–6 years. In V. Reynolds & N. G. Burton-Jones (Eds), *Human behaviour and adaptation*. London: Taylor & Francis.

Pearce, J. & Newton, S. (1969). *The conditions of human growth*. New York: Citadel.

Rosensweig, M. R., Bennett, E. C., & Diamond, M. C. (1972). Brain changes in response to experience. *Scientific American, 226*, 22–29.

Vaillant, G. (1977). *Adaptation to life*. Boston, Mass.: Little, Brown & Co.

2 Social Geometry and Cohesion in Three Primate Species

Gary R. Emory[1]
Institute of Ecology, University of California (Davis), U.S.A.

The structure and function of the geometrical organisation of animal groups has long been one focus for ethologists, ecologists, and other evolutionary biologists interested in the adaptive significance of animal sociality (Brown, 1975; Darwin, 1874; Deag, 1980; Dimond & Lazarus, 1974; Edmunds, 1974; Hamilton, 1971; Treisman, 1975; Williams, 1966; Wilson, 1975; Wittenberger, 1981). Studies of social geometry have involved a broad range of vertebrate taxa including fish, birds, ungulates, and especially primates[2]. Such work has led to a proliferation of inferred or proposed, plausible functions for specific forms of geometrical organisation including increased effectiveness of foraging and locomotion, as well as enhanced detection of and defence against predators.

These studies have dealt predominantly with certain vectoral and spatial aspects of group geometry in relation to ecological selection pressures. However, an equally important area of investigation has been relatively neglected. This concerns the vectoral and spatial aspects of group geometry in relation to social selection pressures, and particularly pressures

---

[1]Address for correspondence: 6050/15 Henderson Drive, La Mesa, CA 92041, U.S.A.

[2]References to studies addressing various aspects of social geometry in vertebrates include: fish (Brock & Riffenburgh, 1960; Williams, 1964); birds (Bertram, 1980; Goss-Custard, 1970; Horn, 1968; Jennings & Evans, 1980; Murton, 1968; Tinbergen, 1951; Vine, 1971); ungulates (Altmann, 1956; Darling, 1937; Galton, 1871; Kruuk, 1972); and primates (Altmann, 1974; 1979; Altmann & Altmann, 1970; Busse, Note 2; Crook, 1966; 1972; Deag, 1974; 1978; DeVore & Washburn, 1963; Hall & DeVore, 1965; Harding, 1977; Rhine, 1975; Rhine & Owens, 1972; Rhine & Westlund, 1981; Robinson, 1981; Rowell, 1966; 1972; 1979; Wilson, 1972).

deriving from social dynamics (in the sense of term used by Crook & Goss-Custard, 1972; Crook, Ellis, & Goss-Custard, 1976). The small amount of work that has been done concerns mainly primates (Anderson, Note 1; Kummer, 1967; 1968; Nagel, 1979). In this chapter, research relevant to the geometry of social groups will be discussed. The topics examined are social geometry as a branch of biological inquiry, including the empirical aspects of recently completed research, some of the implications of this work for theoretical ethology, and a possible role for dynamic social geometry as a mechanism for maintaining social cohesion.

PHENOMENA OF SOCIAL GEOMETRY

The literature of biological research relevant to social geometry is vast, as shown in my introductory text. However, the conceptual basis of this area of inquiry as a subdiscipline of social biology has not been thoroughly explored, only given limited consideration by a few authors (Altmann, 1979; Hamilton, 1971). This chapter will not attempt a definitive explication of the topic but a brief indication of the nature and range of phenomena of interest in social geometry is appropriate.

Social geometry in the broad sense can be considered to be the biological study of all vectoral (directional) and spatial (distributional) relations among individuals and social units in animal societies. As a biological study, it includes both structural and functional approaches, as well as consideration of the traditional issues of proximate and ultimate causation. It can be investigated at the level of the individual as well as at the higher levels of populations, groups, and subgroups. These multiple levels can interact in a hierarchical fashion, with processes of geometrical causation proceeding both up and down the hierarchy. Only the levels of individuals and groups will be further considered.

Table 2.1 gives a broad overview of the conceptual basis for the study of social geometry. Examples are provided for the various levels and topics of analysis.

SOCIAL GEOMETRY AND SOCIAL DYNAMICS

As group-living animals behave and move in three-dimensional space they necessarily and continuously alter the geometrical organisation of their social groups. Thus, even when any given individual remains absolutely static for a given period of time, its geometrical relationships with other group members may be highly dynamic due to their own behaviour and movements. The geometrical organisation of animal groups is thus a dynamic, not a static, phenomenon, so it may be heuristic to investigate it in relation to social dynamics.

TABLE 2.1
Conceptual Overview of Social Geometry

	Vectoral Relations	
	Orientational Factors	*Directional Factors*
Individual Level	Examples: head orientation body orientation	Examples: inter-individual interaction pathways directional location
Group Level	Examples: overall group head orientation overall group body orientation	Examples: inter-individual interaction network
	Spatial Relations	
	Metrical Factors	*Configurational Factors*
Individual Level	Examples: inter-individual distance	Examples: nearest neighbours
Group Level	Examples: overall group dispersion[a] local density[b]	Examples: overall group shape[c] overall group deployment[d]

[a]area, volume, minimum/maximum diameters. [b]relative concentration at centre and periphery. [c]circle, ellipse, line. [d]relative orderliness of positional arrangement of individuals or age/sex classes.

Such a research strategy was followed in the studies described in the following sections. A structural approach was adopted to investigate the relations between social geometry and social dynamics in a range of cercopithecine taxa.

SCOPE OF THE EMPIRICAL STUDIES

The purpose of these studies was to explore a series of relationships between social geometry and social dynamics. For social geometry, selected topics of both vectoral and spatial relations at the individual level of analysis were studied. Vectoral relations[3] comprised head orientation, body orientation, and directional location, spatial relations were inter-individual distances. Social geometry was related to social dynamics through analysis of status hierarchy and age/sex class categorisation.

Of these aspects of social geometry and social dynamics, not all were examined in each species. The cercopithecine monkey species that were studied include: mandrills *Mandrillus sphinx*, gelada baboons *Theropithecus gelada*, Guinea baboons *Papio papio*, and long-tailed macaques *Macaca fascicularis*.

REVIEW OF THE EMPIRICAL STUDIES

Three separate studies were conducted: These are presented in chronological order as research environment, study population, topics investigated, methods of data acquisition, specific relationships analysed, and results.

Captive Mandrills and Gelada Baboons

This study was conducted at the San Diego Zoological Gardens, California, U.S.A. during 1972 and 1973 (Emory, 1975; 1976a; 1976b; 1976c). Two identical rectangular enclosures measuring approximately 6.2m × 5.5m × 3.5m housed the separate groups. Each enclosure had a multilevel

[3]It is to be appreciated that vectoral (directional or orientational) motions of the anatomical components of the bodies of organisms (e.g., the eyes, head, trunk, limbs of vertebrates) are capable of individual independent action to an extent influenced by the general body plan, muscular relationships, and associated innervation pathways. In other words, the extent to which any two bodily parts can move independently of each other depends on overall morphology and physiology. Relevant research by Jung (summarised by Chance, 1975; 1976) and by von Cranach (1971) on humans suggests that, in our species at least, head and body orientational movements tend to be highly correlated. In these monkey studies, on the other hand, the degree of interdependence between these parameters was an empirical issue to be resolved by the data.

structure, providing a degree of environmental complexity for the animals as well as excellent visibility for the observer.

The study population for each species comprised one social group in which the individual identities were known. The Mandrill group was two males (adult, juvenile) and three females (adult, juvenile, infant); the gelada baboon group was two males (adult, infant) and three females (adult, two subadults). The status hierarchy of the mandrills was: adult male, adult female, juvenile male, juvenile female, infant male. That of the gelada baboons was: adult male, adult female, subadult female, subadult female, infant male. Because of the identical enclosures and group sizes, near identical group compositions, and similar age/sex class positioning in the status hierarchies, the study of these two species was comparative.

The aspects of social geometry investigated included head orientation, body orientation, directional location and inter-individual distance. The social dynamic was status hierarchy, as discussed later. Social geometry records were made from all individuals simultaneously, through instantaneous sampling of discrete states. Data, in the form of arrows representing each individual of each group, were recorded directly onto maps of the enclosures (drawn to scale for direct vectoral and spatial analysis). From these raw data the four aspects of social geometry were extracted; details of the extraction methods are shown in Fig. 2.1. The status hierarchies were multifactorially derived on the basis of the dyadic distribution of these behavioural criteria: supplantation and priority of access to limited food sources.

The specific relations between social geometry and social dynamics included: (1), head orientation of each individual in relation to the dominant male; (2), head orientation of each of two middle-ranking individuals in relation to all other individuals; (3), head orientation of the dominant male in relation to all other individuals; (4), body orientation of each individual in relation to the dominant male; (5) body orientation of each of two middle-ranking individuals in relation to all other individuals;

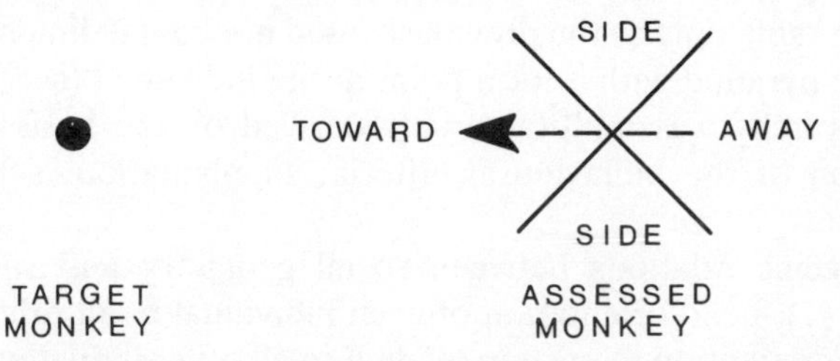

FIG. 2.1 Identification of directions used in comparing visual attention and body orientation for any two individuals. The arrow indicates in which quadrant the body or gaze of the individual, being assessed, is oriented.

(6), directional location of each individual in relation to the dominant male; (7), inter-individual distance from the dominant male.

The data showed the following patterns. Head orientation of each individual to the dominant male was less frequent in the mandrills than in the gelada baboons, with the same pattern prevailing for the head orientation of the dominant male to all other individuals. Body orientation of each individual to the dominant male was less frequent in the mandrills than in the gelada baboons. Directional location of each individual in relation to the dominant male was essentially random in both mandrills and gelada baboons. Inter-individual distance from the dominant male was greater in the mandrills than in the baboons.

Captive Long-tailed Macaques

This study was conducted at the Ethology Laboratory, Uffculme Clinic, Birmingham, U.K. during 1975 and 1976 (Emory & Harris, 1978; 1981a). Two rectangular enclosures measuring approximately 2.8m × 3.8m × 2.4m and 2.7m × 2.9m × 2.6m, connected by a tunnel, housed the group. Each enclosure had a multilevel structure.

The study population comprised one social group in which the individual identities were known: three males (adult, subadult, juvenile) and six females (adult, subadult, four juveniles). Only the five oldest individuals of the group served as a source of data. The status hierarchy of these individuals was: adult male, adult female, subadult male, subadult female, juvenile male.

The aspects of social geometry were head orientation and body orientation, and the social dynamics were status hierarchy and age/sex class of the individuals. Social geometry records were sampled from individuals in dyads. These data were collected by means of continuous sampling of discrete events over time (head orientation), and by instantaneous sampling of discrete states (body orientation). These sampling methods were derived from those used in the mandrill and gelada-baboon study, with a modification of the directional grid that incorporated the vertical dimension at the same time as the previously used horizontal dimension to form a four-sided pyramid with a focal point at the individual (see Fig. 2.1). The status hierarchy was multifactorially derived on the basis of the dyadic distribution of the behavioural criteria: supplantation, submission, and aggression.

The specific relations between social geometry and social dynamics included: (1), head orientation of each individual to all other individuals; (2), body orientation of each individual to all other individuals.

The data revealed the following patterns. Head orientation of each individual to all other individuals was distributed such that high-status individuals directed their head orientation toward others at relatively low frequency, whereas low-status individuals did so at relatively high frequency. Further, high-status individuals received head orientation toward them from other individuals at relatively high frequency but low-status individuals did so at relatively low frequency. The direction of body orientation of each individual to all other individuals had no clearly discernible pattern with respect to status. However, high-status individuals received body orientation toward them from others at relatively high frequency, and low-status ones at relatively low frequency. The orientation of female body presentations (presents) was such that males primarily received standard, hind-end presents, whereas females received mainly sideways presents.

Free-ranging Guinea Baboons

This study was conducted in an arid valley on the southern fringe of the Sahara Desert in the Islamic Republic of Mauritania, West Africa during 1980 and 1981 (Emory, Note 3). The valley floor was surrounded on three sides by the escarpment cliffs of the Assaba Mountains. The valley floor was predominantly sand, gravelly pans, and boulder-strewn flats, and the surrounding escarpments comprised vertical cliff faces and scree slopes. Vegetation was adapted primarily to the arid desert conditions, except for the plants in the wetter environment in and around the water courses that originated from the scree slopes. Visibility for the observer was generally good.

The study population comprised several social groups in which the individual identities were not known. Groups were classified by their composition, according to the presence of individuals in sex-specific demographic classes: adult, subadult, juvenile, infant. Different combinations of individual age/sex classes thus constituted different group-compositional types. As there were no identifiable individuals, status hierarchies were, of course, unknown.

Investigated were the social geometry of head orientation, body orientation and inter-individual distance, and the social dynamic of age/sex class of individuals. Social geometry records were sampled from individuals in dyads. These data were collected by continuous sampling of discrete events over time (head orientation), and by instantaneous sampling of discrete states (body orientation, inter-individual distance). These methods were essentially the same as in the study of long-tailed macaques.

The only data so far analysed from this study are the body orientations of female presents. The pattern, similar to that of the long-tailed macaques, was that males primarily received standard, hind-end presents, and females received mainly sideways presents.

IMPLICATIONS FOR THEORETICAL ETHOLOGY

The empirical studies reviewed in the previous section were largely stimulated by the theory of social attention structure in primates (Chance, 1967; 1975; 1976; Chance & Jolly, 1970; Emory, 1975; 1976a; 1976b; 1976c; Emory & Harris, 1981a; 1981b). Some implications of these empirical results for this theory will now be considered briefly.

At its most fundamental level, the theory[4] postulates a differential distribution of social attention with respect to relative status among individuals of primate groups. It is suggested that this differential distribution may involve a negative association between the giving of attention and status, and a positive association between the receiving of attention and status. In other words, low-ranking individuals should tend to monitor high-ranking individuals more so than vice versa.

The theory further postulates a continuum of modes of social structure ranging from centripetally to acentrically organised groups. The conceptual basis for this is that increasing degrees of centripetal group organisation are positively associated with increasing degrees of congruence in the proposed differential distribution of social attention, whereas increasing degrees of group acentricity are negatively associated with increasing congruence. Thus, the more that the overall social attention of a group is focused on one or more central individuals, the more that group can be expected to travel, forage, interact, and cope with predators as a single coherent social unit.

Independent evidence, not reviewed here (Emory, 1975; 1976b; Emory & Harris, 1981b), suggests that all four species studied had centripetal organisation to varying degrees, with this order of increasing centripetalism: mandrills, long-tailed macaques, Guinea baboons, gelada baboons. Hence, according to the theory, increasing degrees of congruence in the proposed pattern of differential distribution of social attention would be expected to follow this order of increasing centripetalism.

Do the results of these studies support this particular aspect of the theory? If head orientation indicates potential visual attention, then analysis of head orientation could address this question. The mandrill and gelada-baboon study examined this aspect of the theory. The results do indeed appear to support it in that the strongly centripetal gelada baboons directed a larger amount of social attention to the dominant male than the weakly centripetal mandrills.

The study of long-tailed macaques considered a slightly different aspect of the theory. It focused on analysis of the proposed negative association

[4]A theory of social attention structure is considered as distinct from a theory of environmental attention structure (Emory, 1976a, in preparation; Emory & Harris, 1981a). The environmental theory considers the distribution of attention in relation to biologically significant, environmental features, such as predators, competitors, food, water, shelter, and substrate.

between status and the giving of social attention, and the positive association between status and the receiving of social attention. Again, the head-orientation data appear to support this aspect of the theory.

In the Guinea baboons head-orientation data have yet to be analysed but these would not be directly relevant to this aspect of the theory because lack of identifiable individuals precludes this type of status analysis. Instead, the data should reveal the structure of social and environmental attention within groups of differing compositional types.

It is appropriate to consider all relevant studies on primate species in relation to the theory of social attention structure. The issue in question specifically concerns the *theory* as opposed to the *existence* of a social attention structure. Its existence implies only consistency or regularity of social attention in any form, whereas the theory implies a specific form of regularity of social attention, namely the relationship of social attention with status.

There have been many studies of primates in which elements of social geometry have been used to investigate the theory. Varying degrees of support for it have been found in the recorded behaviour of a wide range of captive South American, Asian and African species including: capuchin *Cebus apella* (Assumpcao & Deag, 1979); rhesus macaque *Macaca mulatta* (Haude, Graber, & Farres, 1976; Hinde & Detweiler, 1976; Mitchell, 1972; Virgo & Waterhouse, 1969; Waterhouse & Waterhouse, 1976); long-tailed macaque *Macaca fascicularis* (Emory & Harris, 1981a; Low, personal communication; Pitcairn, 1976; 1979); talapoin monkey *Miopithecus talapoin* (Dixon, Scruton & Herbert, 1975; Keverne, Leonard, Scruton, & Young, 1978; Scruton & Herbert, 1970; 1972); vervet monkey *Cercopithecus aethiops* (Assumpcao & Deag, 1979); hamadryas baboon *Papio papio* (Rijksen, 1981); mandrill *Mandrillus sphinx* (Emory, 1975; 1976a; 1976b; 1976c; gelada baboon *Theropithecus gelada* (Emory, 1975; 1976a; 1976b; 1976c); and chimpanzee *Pan troglodytes* (Reynolds & Luscombe, 1969; 1976).

Lack of support for the theory has been found in two South American and African species, including the squirrel monkey *Saimiri sciureus* and diana monkey *Cercopithecus diana* (Assumpcao & Deag, 1979).

In the totality of these studies, there is general support for the theory, although it must be appreciated that different aspects were examined in individual studies. Such broadly based support, given the wide variation in measures, methodology, environment, population characteristics and species, argues strongly for the robustness of the theory.

SOCIAL GEOMETRY AND SOCIAL COHESION

I will conclude this chapter with a consideration of the means by which dynamic use of appropriate patterns of social geometry might influence or act as a mechanism for the maintenance of social cohesion. On this complex topic, two biologically significant behavioural pathways will be considered.

One such pathway concerns the relationships among social geometry, social dynamics (specifically, intra-group aggression), and social cohesion. Consider that inter-individual aggression, if intense or prolonged, can act to disperse the members of a social group. Therefore, if such aggression can be reduced, the tendency for group dispersal may then be reduced. Certain components of dynamic and situationally appropriate social geometry can achieve this end by minimising inter-individual aggression, which curtails dispersal of the group, and so in turn minimises the disruption of social cohesion.

For illustrative purposes, consider the following scenario. Dynamic head orientation in social attention allows the location and behaviour of conspecifics to be monitored. By taking advantage of and acting upon the social information that may be gained through this, individuals can perform situationally appropriate and responsive vectoral and spatial equilibration in relation to aggressive group members. Such reactive behaviour can minimise the provocation of aggression.

An appropriately responsive vectoral equilibration would thus be dynamic head orientation. Prolonged direct head orientation to potentially aggressive individuals, which might be interpreted as a threat, could be avoided. A parallel example is dynamic body orientation, with avoidance of prolonged direct body orientation (which might also be interpreted threateningly) to potentially aggressive individuals. Appropriately responsive spatial equilibration would be the use of dynamic inter-individual distance: avoidance of being too close to or too far from potentially aggressive individuals.

The second behavioural pathway concerns the relationships among social geometry, social co-ordination (specifically intra-group response to predatory attack), and social cohesion. Consider that a co-ordinated group response to predatory attack has high value in ecological selection. Therefore, the maintenance of social cohesion and social co-ordination would also have high selective value. Dynamic and contextually suitable social geometry can serve this purpose; certain components of social geometry can maintain cohesion and co-ordination, allowing a co-ordinated group response to predatory attack.

Again, for illustrative purposes, consider this scenario. Dynamic head orientation in the form of environmental attention allows the location and behaviour of predators to be monitored. By taking advantage of and acting upon the information that may be gained through such orientation, individuals can perform contextually adaptive and responsive vectoral and spatial manoeuvres in relation to other individuals. Such reactive behaviour can enable a co-ordinated group response to the presence of potential predators.

An example of adaptively responsive vectoral manoeuvres once again involves dynamic head orientation. Individuals could influence other individuals to perform similar head orientation so that they might gain similar environmental information about the predator. Another example is of dynamic head orientation in the whole group; a co-ordinated pattern of outwardly radiating head orientation might allow vigilance in all directions. Similarly, with dynamic body orientation in the whole group, a co-ordinated, outward radiating pattern might allow a defensive front in all directions.

An adaptively responsive spatial manoeuvre is dynamic dispersion of the group. Co-ordinated interplay between intra-group and extra-group movements might allow effective alternation between cohesion-decreasing individual foraging or socialising manoeuvres and cohesion-increasing predator-response manoeuvres. Another example is dynamic overall deployment of the group. A co-ordinated pattern of deployment, assuming there are individual and age/sex class variations in inherent ability to defend against predators, might allow more effective protection of those individuals at high risk.

These are only a sample of the many and varied behavioural pathways by which dynamic and situationally adaptive social geometry may influence social cohesion. It is hoped that this chapter will stimulate further research into the geometry of animal sociality.

REFERENCES NOTES

1. Anderson, J. *The spatial structure of a yellow baboon troop*. Unpublished manuscript.
2. Busse, C. D. *Geometry of baboon troops: A predation model for measuring sociospatial patterns*. Unpublished manuscript.
3. Emory, G. R. *On the directional orientation of female presents in* Papio papio. Unpublished manuscript.

REFERENCES

Altmann, M. (1956). Patterns of herd behavior in free-ranging elk of Wyoming, *Cervus canadensus nelsoni*. *Zoologica*, *41*, 65–71.

Altmann, S. A. (1974). Baboons, space, time and energy. *American Zoologist*, *14*, 221–248.

Altmann, S. A. (1979). Baboon progressions: Order or chaos? A study of one-dimensional group geometry. *Animal Behaviour*, *27*, 46–80.

Altmann, S. A. & Altmann, J. (1970). *Baboon ecology: African field research*. Chicago: University of Chicago Press.

Assumpcao, C. T. de & Deag, J. M. (1979). Attention structure in monkeys: A search for a common trend. *Folia Primatologica*, *31*, 285–300.

Bertram, B. C. (1980). Vigilance and group size in ostriches. *Animal Behaviour*, *28*, 278–286.

Brock, V. E. & Riffenburgh, R. H. (1960). Fish schooling: A possible factor in reducing predation. *Journal du Conseil, Conseil Permanent International pour l'Exploration de la Mer*, *25*, 307–317.

Brown, J. L. (1975). *The evolution of behavior*. New York: Norton.

Chance, M. R. A. (1967). Attention structure as the basis of primate rank order. *Man*, *2*, 503–518.

Chance, M. R. A. (1975). Social cohesion and the structure of attention. In R. Fox (Ed.), *Biosocial anthropology* (Pp. 93–113). London: Malaby.

Chance, M. R. A. (1976). The organization of attention in groups. In M. von Cranach (Ed.). *Methods of inference from animal to human behavior* (Pp. 213–235). The Hague: Mouton/Aldine.

Chance, M. R. A. & Jolly, C. J. (1970). *Social groups of monkeys, apes and men*. New York: Dutton.

Crook, J. H. (1966). Gelada baboon herd structure and movement: A comparative report. *Symposia of the Zoological Society of London*, *18*, 237–258.

Crook, J. H. (1972). Sexual selection, dimorphism, and social organization in the primates. In B. G. Campbell (Ed.), *Sexual selection and the descent of man, 1871–1971* (Pp. 231–281). Chicago: Aldine.

Crook, J. H. & Goss-Custard, J. D. (1972). Social ethology. *Annual Review of Psychology*, *23*, 277–312.

Crook, J. H., Ellis, J. E. & Goss-Custard, J. D. (1976). Mammalian social systems: Structure and function. *Animal Behaviour*, *24*, 261–274.

Darling, F. F. (1937). *A herd of red deer: A study in animal behaviour*. London: Oxford University Press.

Darwin, C. (1874). *The descent of man and selection in relation to sex* (2nd ed). London: Murray.

Deag, J. M. (1974). *A study of the social behaviour and ecology of the wild Barbary macaque Macaca sylvanus*, L. PhD thesis, University of Bristol.

Deag, J. M. (1978). The adaptive significance of baboon and macaque social behaviour. In F. J. G. Ebling & J. M. Stoddart (Eds.), *Population control by social behaviour* (Pp. 83–113). London: Institute of Biology.

Deag, J. M. (1980). Social behaviour of animals. *Studies in Biology*, *118*, 1–92.

DeVore, I. & Washburn, S. L. (1963). Baboon ecology and human evolution. In F. C. Howell & F. Bourliere (Eds.), *African ecology and human evolution* (Pp. 335–367). Chicago: Viking Fund Publications in Anthropology.

Dimond, S. & Lazarus, J. (1974). The problem of vigilance in animal life. *Brain, Behavior and Evolution*, *9*, 60–79.

Dixon, A. F., Scruton, D. M., & Herbert, J. (1975). Behaviour of the talapoin monkey (*Miopithecus talapoin*) studied in groups in the laboratory. *Journal of Zoology*, *176*, 177–210.

Edmunds, M. (1974). *Defense in animals: A survey of anti-predator defenses*. Harlow England: Longman.

Emory, G.R. (1975). Comparison of spatial and orientational relationships as manifestations of divergent modes of social organization in captive groups of *Mandrillus sphinx* and *Theropithecus gelada*. *Folia Primatologica*, *24*, 293–314.

Emory, G. R. (1976a). Aspects of attention, orientation, and status hierarchy in mandrills (*Mandrillus sphinx*) and gelada baboons (*Theropithecus gelada*). *Behaviour*, *59*, 70–87.

Emory, G. R. (1976b). Attention structure as a determinant of social organization in the mandrill (*Mandrillus sphinx*) and the gelada baboon (*Theropithecus gelada*). In M. R. A. Chance & R. R. Larsen (Eds.), *The social structure of attention* (Pp. 29–49). London: John Wiley & Sons Ltd.

Emory, G. R. (1976c). Social structure in mandrills and gelada baboons. *Bios*, **1**, 1–9.
Emory, G. R. & Harris, S. J. (1978). On the directional orientation of female presents in *Macaca fascicularis*. *Primates*, *19*, 227–229.
Emery, G. R. & Harris, S. J. (1981a). Attention, orientation and socioecological systems in cercopithecine primates. I. Observational study. *Social Science Information*, *20*, 259–286.
Emory, G. R. & Harris, S. J. (1981b). Attention, orientation and socioecological systems in cercopithecine primates. II. Taxonomic comparisons. *Social Science Information*, *20*, 537–559.
Galton, F. (1871). Gregariousness in cattle and men. *Macmillan's Magazine*, *23*: 353.
Goss-Custard, J. D. (1970). Feeding dispersion in some overwintering wading birds. In J. H. Crook (Ed.), *Social behaviour in birds and mammals: Essays on the social ethology of animals and man* (Pp. 3–35). New York: Academic Press Inc.
Hall, K. R. L. & DeVore, I. (1965). Baboon social behavior. In I. DeVore (Ed.), *Primate behavior: Field studies of monkeys and apes* (Pp. 53–110). New York: Holt, Rinehart & Winston.
Hamilton, W. D. (1971). Geometry for the selfish herd. *Journal of Theoretical Biology*, *31*, 295–311.
Harding, R. S. O. (1977). Patterns of movement in open country baboons. *American Journal of Physical Anthropology*, *47*, 349–354.
Haude, R. H., Graber, J. G., & Farres, A. G. (1976). Visual observing by rhesus monkeys: Some relationships with social dominance rank. *Animal Learning and Behavior*, *4*, 163–166.
Hinde, R. H. & Detweiler, D. H. (1976). Visual observing by rhesus monkeys: Influence of potentially threatening stimuli. *Perceptual Motor Skills*, *43*, 231–237.
Horn, H. S. (1968). Adaptive significance of colonial nesting in the Brewer's blackbird. *Ecology*, *49*, 682–694.
Jennings, T. & Evans, S. M. (1980). Influence of position in the flock and flock size on vigilance in the starling, *Sturnus vulgaris*. *Animal Behaviour*, *28*, 634–635.
Keverne, E. B., Leonard, R. A., Scruton, D. M., & Young, S. K. (1978). Visual monitoring in social groups of talapoin monkeys (*Miopithecus talapoin*). *Animal Behaviour*, *26*, 933–944.
Kruuk, H. (1972). *The spotted hyena: A study of predation and social behavior*. Chicago: University of Chicago Press.
Kummer, H. (1967). Dimensions of a comparative biology of primate groups. *American Journal of Physical Anthropology*, *27*, 357–366.
Kummer, H. (1968). *Social organization of hamadryas baboons: A field study*. Chicago: University of Chicago Press.
Mitchell, G. (1972). Looking behavior in the rhesus monkey. *Journal of Phenomenological Psychology*, *3*, 53–67.
Murton, R. K. (1968). Some predator–prey relationships in bird damage and population control. In R. K. Murton & E. N. Wright (Eds.), *The problems of birds as pests* (Pp. 157–169). New York: Academic Press Inc.
Nagel, U. (1979). On describing primate groups as systems: The concept of ecosocial behavior. In I. S. Berstein & E. O. Smith (Eds.), *Primate ecology and human origins: Ecological influences on social organization* (Pp. 313–339). New York: Garland.
Pitcairn, T. K. (1976). Attention and social structure in *Macaca fascicularis*. In M. R. A. Chance & R. R. Larsen (Eds.) *The Social Structure of Attention* (Pp. 51–81). London: John Wiley & Sons Ltd.
Pitcairn, T. K. (1979). *The structure of attention and social behaviour in two groups of Macaca fascicularis*. PhD thesis, University of Birmingham.
Reynolds, V. & Luscombe, G. (1969). Chimpanzee rank order and the function of displays. *Proceedings of the Second International Congress of Primatology*, *1*, 81–86.
Reynolds, V. & Luscombe, G. (1976). Greeting behaviour, displays and rank order in a

group of free-ranging chimpanzees. In M. R. Chance & R. R. Larsen (Eds.), *The social structure of attention*, (Pp. 105–115). London: John Wiley & Sons Ltd.

Rhine, R. J. (1975). The order of movement of yellow baboons (*Papio cynocephalus*). *Folia Primatologica*, *23*, 72–104.

Rhine, R. J. & Owens, N. W. (1972). The order of movement of adult male and black infant baboons (*Papio anubis*) entering and leaving a potentially dangerous clearing. *Folia Primatologica*, *18*, 276–283.

Rhine, R. J. & Westlund, B. J. (1981). Adult male positioning in baboon progressions: Order and chaos revisited. *Folia Primatologica*, *35*, 77–116.

Rijksen, H. D. (1981). Infant killing: A possible consequence of a disputed leader role. *Behaviour*, *78*, 138–168.

Robinson, J. G. (1981). Spatial structure in foraging groups of wedge-capped capuchin monkeys *Cebus nigrivittatus*. *Animal Behaviour*, *29*, 1036–1056.

Rowell, T. E. (1966). Forest living baboons in Uganda. *Journal of Zoology*, *149*, 344–364.

Rowell, T. E. (1972). *Social behavior in monkeys*. Kingsport: Kingsport.

Rowell, T. E. (1979). How would we know if social organization were *not* adaptive? In I. S. Berstein & E. O. Smith (Eds.), *Primate ecology and human origins: Ecological influences on social organization* (Pp. 1–22). New York: Garland.

Scruton, D. M. & Herbert, J. (1970). The menstrual cycle and its effect on behaviour in the talapoin monkey (*Miopithecus talapoin*). *Journal of Zoology*, *162*, 419–436.

Scruton, D. M. & Herbert, J. (1972). The reaction of groups of captive talapoin monkeys to the introduction of male and female strangers of the same species. *Animal Behaviour*, *20*, 463–473.

Tinbergen, N. (1951). *The study of instinct*. Oxford: Clarendon Press.

Treisman, M. (1975). Predation and the evolution of gregariousness. I. Models for concealment and evasion. *Animal Behaviour*, *23*, 779–800.

Vine, I. (1971). Risk of visual detection and pursuit by a predator and the selective advantage of flocking behavior. *Journal of Theoretical Biology*, *30*, 405–422.

Virgo, H. B. & Waterhouse, M. J. (1969). The emergence of attention structure in rhesus macaques. *Man*, *4*, 85–93.

von Cranach, M. (1971). Orienting behavior in human interaction. In A. Esser (Ed.), *Behavior and environment: The use of space in animals and man* (Pp. 217–237). New York: Plenum.

Waterhouse, M. J. & Waterhouse, H. B. (1976). The development of social organization in rhesus monkeys (*Macaca mulatta*): An example of bimodal attention structure. In M. R. A. Chance & R. R. Larsen (Eds.), *The social structure of attention* (Pp. 83–104). London: John Wiley & Sons Ltd.

Williams, G. C. (1964). Measurement of consociation among fishes and comments on the evolution of schooling. *Publications of the Museum, Michigan State University, East Lansing, Biological Series*, *2*, 351–383.

Williams, G. C. (1966). *Adaptation and natural selection: A critique of some current evolutionary thought*. Princeton, N.J.: Princeton University Press.

Wilson, C. C. (1972). Spatial factors and the behavior of nonhuman primates. *Folia Primatologica*, *18*, 256–275.

Wilson, E. O. (1975). *Sociobiology: The new synthesis*. Cambridge, Mass.: The Belknap Press.

Wittenberger, J. F. (1981). *Animal social behavior*. Boston, Mass.: Duxbury.

3 Social Attention and Social Awareness

Thomas K. Pitcairn
Department of Psychology, University of Edinburgh, U.K.

In this Chapter I shall be concerned with theories of social cohesion and sociality in general among primates. That most primates are social to the extent of living in fairly large, mixed sex and age groups is not in doubt, but how they can achieve this can only be a matter of wonder, given the limitations in our modern theoretical formulations.

In terms of ethological theory, one of the first mechanisms of social (inter-individual) cohesion posited was that of Lorenz (1966), who described a system of "bonding by aggression". In essence, animals in pair bonds both produced and reinforced this relationship by directing aggression to other conspecifics outwith the pair. Thus the pair both reduced their own aggression to each other, which would arise because of the ever-present aggressive drive and their spatial closeness to each other, and simultaneously defined other conspecifics as out-group members—"not me/us". This, of course, is still used as an explanatory mechanism within social psychology—for example, in Sherif and Sherif (1969), who discuss inter-group competition amongst adolescent boys, or Tajfel (1979), a general discussion of inter-group conflict. The essential sociality of the pair bond is regarded as given, and is allowed expression by reduction in aggression within the pair. The mechanism of cohesion, therefore, is simply the reduction of aggression—even the distinct "love" bond Eibl-Eibesfeldt (1971) arises out of the essential (aggressive) conflict between proximity to others and maintenance of self-identity.

The social concept, "not me/us", therefore relies on an internal state, or rather the reduction of an internal condition, that of aggression, for its expression. Inter-individual (social) relations are expressed in terms of intra-individual behaviour.

This production of inter-individual relations by intra-individual constraints is also an important concept in the area of developmental social psychology, both in the work of social-learning theorists such as Bandura (1977) and Corsaro (1981), and in the very different work of Piaget (1932) and his followers (e.g., Selman, 1981). In brief, the social-learning theorist holds that the child gradually learns to be sociable, operating always on the basis of external or internal rewards for performance—for a Piagetian, the pre-operational child (below and about the age of seven years) is said to be egocentric, unable to take any perspective but their own. It is not until the *logical* structure for such "other perspective" taking is developed that the child can be said to be fully social, aware of others as social partners. The child at this early stage engages in "collective monologues" in a group setting where the expression of action, opinion, etc., is not contingent upon any particular other, or indeed really on any other at all. The child's relationships, friendship structures, and the like are based mainly on propinquity, and the accidental holding in common of a particular desire (for an object, or game) by more than one child.

The point of this discussion is simply to illustrate the nature of the inter- versus intra-individual controversy. These descriptions of social states and relations (*not* processes) may be labelled as "bottom-up" analyses in which the nature of sociality is explained, ultimately and proximally, by a state, or set of conditions, within the individual. Any analysis of social relations that uses these single features, such as aggression, kinship, or social attention in its simplest form, is open to this charge. Of course, many workers in the field would not regard this as a problem—clearly, the scientific method is designed to produce a set of simplifying factors to explain seemingly complex processes. However, although the basic events are not in question (i.e., aggression *does* occur), to concentrate on them entirely denies the importance of their organisation into higher-order emergent properties, which exert a downward causal control without interfering with the interactions of these elements at their own level. This latter position has been adduced within the brain sciences by Sperry (1964; 1969), recently criticised by Klee (1984). Klee's view is that what he calls the "micro-determinist" position, which is equivalent to the reductionist view within biology, is sufficient to explain all higher-order interactions and properties—the "new relational structure" is a direct product of the properties of lower-order events and materials. Sperry's macro-determinist theory, however, Klee finds more difficult to deal with, and he dismisses it eventually as based only on analogy or metaphor, and therefore weak. As Sperry (1964, p. 77) states, however, in the brain, ". . . the simpler electric, atomic, molecular, and cellular forces and laws, though still present and operating, have been superseded by the configurational forces of higher-level mechanisms. At the top, in the human brain, these include the powers of perception, cognition, reason, judgement, and the like, the operational causal effects and forces of which are equally or more potent in brain dynamics than are the outclassed inner chemical forces".

Sperry (1985) makes clear in his reply to Klee that the models of downward causation are clearly strong forces that control the underlying components through their own spatial–temporal relationships in the same way that certain aspects of the behaviour of the D.N.A. molecule are predicated by its helical shape and not by the properties of its component parts, such that the fate of these parts is then determined or controlled by these higher-order properties. Similarly, Trevarthen (1980) points out that processes of brain development in human infants are determined by the higher-order property of human sociality.

Thus in the field of neuroscience there is a strong case to be made for the importance of higher-order processes. In the field of human and animal behaviour such a case can also be made. For example, the search for a single trait, or combination of traits, to explain leadership, has been a common game since the 1930s (see Mann (1959) for a review). None of the studies has come up with a correlation coefficient of more than 0.5 for any trait, and indeed for any single trait the correlation has varied in different studies from 0 to 0.5, a condition in which the reliabilities are very low indeed. However, we know from the pioneering work of Landau (1951) on the determinants of rank orders in animals, that very high and reliable coefficients for a determining trait are necessary to produce the stability and transitivity to be found in animal social relations. Recent work, such as that of Chase (1974; 1982; 1984) has moved toward looking at the nature of the process itself in the determination of such hierarchies. What Chase was able to determine, as an important controller of the eventual rank order in groups of three and four chickens, was the strategies of attack themselves, which proceeded in such a way as to guarantee transitivity in the final rankings; i.e., that A defeats B who defeats C etc., with no reversals. For example, in a triad of animals, if A attacks and defeats B, four possible further developments remain:

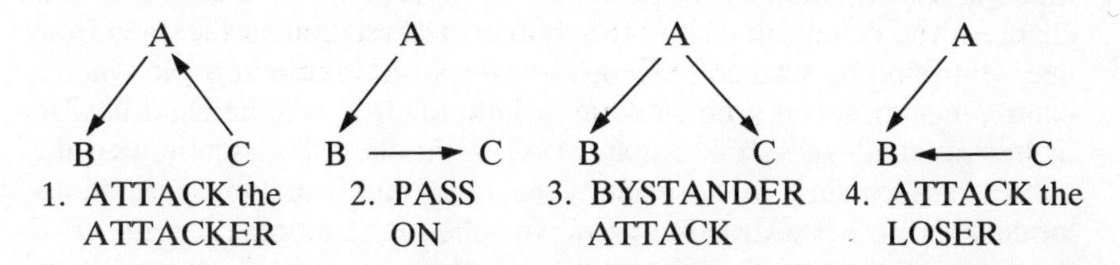

Only strategies 3 and 4 guarantee a linear rank order, and these occurred in about 90% of the cases observed by Chase. Thus the establishment of hierarchies could be a product of a multiplier principle, in which initial small differences among a population of animals would give an individual increasing advantage through the different strategies of rank-order determination.

However, even here the individual can be operating on "automatic". We could design a simple set of constraint principles and determinants of action that would guide *the individual* to act socially. What we are looking for is evidence of, and a theoretical approach to, the concept of the animal acting *socially*, that is *with awareness of self and other*. Interestingly, Humphrey (1980) thinks that it is precisely in the social context that it is most likely that such self-awareness in general is to be found and to have been evolved. Crook (1983) points out that it was J. B. Watson (1913), in his manifesto for modern psychology, who argued that all words referring to unobservable subjective states, in both humans and animals, should be removed from the realm of subjects fit for psychological investigation. Thus consciousness became irrelevant, and the animal awareness a dirty word, despite William James' (1890) important distinction between "me" and "I". The latter "I" concerns the individual's central awareness of sensation and feeling, and as a result of psychological research concentrating on James' "me", social awareness and action is now thought to be a product of the external world, derived from pressures and rewards from other individuals. This is perhaps most clearly seen in work on the social relations of young children—for example, Lewis and Brooks-Gunn (1979) describe the emergence of self-awareness of young children through the relationship to others in the social environment. One-year-olds' interest, particularly in children of the same sex, whom they can identify and preferentially pay attention to even when the stimulus provided is no more than a series of light dots attached to the other child, is seen as an example of the child's self-awareness in relation to gender. Similarly, when a younger infant (around 7–8 months) shows little stranger-fear to an unknown child entering the room, this is also part of the infant's self-awareness as a child. This self-awareness arises from the contingencies of the response of others to one's own actions. This research follows very much in the tradition of the social-learning theorists cited earlier in this chapter, who believe that all of the child's social relations are learned from and controlled by such contingencies of response from others. So when a child is having social problems, the solution is to teach the child how to make friends (Asher & Renshaw, 1981)—the fault lies usually with the child's performance as a social being, and thus can be corrected by modelling or other instruction. However, the social awareness of the child does not end with the knowledge of these response contingencies: Chisholm (1984) has shown that children are interested in the formation and maintenance of relationships *per se*. She has analysed the semantic content of children's interactions and found that children make selective use of, for example, subgroups' inclusion and exclusion statements, which relate only to individual other children, and not to the child's social ability as an individual or to the nature of the game proposed or in progress. In other words, the children are aware of and respond to the others in the groups in terms of their likes and dislikes, and the relationships within the group as a whole. Similarly, Maxwell (1983), and Maxwell and Pitcairn (1984, Note 1), have shown that the goals to which a child aims are an important determinant of that child's friendship pattern—for example, the child who forms a one-sided (i.e., unreciprocated) relationship with another in the group may be doing so in an attempt to establish a long-term recriprocal friendship, or may be merely using that child as a means of entry into some specific group activities. Children, then, are very aware of the social actions of others within their subgroup. They are not as limited in their abilities, either cognitively or socially, as much of the research in the area would suggest.

In non-human primates, too, it has been known for some time that individuals are aware at least of paired relationships between others in the troop. Kawai (1958) with his dependant-rank hypothesis, showed that rank relations between two animals may depend upon the presence or absence of a third party who will support one of the antagonists. These relations can also be seen in hamadryas baboons (Kummer, 1968), where a low-ranking individual may use the presence of the dominant male to threaten a third party, positioning themselves in such a way that any return threat may be perceived by the male to be directed at himself. The animals are clearly using a strategy that relies on their own awareness of the relations existing among other troop members. It is clear that, in some sense, the troop members understand these relationships—it may be argued that such an awareness is "merely" the recognition of a rank position, but the animal must at the very least know not only where it stands in such a system, but also the relative position of others.

The assessment of triadic relations, then, can be seen as an intermediate theory between a bottom-up view of social relations as being totally the product of, let us say, the management of aggression, and a fully-fledged, top-down or macro-deterministic position of sociality and social cohesion, which would stress that social relations in themselves control the lower-level elements such as the distribution of aggression. This does not deny the importance of aggression as a social behaviour important to the formation of bonds—indeed de Waal (Chapter 5) explicitly makes use of this Lorenzian model in his description of "chimpanzee politics", or bonding by reassurance. It is simply to state that there must be a *higher-order organising level* in operation, which reflects the complex social structure which results. Let us look at one principle of social organisation found in many primate troops, that of rank order.

Rank orders, or dominance orders, have been the source of much discussion in the primatology literature. It is not my wish to review this, but it is reasonably well-accepted now that such rank orders, based primarily on agonistic relations, do exist in many primates, and that they are at least a candidate principle for social organisations. Such rank orders are (usually linear) status differences amongst the troop members, according various privileges in accord with rank, such as for males access to oestrous females, or to preferred foods. However, these orders have also been seen to relate to differences in various social behaviours. Higher-ranking animals tend to receive more grooming from other troop members (Fairbanks, 1980; Sade, 1967; Seyfarth, 1976); they receive more visual attention from lower-ranking individuals (Chance, 1967; Pitcairn, 1976; Scruton & Herbert, 1970; Virgo & Waterhouse, 1969); and animals tend to space themselves out (see Chapter 2) with reference to the position of dominant animals (Alexander & Bowers, 1969; Chance, 1956; Imanishi, 1957; Itani, 1954).

Thus various features of the social life in primate troops assort in relation to the ranking structure. Indeed, Seyfarth (1977; 1980) has developed a model of female attractiveness, relating directly to rank, which explains the amount of grooming each female receives. Females with infants are also particularly attractive, but the main element in the model is the competition for higher-ranking females. These females are attractive because, in Seyfarth's functional interpretation, they may provide a supporting role or service to the groomee—in other words, the lower-ranking females attempt to form alliances with higher-ranking ones in order to protect their own interests. This conforms to Trivers' (1972) model of reciprocal altruism. Competition enters into the model, in that the higher-ranked a female may be, still higher-ranking is the female she may groom. Thus there is a tendency for the females to groom the next highest female in the rank order. However, Fairbanks (1980), although supporting the attractiveness model in general, found no evidence for it in terms of the interventions and support provided by high- to low-ranking females, independent of kinship, and Chapais (1983), in a different species, found little evidence for the specific "next-in-rank" effect proposed by Seyfarth.

Another model for these relationships is offered by Colvin (1983), which states that attractiveness is centred around similar rank, rather than higher rank. The similar-rank model works from various basic assumptions that Colvin makes—that similar-ranked individual needs are more likely to be reciprocal, that they will incur similar costs, and that the individuals are more familiar with one another. This familiarity is a product of prior exposure, and thus related to kinship, and further allows each individual to predict more accurately what the other will do, and so monitor the partner more effectively.

Monitoring the other is, of course, a central concept in the attentional theory of Chance (1967; Pitcairn, 1976). Thus what is being posited is that sociality in primates necessitates an awareness of others versus self, at least to the degree of knowledge of others interrelations (Hinde, 1979). However, these effects are only seen as occurring in one direction. For both Seyfarth and Colvin, the nature of the relationship between individuals is defined by the external pressures and rewards that those others will provide, in a reciprocal or delayed-reciprocal manner. But these relationships, of attraction or repulsion, should carry also a reference principle (Pitcairn, 1976). This means that not only are the individuals aware of the presence and actions of others, and these others' relations to third, fourth etc. parties, but also are directly influenced by these others in many ways other than direct dominance, and who may act as models for their behaviour (Hall & DeVore, 1965; Kummer, 1968). These reference individuals may have a positive or a negative valency—at the behavioural level they may attract or repel—and changes in the reference values are responsible for the flexibility in relationships noted by Colvin (1983).

A concept of reference groups and individuals is well-established in social psychology (e.g., Forsyth, 1983), referring usually to pressures put on an individual to conform—again, it seems as an external event that produces a social change—but this is merely to posit the mechanisms behind the event. The idea of reference acting at an individual level comes from attention structure as an emergent property of the social system (Pitcairn & Strayer, 1984). Negative reference individuals are directly rank-related, that is they are the product of an agonistic relationship, which is reflected in the tendency for visual orientation to go up the rank order. Figure 3.1 (modified from Pitcairn (1976)) shows the distribution of attention by rank difference in a group of monkeys, *Macaca fascicularis*. The clear trend is for lower-ranking individuals to look more at the higher-ranking, and the greater the rank difference the larger the amount of gaze. There is no tendency to observe next highest-ranking animals disproportionately. The amount of social contact (contact in huddling groups) is also rank-related, but the trend goes in the opposite direction (Fig. 3.2).

In other words, social contacts, at least in terms of body contact in these huddling groups, *is* directly related to rank, which in itself is partly at least a product of kinship (Kurland, 1977; Sade, 1967). A negative relationship is most clear in those animals who may be seen frequently to be a part of the same contact group, but who themselves are never, or only very rarely, in direct body contact with each other. The assumption here is that they both have a positive relation to a third party, who links the two together.

However, it is clear that *neither* attention (Chance, 1967) *nor* attraction (Colvin, 1983; Seyfarth, 1980) will explain the social organisation. Both reflect merely parts of it at an intermediate level. Attraction is related to

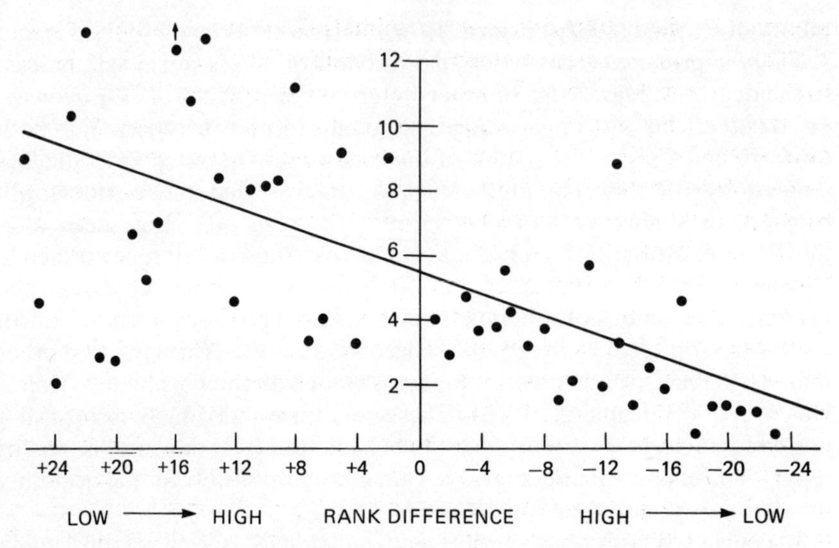

FIG. 3.1 Mean number of looks-at by rank difference: positive values—looks from low- to high-ranking animals; negative values, vice versa.

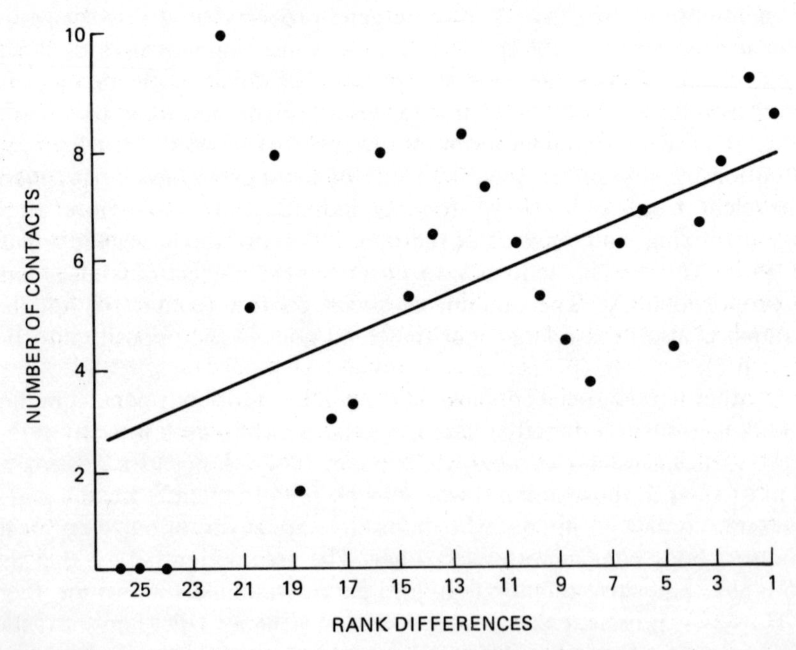

FIG. 3.2 Number of contacts versus rank differences in terms of number of places apart in the rank order.

attention by the monitoring that the animals do of their partners (Colvin, 1983). Attention structure emerges (Pitcairn & Strayer, 1984, p. 373) from, ". . . the set of biases which operate on an individual assessment of the relative importance of others within the group". These biases are rank-related in at least three ways: firstly, the negative reference provided by high-ranking individuals; secondly, the relationship of rank to kinship, which produces familiarity and hence attraction (Colvin, 1983); and thirdly, the information provided to, and received by individuals from, other troop members is related to a behavioural style, which is in itself rank-related. This latter behaviour can be seen most clearly in the response of females to the quiet, non-aggressive approach of a dominant male. Those females who sit unconcernedly as the male approaches, perhaps to move quietly off if the male sits down nearby, provide more information for onlookers than those (usually low-ranking) who move off grimacing and screaming, because their behaviour is differentiated from their response to the male when highly active or aggressive. Similarly, if a female *always* retreats however the male approaches, she has little opportunity to modify her behaviour as she does not receive different information from his various manners of approach—she lacks flexibility.

What is interesting is not that attention or attraction fail as a model to explain social organisation, but that there *are* large, individual differences. If we look at an attentional sociogram of a *M. fascicularis* monkey group (Fig. 3.3), what is important is not that the group members do not all devote most of their attention to the dominant male, as Assumpcao and Deag (1979) say is necessary for attention to be an organising principle, but that there are large individual variations within the trend for attention to pass up the rank order. Thus female (Bluntie) is looked at frequently by higher-ranking female (B.B.1) because they had a particularly poor relationship. They came into contact fairly often as they were members of the same female sub-cluster, with females (Fatty) and (B.B.2). This exemplifies a negative reference. Similarly, B.B.2 has a positive relation with many higher-ranking females, possibly because of her young infant. It is not sufficient to say that she rises in apparent rank because of her baby, for in many other ways she remains low-ranking—she still defers to females higher in rank although she is allowed to join their sub-cluster.

Therefore attention and attraction are both mechanisms that play a part in, but do not constitute entirely, a higher-order social function of the awareness of others within the group. Attractiveness and attention (or the attention-focusing quality; Chisholm, 1976) are differentially distributed within the group, but essentially each individual must be aware of the others in the group. This is a very vague statement, but what it means is that both attention and attraction are general properties of any bond or relationship, which, as Crook (1983, p. 13), has pointed out, is composed

	NEPHEW	BROTHER	ALPHA FEMALE	OTHERS
ORPHAN	■	□		
RED SPOT	□		■	
B.B.1	□			
BLOODY TAIL	■	□		
PARSNIP			■	
DOGFACE	■		□	
STUMP TAIL	□	■		
IRREG. CREST	□	■		
WHISKERS	■		□	
PATCHY	□			
KOSI	□			
FATTY	■		□	
B.B.2			□	
BLUNTIE	■		□	
FLOPPY			□	
GREYFACE				
GREYBACK	□		■	
LEFTIE			□	
bb			■	
SLITEAR	■			
GREYDOT	□			
CURLY			■	
SLICKER				
MOTHER				

Attention - 1st choice ■ or ━━

- 2nd choice □ or ──

Attention down heirarchy ➔

of "... rules of salience which would provide a hierarchy of criteria determining priorities for attentional 'screening' ". Let me give an example in a brief review of the work on human infant attachment.

Attachment is an "affectional bond" (Harlow & Harlow, 1976), acting as a secure base for exploration (Ainsworth, Blehar, Waters, & Wall, 1978), because attachment as a relationship is incompatible with fear. Fear, which is equivalent in some sense to high arousal, produces a need for comforting or conciliation (Mason, 1964; de Waal, Chapter 5). The relationship between attachment and disturbance is:

(a) Mother ~ Infant —Infant may move away, explore.

(b) Mother ~ Infant —Infant moves to mother, dependent upon the degree of disturbance.
↑
External disturbance

(c) Mother # Infant —Infant attaches less securely, moves away less, less exploration.

~ = attachment; # = insecure attachment

Therefore, when an external disturbance produces fear or stress, the infant moves to the attachment figure for comforting (b). When the attachment figure itself produces fear, comforting becomes equivalent to conciliation (de Waal, Chapter 5). The social nature of the bond is most clearly demonstrated when it is disturbed. An insecure attachment (c) leads to less movement away from the mother or other care-giver. This lack of exploration, both social and physical, leads to poorer social relations in the peer groups later on (Easterbrooks & Lamb, 1979; Lieberman, 1977; Sroufe, Fox, & Pancake, 1983). This condition of insecure attachment is equivalent to reverted (Mason, 1964) or reflected (Chance & Jolly, 1970) escape. For example, Kummer (1957) has shown that a female hamadryas' response to male neck bites is not to flee away from the male, but rather to flee towards him. The poor social relations result from the infant's lack of establishment (or knowledge about or understanding of) these social relations. It is *not* that the child does not have a suitable

---

FIG. 3.3 Diagram of Look At. Females in rank order with the first and second choices for their attention marked by a symbol. When the attention is not to the first or second ranked male, or to the first ranked female, it is indicated by the black line, the ends of the line connecting the two animals involved.

repertoire of social skills, as the social-learning theorists would have us believe, but that the child's awareness of social relations is limited.

Thus the subsequent behaviour of the infant is dependent upon the nature of the attachment. Different care-givers provide different attachment qualities (e.g., fathers are different from male acquaintances—Lewis & Weinraub, 1976). So, the nature of the bond can be assessed by the degree of comfort that it affords, and the infant responds differently to different care-givers and to different qualities of the attachment.

Attachment is therefore one example of a bonding system, the mechanisms of which are fairly evident. The security of attachment can be seen as equivalent to the interaction of attractiveness and attention within the social structure. It is a lower-order phenomenon, which relates to the nature of the bond, but is not and cannot be the bond itself. In the same way, primate cohesion and sociality comprises more than the sum total of dyadic or triadic relations. Each animal has a social awareness of their relation to the others, and to the relations among the others. The rules of social cohesion arise from the mechanisms of attention and attraction, through rank order and kinship. The social awareness is, in the terms of Crook (1983), "the subjective aspect of a [social] representational process, and there is no 'ghost in the machine' [Ryle, 1949] additional to this".

REFERENCE NOTES

1. Maxwell, W. S. & Pitcairn, T. K. (1984). *Individual differences in patterns of friendship in young children*. Paper to International Symposium on Child Development, Amsterdam.

REFERENCES

Ainsworth, M. D., Blehar, M. C., Waters, F., & Wall, S. (1978). *Patterns of attachment*. Hillsdale, N.J.: Lawrence Erlbaum Associates Inc.

Alexander, B. & Bowers, J. (1969). The social organization of a troop of Japanese monkeys in a two acre enclosure. *Folia Primatologica*, *10*, 230–242.

Asher, S. R. & Renshaw, P. D. (1981). Children without friends. In S. R. Asher & J. M. Coltman (Eds.), *The development of children's friendships*. Cambridge, England: Cambridge University Press.

Assumpcao, T. de & Deag, J. (1979). Attention structure in monkeys. *Folia Primatologica*, *31*, 285–300.

Bandura, A. (1977). *Social learning theory*. Englewood Cliffs, N.J.: Prentice-Hall.

Chance, M. R. A. (1956). Social structure of a colony of *Macaca mulatta*. *British Journal of Behaviour*, *4*, 1–13.

Chance, M. R. A. (1967). Attention structure as the basis of primate rank orders. *Man*, *2*, 503–518.

Chance, M. R. A. & Jolly, C. J. (1970). *Social groups of monkeys, apes and men*. London: Cape.

Chapais, B. A. (1983). Dominance, relatedness and the structure of female relationships in Rhesus monkeys. In R. A. Hinde (Ed.), *Primate social relationships*. Oxford: Blackwell.

Chase, I. D. (1974). Models of hierarchy formation in animal societies. *Behavioural Science*, *19*, 374–382.
Chase, I. D. (1982). Behavioural sequences during hierarchy formation in chickens. *Science*, *216*, 439–440.
Chase, I. D. (1984). The sequential analysis of aggressive acts during hierarchy formation. *Animal Behaviour*, *33*, 86–100.
Colvin, J. (1983). Familiarity, rank and structure of rhesus male peer networks. In R. A. Hinde (Ed.), *Primate social relationships*. Oxford: Blackwell.
Chisholm, J. (1976). On the evaluation of rules. In M. R. A. Chance & R. Larsen (Eds.), *The social structure of attention*. London: John Wiley & Sons Ltd.
Chisholm, V. A. (1984). *Preschool children within their social structure*. Ph.D thesis, University of Edinburgh, Scotland.
Corsaro, W. A. (1981). Friendship in the nursery school. In S. Asher & J. Gottman (Eds.), *The development of children's friendships*. Cambridge, England: Cambridge University Press.
Crook, J. (1983). On attributing consciousness to animals. *Nature*, *303*, 11–14.
Easterbrooks, M. & Lamb, M. (1979). The relationships between the quality of infant–mother attachment and infant competence in initial encounters with peers. *Child Development*, *50*, 380–387.
Eibl-Eibesfeldt, I. (1971). *Love and hate*. London: Methuen.
Fairbanks, L. A. (1980). Relationships among adult females in captive vervet monkeys. *Animal Behaviour*, *28*, 853–859.
Forsyth, D. R. (1983). *An introduction to group dynamics*. Monterey, Calif.: Brooks/Cole.
Hall, K. R. L. & DeVore, I. (1965). Baboon social behaviour. In I. DeVore (Ed.), *Primate behaviour*. New York: Holt, Rinehart & Winston.
Harlow, H. F. & Harlow, M. K. (1976). The affectional system. In A. Schrier, H. Harlow, & F. Stollnitz (Eds.), *Behaviour of non-human primates* (vol. 2), New York: Academic Press Inc.
Hinde, R. A. (1979). *Towards understanding relationships*. London and New York: Academic Press Inc.
Humphrey, N. (1980). In B. D. Josephson & V. S. Ramanchandran (Eds.), *Consciousness and the physical world*. Oxford: Pergamon.
Imanishi, K. (1957). Social behaviour in Japanese monkeys. *Psychologia*, *1*, 47–54.
Itani, J. (1954). *The monkeys of Takasakiyamo* (Translated by S. L. Washburn). New York: Ford Foundation.
James, W. (1890). *The principles of psychology*. New York: Dover.
Kawai, M. (1958). On the system of social ranks in a natural troop of Japanese monkeys. *Primates*, *1*, 111–130.
Klee, R. L. (1984). Micro-determinism and concepts of emergence. *Philosophy of Science*, *51*, 44–63.
Kummer, H. (1957). *Social behaviour of hamadryas baboons*. Bern: Huber.
Kummer, H. (1968). *Social organization of hamadryas baboons*. Basel: Karger.
Kurland, J. (1977). Kin selection in the Japanese monkey. *Contributions to Primatology* (vol. 9). Basel: Karger.
Landau, H. G. (1951). On dominance relations and the structure of animal societies. *Bulletin of Mathematical Biophysics*, *13*, 1–19.
Lewis, M. & Brooks-Gunn, J. (1979). *Social cognition and the acquisition of self*. New York: Plenum Press.
Lewis, M. & Weinraub, M. (1976). The fathers role in the infant's social network. In M. Lamb (Ed.), *The role of the father in child development*. New York: John Wiley & Sons Ltd.
Lieberman, A. F. (1977). Preschoolers competence with a peer. *Child Development*, *48*, 1277–1287.

Lorenz, K. (1966). *On aggression*. London: Methuen.
Mann, R. D. (1959). A review of the relationships between personality and performance in small groups. *Psychological Bulletin*, *56*, 241–270.
Mason, W. (1964). In L. Berkowitz (Ed.), *Advances in experimental social psychology* (vol. 1). New York: Academic Press Inc.
Maxwell, W. S. (1983). *Patterns of friendship and interaction style in young children*. Ph.D. thesis, University of Edinburgh, Scotland.
Piaget, J. (1932). *The moral judgement of the child*. London: Routledge & Kegan Paul.
Pitcairn, T. K. (1976). Attention and social structure in *M. fascicularis*. In M. R. A. Chance & R. R. Larsen (Eds.), *The social structure of attention*. London: John Wiley & Sons Ltd.
Pitcairn, T. K. & Strayer, F. F. (1984). Social attention and group structure. *Journal of Social and Biological Structures*, *7*, 369–376.
Ryle, G. (1949). *The concept of mind*. London: Hutchinson.
Sade, D. S. (1967). Determinants of dominance in a group of free-ranging rhesus monkeys. In S. H. Altman (Ed.), *Social communication among primates*. Chicago: University of Chicago Press.
Scruton, D. & Herbert, J. (1970). The menstrual cycle and its effect on behaviour in the talapoin monkey. *Journal of Zoology (London)*, *162*, 419–436.
Selman, R. L. (1981). The child as a friendship philosopher. In S. R. Asher & J. M. Goltman (Eds.), *The development of children's friendships*. Cambridge, England: Cambridge University Press.
Seyfarth, R. M. (1976). Social relationships among adult female baboons. *Animal Behaviour*, *24*, 917–938.
Seyfarth, R. M. (1977). A model of social grooming among adult female monkeys. *Journal of Theoretical Biology*, *65*, 671–698.
Seyfarth, R. M. (1980). The distribution of grooming and related behaviours among adult female vervet monkeys. *Animal Behaviour*, *28*, 798–813.
Sherif, M. & Sherif, C. (1969). *Social psychology*. New York: Harper & Row.
Sperry, R. W. (1964). Problems outstanding in the evolution of brain function. *James Arthur Lecture*. New York: American Museum of Natural History.
Sperry, R. W. (1969). A modified concept of consciousness. *Psychological Review*, *76*, 532–536.
Sperry, R. W. (1985). Discussion: Macro- versus micro-determinism. *Philosophy of Science*, *52*.
Sroufe, A. L., Fox, N. E., & Pancake, V. R. (1983). Attachment and dependency in developmental perspective. *Child Development*, *54*, 1615–1627.
Tajfel, H. (1979). *Differentiation between social groups*. London and New York: Academic Press Inc.
Trevarthen, C. (1980). Neurological development and the growth of psychological functions. In J. Sants (Ed.), *Developmental psychology and society*. London: Macmillan.
Trivers, R. L. (1972). Parental investment and sexual selection. In B. Campbell (Ed.), *Sexual selection and the descent of man*. Chicago, Ill.: Aldine.
Virgo, H. & Waterhouse, M. (1969). The emergence of attention structure among rhesus macaques. *Man*, *4*, 85–93.
Watson, J. B. (1913). *Psychological Review*, *20*, 158–177.

4 The Cohesive Foragers: Human and Chimpanzee

Margaret Power
Department of Sociology and Anthropology, Simon Fraser University, Vancouver, Canada

A number of scientists have commented on the many similarities between the organisation and social behaviour of wild chimpanzee groups and the least complex human foraging or hunting–gathering[1] groups (Lee, 1968; Reynolds, 1966; Sugiyama, 1972; Turnbull, 1968b). Both follow a foraging pattern of fission and fusion (Itani & Suzuki, 1967; Kummer, 1971; Lee & DeVore, 1968). However, despite more than two decades of field study of free chimpanzees and human foraging groups, primatologists and anthropologists do not yet entirely understand the social organisation of either the chimpanzee or the human foragers.

Rereading and reassessment of the field-study publications on the social behaviour of wild[2] chimpanzees led to the recognition of a previously undetected form of social organisation. This organisation is based on a network relationship of *mutual dependence* rather than on a hierarchy of

[1]As a whole way of life is the focus of the study and not the technology of humans who follow it, the general term "foragers" is used throughout in preference to the commonly used term "hunter–gatherers". As used here, by foragers is meant groups without one permanent leader that follow a nomadic pattern of fission and fusion, and live off wild food. This definition includes both some human small-scale societies and wild chimpanzee groups, and excludes troop-organised primates and one-leader groups.

[2]Rowell (1972) distinguishes between "wild" and "free-ranging" primates. She considers only undisturbed primates as wild, but includes in this category natural groups that have been fed very minor amounts of bait. Groups of animals that are not confined but that are fed a large proportion of their food by humans are referred to as free-ranging. This distinction is imperative in moving toward an understanding of the natural social order of chimpanzees.

dominance.[3] The theoretical model of this form of social organisation is applicable to our understanding of the structure of both undisturbed food-gathering chimpanzee society and the most simply organised of the few remaining food-gathering human societies. These are the small hunting–gathering groups who live by an "immediate-return" system of foraging (Woodburn, 1982).

According to Woodburn, an immediate-return foraging system is one in which the members gather and hunt and, typically, eat the food they gather the same day. These foragers do not store food. Their tools and weapons are few, simple and utilitarian; easily acquired and easily replaced.

When human foragers or foraging societies are referred to in this paper, it is, in every case, only those who live by the fission and fusion, immediate-return foraging mode. Woodburn produced this category solely in reference to human groups, but chimpanzees also fulfil the criteria of an immediate-return foraging society.

Woodburn is impressed by the high state of egalitarianism that is characteristic of the autonomous humans who live by the immediate-return system of foraging. He suggests (Woodburn, 1982, p. 431) that among these peoples ". . . there is the closest approximation to equality known in any human societies". Leadership is not single or structured. No individual has formal authority, but in these groups there are both men and women whose opinions are respected, whose presence is sought. Service (1966, p. 51) calls them "people with influence" based on attractive qualities of personality. Most anthropologists consider these individuals to be informal "charismatic" leaders (Service, 1966; Turnbull, 1968a).

Like the human foragers, wild (i.e. undisturbed, not artificially fed) chimpanzees lack a rank order and have high individual freedom of choice amounting to autonomy. However, it seems unlikely that these apes have the concept of egalitarianism as do humans, even though they use essentially the same positive and negative sanctions toward any behaviours that enhance or threaten the egalitarian social order. (This shall be taken up shortly.) The wild chimpanzees are, nevertheless, functional egalitarians.

Both Turnbull (1968a) and Woodburn (1982) make the same important point; both emphasise the *necessity* of egalitarianism and cooperation in human immediate-return foraging societies. They agree that if lacking these positive characteristics, the loosely linked foraging societies would collapse. Accordingly, Turnbull suggests (1968a) that it is more illuminating to view these attributes as essential components of the social order,

---

[3]The development of this form of mutual-dependence organisation is a main theme of my larger study (recently completed and publication pending). The points and arguments in this chapter are more fully developed in the larger work.

rather than as norms and values. By extending Turnbull's highly important insight, I will argue herein that *positive relationships and attributes are structural and functional essentials in both human and wild chimpanzee groups*; both of which live by the immediate-return mode of foraging.

Social relationships are the basis of the social structure. Many aspects of behaviour have become structural parts of the social organisation. The nature of the mutual dependence system of itself generates the necessity of certain relationships, behaviours, constraints, and principles in any group so organised, be it human or chimpanzee. The point that structural roles and relationships must be positive in emotive tone or nature cannot be overemphasised. This is not merely desirable, but necessary, if the fluid social order is to hold, co-existent with a great deal of individual autonomy. There must be a strong attraction toward identifying and reuniting as a larger social group, which is not necessary in more rigidly structured, ranked groups.

As indicated earlier, two of the major hypotheses in my larger work are the proposals that:

1. These two primate species, though human and ape, share the same socio-ecological adaptation to a foraging life.
2. Both are organised around a system of *mutual dependence*.

This social system is based on a fluid, mutualistic, status and role relationship between many charismatic leaders and dependent followers. This chapter is a consideration of some of the aspects of this mutual-dependence system, which forge powerful cohesive bonds between all members of what appear to be very loosely organised social groups.

Anthropologists who study human foraging societies do not regard these groups as particularly cohesive. At the same time, they realise that the traditional foraging societies have an extraordinary sense of unity, despite a fluid, seemingly unstructured organisation which lends them a confusing appearance of disarray. Turnbull (1968a, p. 25) suggests that the "amazing strength of unity", characteristic of this type of human society, testifies to the efficiency of this fundamental socio-ecological adaption.

The cohesion of foraging groups is unusually powerful because it is based on and gained from two main organising principles. The first hypothetical principle is a mutualistic status/role relationship of two modal personalities, the charismatic leaders and dependent followers. There are many confident, charismatic individuals who possess, to varying degree, personality qualities that both attract others and facilitate the carrying out of the supportive/protective role behaviours that are required of leaders in the mutual-dependence orders. This property influences other, more nervous and hence more dependent, members to seek their company, to follow their lead. The less confident members then take on the interlocking supportive/dependent status/role.[4]

The second hypothetical basis of cohesion is the pattern of fission and fusion. The parting of subgroups facilitates constantly recurring, larger group reunions. Among wild chimpanzees these larger group gatherings may occur several times in one day. Because of the highly positive tone of these reunions, Reynolds (1965), like Garner (1896), refers to them as "carnivals".

There is considerable evidence in the field-study literature concerning these two foraging species that gives support to these hypotheses. In the case of the chimpanzee, empirical evidence is found in the publications of those observers who studied wild chimpanzees by various naturalistic (non-provisioning) methods. Their work has been undervalued and largely ignored.

Most scientists believe that as well as following the same organisational pattern of fission and fusion as human foragers, chimpanzees share with humans the same spectrum of emotions, used in similar contexts. Hinde and Stevenson-Hinde (1976) suggest that in striving to understand non-human primate relationships we can profit greatly by using some general and basic principles from the literature on human relationships. In this chapter, this suggestion is followed. The use of human-based concepts is restricted to organisational and emotive theory.

CURRENT UNDERSTANDING

Field study of free chimpanzees began in the early 1960s. The pioneers who undertook this difficult task during that decade include Adriaan Kortlandt, the team of Vernon and Frances Reynolds, Yukimaru Sugiyama, and Jane Goodall.[5] All of these field observers used some form of relatively "naturalistic" or unobtrusive methods. They did not feed the apes they studied and generally interfered with the animals' daily rounds as little as possible. Goodall began quite early to feed the Gombe apes with bananas but, before 1965, this was simply by placing heaps of the fruit here and there

---

[4]The term status/role is used in preference to "status and role" to emphasise that the two are inseparable in a mutual dependence system. The statuses ascribed to individuals who act as charismatic leaders, and as dependent followers, are contingent on their carrying out certain necessary, "expected behaviours", i.e., roles. In our society, status and role are sometimes more separable. Titles such as Doctor or General retain a certain status, whether or not the possessor is carrying out the attached role.

[5]For convenience, throughout this study I refer to this scientist as Goodall, although some of her publications are under the name Van Lawick-Goodall. She has published under both maiden and married names, but she is well known by either. I do not think this simplification will cause confusion.

from which the apes could help themselves. This method apparently did not seriously disturb the natural order, if compared to the disruption inadvertently engendered by a human-controlled, restrictive feeding method introduced at Gombe in 1965, and continued with modifications to at least 1970 (Wrangham, 1974; Teleki, Hunt & Pfifferling, 1976; Power, 1986).[6] Quite without such intent, the post-1965 feeding method fostered direct competition for limited food on to the chimpanzees who normally compete indirectly. Hence, prior to 1965, the Gombe apes are, in the terms suggested by Rowell (1972), *wild* chimpanzees, relatively undisturbed. After 1965, the same apes became what Rowell designates a *free-ranging* group, unconfined but artificially supplied with much food.

Naturalistic methods of studying wild apes present many difficulties. As the habituation of the chimpanzees at Gombe has been taken to be totally successful, naturalistic methods were curtailed for many years. Recently, however, field researchers Ghiglieri (1979) and Baldwin (1979) have again attempted the unobtrusive method.[7]

All of these unobtrusive observers, including Goodall (prior to 1965), quite separately—but with striking unanimity—report that wild chimpanzees everywhere share the same form of social organisation and behaviour. Wild chimpanzees are peaceful, non-aggressive, intensely social primates who move about in what seem randomly formed, ever-changing small subgroups with no sign of a dominance hierarchy, no single permanent leader, and no enforced (closed) territoriality.

The unanimity of the separate findings lends strength to their likely validity and is in accord with the striking similarities between the social organisations of immediate-return human foraging societies, wherever they are found. These unobtrusive field reports are rejected by an influential majority of the academic community. The extremity of the

---

[6]The strengths and weaknesses of depending on the data collected from several short-term pioneering field studies of relatively undisturbed wild chimpanzee groups, as opposed to those from longitudinal studies of habituated, provisioned, free-ranging chimpanzees are taken up in my larger study.

It should be clearly understood that although I am strongly critical of the method of feeding introduced at Gombe in 1965, I am not critical of the researchers for initiating it. In 1965, our knowledge of chimpanzee psychology was little developed and the idea of referring from human to chimpanzee emotional reactions was unthinkable. With serious limitations on research funds and time, Goodall had good practical reasons for introducing the human-controlled banana feeding. Nevertheless, the all-pervading nature of the well-documented negative social change at Gombe compels reassessment of our understanding of wild chimpanzee social organisation.

[7]Baldwin undertook an 11-month study of the behaviour and ecology of a group of wild chimpanzees inhabiting an open arid region of Mt. Assirik, Senegal. She hoped to focus on behaviour, but she found the apes so wary of humans that she had to change the focus of the study to that of ecology. Ghiglieri's, too, is an ecological study.

negative social change that resulted from the directly competitive feeding method at Gombe is not recognised. The prevailing view is that feeding these chimpanzees caused some increase in aggressive interaction, but that their behaviour and organisation is essentially typical of wild chimpanzees in general. Hence, many scientists tend to view wild chimpanzees as being naturally organised into some form of male dominance, in which ranking position is gained and maintained by aggressive means. Yet, many biologists now feel that the structure of social groups varies, even within a single species. Rather than social behaviour being a constant natural trait, it is dependent upon the interaction between the group and its environment. A species that is social in one setting may be asocial in another (Maynard Smith, 1978; Wrangham, 1980). Similarly, a species that is positive in its behaviour in one setting may be negative in another.

Primates did not evolve in a crowded world. According to all reports, the natural world of the chimpanzees was a safe and unthreatening one until pressured by human expansion. There was food even in dry years and an absolute minimum of intrusive events from without. Today, in most places the natural habitat of primates is shrinking; loggers, farmers and herders press in on every side, cutting into the foraging ranges. Now few primate species still have recourse to one natural solution, that of some emigrating to a more food-rich or less congested area.

As suggested, a pall of silence seems to cover any reference to Goodall's unobtrusive pre-1965 observations, whereas the data from post-1965 Gombe studies have been accorded a great deal of scientific authority. Consequently, much current understanding of chimpanzee social behaviour and organisation is based on studies carried out at Gombe since 1965.[8] However, the many Gombe studies, when read in chronological order, clearly document an ever-accelerating, very negative social change, which spread in ripple fashion to distort every aspect of what was formerly a highly positive and smoothly functioning social order.

Under the unnaturally prolonged stress that resulted from the method of feeding at Gombe, the constancy of partner changeover tended to solidify into more permanent preferred-partner alliances, and a trend toward territoriality developed. The more permanent nature of the usually fluid, temporary partnerships cut at the root of the mutual-dependence system,

[8]Because of space limitations, I do not refer to the equally important and influential Japanese studies of chimpanzee groups in Mahale National Park, Tanzania. These animals also have been artificially fed by methods in which control of the apes' access to a present and desired bait food is manipulated by the humans, resulting in frustration behaviour. The extent and serious nature of this totally unnatural situation is not recognised. Hence the similarities between the recent aggressive, fiercely competitive behaviour of both artificially fed groups serves to reinforce the current assumption that the male centred, territorial organisation reported at both centres, is the adapted mode of chimpanzee social organisation. (See Power, 1986.)

which depends for its success on bonds between all members of the group, rather than between specific individuals or within closed groups.

In the 1960s, when many of the unobtrusive field studies were carried out, the paradigm of social dominance was a largely unquestioned analytical model, widely used in analysis of both captive and wild primate groups. Consequently, the pioneer field observers also tried to use this trusted analytical model. However, they experienced great difficulty in reconciling the behaviour of the peaceful, non-hierarchical wild chimpanzees they observed, with the aggression-based hierarchical paradigm. For this reason, I accept and use their empirical data, but reject their analyses. When wild chimpanzee behaviour and social organisation is discussed here, the data drawn upon are not from the large and influential post-1965 Gombe literature, but from early and recent naturalistic field studies.

The model of a mutual-dependence order was constructed on the basis of these ignored, unobtrusive studies of wild chimpanzees. The new perspectives it offers permit very different interpretations of chimpanzee behaviour, a few of which are offered in this chapter.

THE MUTUAL-DEPENDENCE SYSTEM

Reynolds (1966), and Izawa and Itani (1966), were the first to recognise that wild chimpanzees and human foragers share many organisational phenomena. Reynolds suggests that the foragers and the chimpanzees are alike in the openness and size of their groups, their long birth intervals, and their wide recognition of relationships. Both lack organised single leadership, and territoriality other than loosely defined and open home ranges. Both inherit behaviour patterns such as tool making and usage. Both have recurring larger social group gatherings. Most significantly, both human and ape foragers share a nomadic way of life, based on a fluid pattern of constant realignment of group composition, the fission and fusion pattern mentioned previously.

Among humans, the very fiuid group composition of fission and fusion entails a constant changeover of personnel within and between the small local groups, and more frequent shifting of campsite location than is necessary for subsistence (Turnbull, 1968). Recurring large group social reunions are the other essential part of the fission/fusion mode.

Chimpanzees, too, move from feeding area to feeding area much more often than would be necessary if food was their foremost objective. Human foragers move every day, every few days, or every few weeks, but among the chimpanzees the fission and fusion pattern takes the form of very temporary small subgroups, which form by free choice to move together for a time, foraging throughout the forest. As often as several times a day, the separated subgroups are summoned to gather as a larger social group by the vocalisations of some of their fellows who have located a good feeding area. These reunions are highly positive in tone, as Reynolds' use of the term "carnival" suggests.

According to Reynolds (1965), during the larger group carnival gatherings a great deal of highly excited positive display, contact behaviour, greeting, and other social interaction takes place, and the small groups mingle and dissolve. After a period of socialising and feeding, subgroups form once again and wander off to resume their travelling, foraging rounds. Characteristically, the reformed groups are composed of different personnel (Goodall, 1965; Kortlandt, 1962; Reynolds, 1965; Sugiyama, 1972).

In brief, the hypothetical mutual-dependence system among chimpanzees is as follows. There are a great many individuals of two modal types among them and, despite the constant change of personnel in the temporary subgroups, each such group contains at least one or more calm, self-assured, charismatic adult apes of either sex, and one or more nervous—hence more dependent—animals, also of either sex. The temporary subgroups form around the charismatics because the dependent individuals are attracted to follow one or another of them for a time. Thus *the composition of the temporary subgroups changes but the structural form remains the same.* This is also the basis of the much more complex human foraging organisation.

It is generally agreed that leadership in human foraging societies in charismatic and temporary (Service, 1966; Turnbull, 1968a); but it has not been recognised that leadership among wild chimpanzees takes the same characteristic form. In both groups, followership is voluntary. Both leadership and followership have attached mutualistic status/roles. This interdependence, and the factor of free, constantly changing choice of companion based on positive attraction, lends a strongly cohesive influence to the temporary relationships, and to the larger social group.

The evidence is that among chimpanzees, as in human foraging societies, charismatic leaders have influence through example, the carrying out of a certain role and personality-connected behaviours. However, in both societies, these leaders are without any form of power to coerce others into following. This status/role cannot be assumed, it is assigned, as followers may (and do) move freely from one leader to another. Hence, in order to be followed, and so to be a leader, the candidate apes must show the qualities of charismatic attraction: that is, by their demeanour and behaviour, convince the more dependent members of their readiness and suitability to carry out those responsibilities and expected behaviours integral to the leadership status/role. So, if charismatics fail to act in a fashion that attracts followers, they no longer have the leadership status/role. In this way, the behaviour of the charismatic apes is controlled by the more dependent who follow—if they wish to. The charismatics have influence, but the dependent possess a veto.

With the highly fluid group composition and location, individual autonomy, and open groups, and without serious threat from within the group or without a particularly strong bond of cohesion is needed to keep the society from amiably disintegrating. In the following sections of the chapter, I will suggest some of the modes and means of the immediate-return foraging organisation that do weave an exceedingly strong web of cohesion between all members of the loosely organised, larger social groups.

CHARACTERISTICS OF HUMAN FORAGING SOCIETIES

The foraging life mode is considered to be the most successful and persistent adaptation that humans ever made. It accounts for over 90% of the historical existence of humankind. Until approximately 10,000 years ago, humans everywhere in the world lived in foraging groups. Today, there remain only a few scattered pockets of people who live according to this once universal mode, in environments so inhospitable as to be undesired by agriculturalists or industrialists, and even these are being rapidly taken over (Murdock, 1968).

Woodburn (1982, p. 433) suggests that until recently most anthropological research has been of groups living by some form of "delayed return" foraging system in which, ". . . people hold rights over valued assets" (such as nets, boats, traps and so on): these involve binding commitments between people, ". . . a set of ordered, differentiated, jurally-defined relationships through which crucial goods and services are transmitted", and a "delayed return" for labour. Groups living by an immediate-return system of foraging organisation have only recently begun to be properly investigated, Woodburn indicates, and so this social system is still relatively unfamiliar.

According to Woodburn (1982), the social organisation of immediate-return foraging societies has the following basic characteristics:

1. Social groupings are flexible and constantly changing in composition.
2. Individuals have a choice of whom they associate with in camp gatherings in the food quest, travelling, (etc.).
3. People are not dependent on *specific* other people for access to basic requirements.
4. Relationships between people, whether relationships of kinship or other relationships, stress sharing and mutuality, but do not involve long-term binding commitments and dependencies of the sort that are so familiar in delayed-return systems.

So far, six contemporary human societies have been identified as following the immediate-return system of foraging[9] and, according to Woodburn (1982, p. 434), of necessity all are ". . . egalitarian, profoundly . . . egalitarian".

The evidence from the unobtrusive field studies is that social groups of undisturbed chimpanzees, too, are characterised by these organisational principles. For this reason, I analyse the social organisation of wild chimpanzees, in anthropological terms, as being that of foragers who live by an immediate-return fission/fusion system, despite their being nonhuman primates. Although the theory of immediate-return foraging social organisation was produced by anthropologists through the study of human groups, it is broadly a primate theory, and one which need not be restricted in its application to the human private only.

Woodburn (1982, pp. 434–435) emphasises that in immediate-return human societies a mobile, flexible nomadism is fundamental: "Anyone may live, hunt and gather wherever he or she likes without restraint." There are no permanent dwellings, fixed base camps, or territories beyond vaguely defined, permeable familiar "home range" areas. There is no exclusion of outsiders. People move frequently and freely from group to group.

Although a fundamental nomadism is required by the foraging ecology, Woodburn warns that this nomadism should not be interpreted in terms of ecological factors alone. The social—in the broadest sense, political—aspects of the nomadic pattern are of great organisational importance. Simplicity and meagreness of material culture is necessitated by their constant movement but in this, too, the socio-political implications are important. In these highly egalitarian, noncompetitive societies, acquisitiveness of any sort leads to loss of status.

In immediate-return foraging societies, there is an absence of institutionalised leadership, and of specialised or formalised institutions or groups that can be differentiated as economic, political, judicial, religious, and so on. Kinship concepts and terms are a great deal broader than our own, and they designate social relationships rather than actual kin. Leadership is not sought after (Service, 1966; Turnbull, 1968a). It is charismatic in form, essentially a matter of having influence over others on the basis of having admired qualities of personality. However, expected behaviours

---

[9]Woodburn (1982) lists the six societies so far identified as fitting into the immediate-return category as being the Mbuti pygmies of Zaire; the !Kung Bushmen (San) of Botswana and Namibia; the Pandaram and Paliyan of south India; the Batek Negritos of Malaysia; and the Hadza of Tanzania. (References to the literature on these societies are given in Woodburn, 1982.)

(roles, statuses, and a mutualistic relationship between those who lead and those who prefer to follow), are involved. The fluid, positive, mutual-dependence relationship between charistmatic and dependent members is an essential structural and functional basis of the immediate-return foraging system.

Composition of Groups

The composition of subgroups among immediate-return human foragers is not rank- or kin-based, but essentially a matter of people who wish to be together, joining to move about together for a time. The ties are affinity and social kinship, not consanguinity (Gardner, 1972; Turnbull, 1972;).

The researchers who followed naturalistic methods of field study of wild chimpanzee groups agree that among these apes, the small subgroups that constantly form are also a matter of free choice based on personal preference (Goodall, 1965; Kortlandt, 1962; Reynolds & Reynolds, 1965)—without restriction by age, sex, or rank, Sugiyama (1972) adds. The constant rearrangement of the personnel forming the temporary small subgroups among both human and chimpanzee foragers takes place freely and amiably. The naturalistic field researchers agree that the participation of wild chimpanzees in any kind of congregation did not appear to be fixed or controlled in any way. However, evidence can be found in Goodall (1971) that there is a pattern of preference for companions of the complementary personality type—charismatic for dependent, dependent for charismatic. Almost any of these seem acceptable as a temporary companion (Ghiglieri, 1979; Goodall, 1971). Each very temporary companionship alliance thus acts to represent not only the larger social group, but also the essential charismatic–dependent relationship.

The Function of Fission and Fusion

Although there is no consensus as to the adapted social organisation of the wild chimpanzee, there is firm agreement that small subgroups of chimpanzees do constantly form, dissolve, and reform in the very fluid manner described earlier. As suggested, this fundamental mode of social organisation has important repercussions on social relations in both human and chimpanzee groups.

Anthropologists generally view the primary function of the fluid associative pattern among human foragers as a highly efficient social mechanism for control of tension that might lead to aggression. But they also recognise that the constant reuniting and regrouping also acts to maintain social contacts and friendly relations between all members of the larger social group. In terms of cohesion, both enhance positive emotive bonds between all members of the usually separated, larger social group.

From a biological perspective, the indirect food competition (separate, simultaneous seeking of the same foods; Mayr, 1970), which is inherent in the fission and fusion mode of foraging, affords a solution to the problem of members coming into direct competition for resources. Indirect competition for resources eliminates interest in controlling the access of others. It also eliminates a conflict of interest between the benefit of the group as a whole and that of each member as an individual. Direct competition for the necessities of life and reproduction jeopardises both the benefits of cooperation and the group. From a social, anthropological perspective, the frequent changeover of personnel in the subgroups prevents the temporary coalitions from becoming alliances or power blocs, and the assigned influence of the leaders from becoming coercive power. In this way, stratification and division are avoided. It is through the maintenance of wide friendly social contacts that cohesion is obtained in the fluid foraging groups.

Charismatic Authority

Some form of authority is present in every society, animal or human, even when it is so informal as to create the illusion of non-existence, an impression one does get from both human foraging and wild chimpanzee societies. Even carrying out role-connected behaviours involves an understood pattern of informal authority (Nisbet & Perrin, 1977). They indicate that informal authority arises and assumes its specific nature from concrete, inter-personal events in which the established rules are so few and so flexible that they may appear not to exist at all, and yet such authority is an inseparable part of the situation in which it is manifest. This very informal, almost invisible mode of authority is most commonly found in small-scale, simple social unions such as human foraging groups—and in chimpanzees.

Nesbit and Perrin also suggest that there is an element of charisma in this personal form of authority. Informal authority tends to be characterised both by attractive qualities, such as good nature and charm, and by its spontaneous, relaxed, usually face-to-face nature. For example, charismatic chimpanzees characteristically reassure and calm their more nervous companions through positive physical contacts: friendly touches, pats, embraces, and so on, and the reassurance gained is often mutual. They further suggest that it is when a group's objectives are vague, and not directed toward any productive output, that the authority structure tends to be strongly coloured by the personalities of those participating. In other words, it is when a group is organised on the basis of positive social relations, as are the human foragers and the wild chimpanzees, that the charismatic–dependent relationship can function as a main organising principle, and the mutual dependence order hold.

Charismatic Qualities among Chimpanzees

In general terms, charisma among humans is defined as being that quality which is imputed to individuals because of their presumed connection with fundamental, order-determining powers and/or their demonstration of an exemplary inner state (Shils, 1968). Charisma among chimpanzees (and human foragers) is not of the intense form theorised by Weber (1947), but closer to the dispersed and diluted form that, as Shils indicates, is found throughout all human societies. In this "normal" form of charisma, possession of an exemplary inner state does not imply some degree of sacredness, but simply an exemplary adherence to, or personification of, the norms and values of the group. Through their personalities, attitudes, and role behaviours, the charismatic chimpanzees, demonstrate a state that is both exemplary and necessary in a society which is organised around a system of mutual dependence.

Shils (1968, p. 390) vouches that charisma in humans is a quality that can be cultivated, ". . . so prized that individuals are encouraged to allow it to come forward in their sensitivity". Any quality based on personality can be culturally encouraged or discouraged. Dependence, too, can be and is encouraged among both chimpanzees and human foragers.[10]

Those who theorise about charisma suggest that the presumed connection of charismatic humans with vital elements is observable through their bearing, demeanour, and actions. Subjective experience of a person's possession of charismatic qualities is manifested in the demeanour and actions of other individuals toward them (Shils, 1968). The subjective expression of the charisma of chimpanzee leaders is also revealed through the attention paid to the charismatics by the less self-assured.

Possession of confidence by the charismatic apes corresponds to sureness, to lack of hesitation, to trust, and to self-confidence. Confidence and not being easily roused to aggression are necessary to the leader status/role, which is essentially one of peacekeeper as well as protector. Confi-

---

[10]John Bowlby (1973) suggests that there are misleading negative connotations to our conceptualisation of dependence. Dependence is not a state of weakness, but a reliance on some other for support of some sort. Based on data from studies of both human and non-human primates, Bowlby argues (Pp. 23–26) that the essential ingredients of a "healthy personality" in a mature adult are a capacity to rely trustingly on others at times, and to be able to assume the other supportive role when the situation changes, and to provide a secure base from which their companion can operate, at other times.

Among foraging humans and, the evidence suggests, wild chimpanzees, taking the role of follower (dependent) is not viewed as a step downward in status or rank. This egalitarian attitude is necessary in order to facilitate the dynamic nature of the two status/roles. The "average" individual then willingly moves from one status/role to the other, according to the requirements of the situation. This has high adaptive value.

dence is in part postural, as also is the lack of assurance. Hence, the *most* confident and the *most* nervous chimpanzees can be distinguished through their bearing, as well as their behaviour. In more average individuals the definition of the qualities that constitute either of the ideal-type personalities is not so pronounced. The adaptive value of this is suggested in my larger study.

Nervous, potentially dependent chimpanzees may also be distinguished by their behaviour. They may have opposite personalities and temperaments to the calmer, more assured chimpanzees. They tend to be unsure, hesitant, less trusting, and more quickly roused to aggression. Yerkes (1925, pp. 246–248), writing of the contrasting temperaments of a pair of captive infant chimpanzees, reports that one was bold, alert and eager for new experiences, even-tempered, good-natured, and remarkable for his observational ability, intelligence and great delight in "showing off". The other was quite the opposite, being "timid, nervous, hesitant before anything novel or new". In unfamiliar situations, the confident ape took the lead and the timid one followed. The contrasting temperaments and/or personality qualities are functionally complementary in the mutualistic charismatic–dependent role relationship.

In any relationship all participants have roles. There are also certain expected supportive behaviours attached to the dependent status/role. These too are exemplary and necessary to complete the relationship, which facilitates the social system.

Clearly, two outstanding chimpanzees at Gombe were the mature male, whom Goodall dubbed David Greybeard, and an elderly and frail-appearing female, Flo. Flo was a highly charismatic leader: her courage and character—of which she had as much as "a whole platoon of chimpanzees"—gave her high status in the society (Goodall, 1965; 1971). However, the data on "the dauntless Flo", as Goodall (1971) sometimes called her, are largely in connection with her sexual and child-rearing roles. The data on the equally charismatic male, David Greybeard, are more abundant. He did not lose charismatic status in the increasingly aggressive society, so these data are more often used in identifying the qualities that compose charisma among chimpanzees.[11]

According to Goodall (1971), David was a gentle and exceptionally calm chimpanzee, although he was, in her opinion, potentially the most danger-

[11]Goodall's data indicate that in the early years Flo was a relaxed and tolerant individual, and that later (under stress) she became aggressive. Being frail, and having dependent young, the old female could not gain position in the dominance hierarchy that formed on the basis of aggressive competition. Indeed, few females could, in competition with the larger, non-childrearing males. Neither does aggressive interaction fit with the female parental (mothering) role (see Footnote 12).

ous animal when roused. He was quick to respond in a friendly, reassuring way to others' approaches. He was deliberate in manner, unhurried and trusting, with "an air of natural dignity". David moved about in leisurely fashion, often pausing to greet, reassure, and calm with a touch, other apparently less assured chimpanzees. Goodall indicates that in the early days before the wary (unhabituated) apes had become accustomed to her presence, she was always glad to see this ape among a group, for with him present she had a better chance of approaching closer. Her impression was that David's lack of fear soothed the more nervous apes, a judgement with which most unobtrusive observers would agree. In a Guiana study group (Albrecht & Dunnett, 1971, p. 13), the charismatic leader about whom virtually the same statement is made was a mature childless female, Pandora. She, and ". . . not mature ♂♂, as might be expected", was first to approach their hide. After which the rest of the group approached and were ". . . noticeably much tamer". Like most observers, Albrecht and Dunnett attempt to establish the notion of a hierarchy of male dominance among the wild chimpanzees. Not being familiar with the concept of charismatic leadership among chimpanzees, they do not recognise this female's confidence and influence on the behaviour of others as attributes and indications of charismatic leadership.[12]

There are both male and female charismatic leaders in the human groups that live by an immediate-return system of foraging (Lee, 1979; Service, 1966; Turnbull, 1968a). There is an accepted but not rigid utilitarian division of labour—childfree men hunt and child-nurturing women gather—but this involves no inequality of status or role (Turnbull, 1968a). These societies are profoundly egalitarian.

There is no division of labour among wild chimpanzees. However, the two parental strategies—the female's one of retreat when mothering because her parental responsibility is the intimate nurturing and protection of her own biological offspring; and the male's widely generalised, protective, "social fatherhood"—tend to separate the sexes into active mobile

---

[12]Pandora's behaviour is typical of childless charismatic leader chimpanzees. Childless leaders are, of course, typically and in very large majority, male. It is part of the protective mothering role (i.e., the parental strategy of the female) that a female with a dependent child will retreat from potentially dangerous situations in order to protect her offspring; but this is role- not personality-based behaviour. For example, according to de Waal (1982), the senior female, Mama, in the Arnhem zoo colony took the active (generalised protective) role until she gave birth to an infant. Then she took on the role behaviour necessary to a mothering chimpanzee, i.e., concentration on nurturing and protection of her own dependent offspring. Mama remained a powerful and influential charismatic leader, but her leadership, in accord with the retreative female parental strategy, took a less physically involved, less easily discernible form. Clearly, her fearless and confident personality did not change, nor did her charismatic attraction for others. Her role changed.

groups of far-ranging males and the more sedentary, less far-ranging, mothering females. As Kortlandt (1962, p. 132) hypothesised more than two decades ago, a main social division in undisturbed chimpanzee groups, "... is between childless and childbearing adults rather than between males and females".

Carrying out either easily observed, physically active, or a sedentary less easily discerned charismatic leadership role is based on suitable personality qualities and behaviour plus availability and personal inclination, as Mama's actions indicate. Awareness that this division is according to parental role rather than by sex as such, and that it involves a difference in the expression or form of leadership rather than an inequality in the status and roles of males and females, is essential for true understanding of the egalitarian social orders of both undisturbed chimpanzees and human foragers.

Complementary Personality-type Preferred Companions

That chimpanzees have highly distinctive individual personalities is well known. Most field observers report two distinct personality types among these apes; that is, individuals who are calmer and more confident than others, and those who, through their actions, appear to be more nervous (Goodall, 1965; Reynolds, 1965; Ghiglieri, 1979; Sugiyama, 1972).

That chimpanzees prefer companions of the complementary personality type became readily apparent under conditions of stress at Gombe. The former, very temporary, preferred partnerships between individuals of complementary personalities tended to solidify and acquire aspects of a power bloc in the aggressively and directly competitive society.

This solidification of formerly fluid alliances permitted Goodall to identify a number of partnerships between chimpanzees who, she realised, were remarkably different in temperament, yet tended to spend much time together. In her book, *In the Shadow of Man* (Goodall, 1971), which she wrote for a general readership, she gives example after example of these opposite personality-type preferred alliances. By far the best documented of these combinations is that of the three males who, lured by bananas, were for over a year the only chimpanzees to venture into Goodall's camp.

This trio was composed of the confident David Greybeard, an extremely timid, nervous–dependent male, William, and Goliath, a powerful nervous–aggressive animal, whom only David seemed able to calm and tolerate. The charismatic Flo and an elderly, highly nervous female, Olly, constituted another prominent combination. These temperamentally dissimilar females often travelled together in the forest and their children were each other's most constant playmates.

Goodall (1971) realises that there is benefit in the interaction of these pairs. In the strained and explosive atmosphere of the feeding station, the nervous, less assured members of the charismatic–dependent alliances were treated with more restraint by aggressive chimpanzees when the nervous individual's protective and confident friend was present. In turn, the dependent chimpanzees took a supportive role when the more charismatic member of an alliance needed assistance.

FACTORS OF SOCIAL COHESION

If there is no inside power to coerce, and no real threat from outside, why then do the individuated, autonomous chimpanzee foragers take on the leader and follower status and perform the attached roles? Why, too, do the human foragers?

The charismatic–dependent relationship is adaptive toward keeping subgroups together as they range, foraging. If all members of a subgroup were equally self-assured, there would be more of a tendency for each individual to go its own way, and a lone animal or human is usually more vulnerable to attack. Hence, this relationship may also be viewed as an efficient survival mechanism in terms of structure, function, and strategies.

This is most easily seen among the chimpanzees, whose social organisation is complex, but much less so than that of the most fundamental of human foraging groups. The more nervous, easily roused, dependent chimpanzees, through their quick response to disturbance in the group or to danger from without, act to alert the less easily aroused, less observant, charismatic members. One or more of the confident individuals then respond to the situation, backed by the support of more dependent members. Thus the response is more by group consensus than leader-initiated.

Affiliation theorists suggest the strength of social cohesion that draws and binds a human group together varies with the attractiveness of the group for its members. The more the members want or need the group, the stronger the cohesive bond. There seems no reason to restrict this concept to the human primate only. As a result of studying social behaviour in chimpanzees, Mason (1965) suggests that social activities may serve as rewards for certain learned responses among these apes. I suggest that among the reasons for the social cohesion of both human foragers and chimpanzees are the intangible but very real emotive rewards, which are an integral part of the charismatic/dependent relationship (the authority structure), and the pattern of amiable parting and positive reunion of the larger social group.

Writing on the psychology of affiliation among humans, Schachter (1959, p. 2) hypothesises that people ". . . in and of themselves, represent goals for one another", as they have powerful needs for approval, support, friendship, prestige and the like, which can be satisfied only through inter-personal relations. Such gratifications, obtainable only through association with others, are both reward and goal. He suggests that in non-acquisitive societies (such as immediate-return foragers), such intangible rewards are particularly effective.

Homans (1961) views human social behaviour as an exchange between at least two people in which the actions of one are sanctioned, rewarded, or punished by the actions of the other. He theorises that the more often a particular action of an individual is rewarded, the more likely they are to perform the action. This results in further reward for the performer. Homans, like Schachter, cites the same type of intangible but highly effective rewards as the return for the assumption of an individual's role. This suggests that the reward for carrying out the role behaviours attached to the two interacting charismatic–dependent status/roles can be expected to encourage individuals to move toward taking on the appropriate role and status. Each role rewards as well as supports the other.

That chimpanzees share the same spectrum of emotions as humans (Hebb, 1946), and many of the same psychic capabilities and motives (Shafton, 1976), is now widely accepted. Hence, it seems reasonable to assume that Schachter's and Homans' insights may be useful in understanding chimpanzee behaviour as well as that of the human primate. It also seems reasonable to assume that the positive sanction or reward of the affection-based deference shown by the dependent chimpanzees can be expected to encourage development of the charismatic qualities of leadership and a readiness to take on the role among the calmer, less easily roused chimpanzees. On the other hand, the investment of the proffered attention, support and friendship of these respected individuals surely encourages and rewards the more nervous members for taking the equally essential dependent role. Through this individual experience of reward the charismatic–dependent relationship is strengthened.

In societies organised around a system of mutual dependence, leadership and followership are statuses and roles, not ranks. Although respect (shown through deference) is assigned to the charismatic leaders by the dependent members, it is in return for the charismatics' enactment of the attached, protective/responsive, role responsibilities toward the dependents.

In this way, individuals of both personality types must look to the other for reciprocal reward. It is the interaction, and the interdependence of the two personality types within the mutualistic roles that structures the positive system of mutual dependence. This is certainly where the emphasis must lie in order to understand the social organisation and cohesion of immediate-return fission/fusion foragers, whether human or chimpanzee.

This interdependence gives the fluid charismatic–dependent relationship great strength, despite the constant change of component personnel. Among chimpanzees as among human foragers the constant change of companions assures that the strongest commitment is to the relationship and the group, more than to the individuals who enact it.

The commitment of chimpanzees to the group is most clearly demonstrated in the highly positive, frequent, larger group gatherings. In these ritualised carnivals of display, socialising, and feeding, the subgroups dissolve and reform with new personnel before resuming their separate foraging rounds.

Sugiyama (1972), who likens the psychological release and gratification apparently experienced by chimpanzees in carnival reunion to that obtained by !Kung San foragers through their trance dance gatherings, suggests that the chimpanzee reunions facilitate communication among the larger regional population, which strengthens wide social bonds. Lorna Marshall (1965, p. 271) comes to a similar conclusion regarding the function of the !Kung San dance ceremonies: she comments that these draw the larger society together ". . . as nothing else does", bringing the people ". . . into such union that they become like an organic being". Coming together, they celebrate the group. The positive societal atmosphere is the basis of an "elastic" emotive bond, which brings back together the usually scattered chimpanzees, and the human foragers, in periodic, positive reunion, and which lends the seemingly fragile groups the amazing strength of unity of which Turnbull writes.

Ritualisation is a standard term in ethology. Goffman (1971, p. 62), in writing on the relations and face-to-face interactions between people in public, indicates that ritualisation, as used by ethologists, is usually understood to refer to, ". . . a physically adaptive behavior pattern that has become removed somewhat from its original function, rigidified as to form, and given weight as a signal or 'releaser' to conspecifics". Among humans, Goffman explains, ritual is ". . . a perfunctory, conventionalized act through which an individual portrays his respect and regard for some object of ultimate value to that object of ultimate value or to its stand-in".[13] Goffman refers to these small rituals as supportive interchanges: in reference to these, he reminds us that Durkheim (1926), in his classic analysis of religion, divided ritual into two classes, positive and negative. Negative rituals involve avoidance, keeping away; the positive rituals license and even oblige approach, hence some type and degree of bond is implied.

Through positive interactions, homage or other reward is paid or ex-

[13]As I understand Goffman's quote, he is saying that ritual is used by individuals to convey respect and regard for someone (or some object), to that person or object.

changed through offerings of some kind. The small, taken for granted, positive rituals that humans produce, such as nodding, smiling, handshaking and so on, act to support the social relationship between the actors involved. From this same perspective, the positive contact behaviour so characteristic of chimpanzees, who hug, pat, and touch as greeting and as reassurance, supports the social relationship between charismatics and dependents, and between the members more generally.

The larger carnival gatherings fit within Goffman's concept of an "arena-related" ritual: such rituals are thc larger, less frequent meetings between socially supportive people who are in contact less often than those working or living near to each other. The example Goffman gives of this ritual is a "maintenance rite". Parties to such a relationship may engineer a coming-together because there has not been a recent reason for one to occur. Coming together is important for the well-being of the relationship for, as Goffman (1971, p. 73) indicates, ". . . the strength of a bond slowly deteriorates if nothing is done to celebrate it".

The wild chimpanzees, through their carnival gatherings, constantly strengthen the social bond between the autonomous subgroups and the larger society. The periodic larger group gatherings, characteristic of human foraging peoples, reaffirm and strengthen the resilient cohesive bond between the autonomous members of their fluxing societies.

Our understanding of the extraordinary strength of cohesive bonding, vital to a social system based on the charismatic–dependent relationship and the recurring larger group reunions, may be further expanded. Consider the human-based findings of Festinger, Pepitone, and Newcomb (1952) regarding the specific needs that only inter-personal contact within a group can satisfy. Two classes of social needs must be met if a group is to be successful. The first was mentioned earlier, as both a reward and a goal of human affiliation; these are the needs for approval, respect, support and so on, which are gained through singling out the group members as individuals, and paying attention to them in a manner that satisfies those needs. The second class of needs, they propose, requires de-individualisation or submersion in the group. Action as a group facilitates a reduction of inner restraints against doing things or performing behaviours that are often inhibited when a member is being paid attention by others as an individual. Groups that provide only one of these two classes of needs are usually unstable and not very satisfying, Festinger and his colleagues maintain. They suggest that although a group cannot satisfy both types of need at the same time, it can provide both types of situation on different occasions; and those which do so are highly attractive to their members.

The individualised attention implicit in the charismatic–dependent relationship, and the release in the de-individualised larger group celebrations combine to make the fission and fusion foraging away of life an attractive and successful one. It answers emotional and social needs of both humans and chimpanzees, as well as solving ecological problems. Fission into small individualised groups is an efficient foraging technique. It at once facilitates maximum use of the foraging range and keeps competition for food indirect. The de-individualised emotive release found in the larger group reunions acts to draw the separated subgroups back together and so reaffirms the social bonds of the larger society. Thus the mutual-dependence system combines a maximum of group cohesion with a minimum restriction of individual autonomy.

CONNECTIONS AND IMPLICATIONS

This new view links with the findings of several contributors to this book. The ideas advanced by Chance (Introduction; Chapter 1) and by Pitcairn (Chapter 3) are based on the study of caged groups of monkeys; and de Waal's on long observation of a captive colony of chimpanzees at the Arnhem zoo (Chapter 5). The studies of Montagner and colleagues (Chapter 10) are of the social behaviour of contemporary, Western pre-school age children in kindergarten and day-care centres. None of these settings is, of course, the natural environment in which the social behaviour of the various species evolved, or for which they are adapted.

My argument has been stated: it is that of necessity the social organisation of foragers living by an immediate-return fission–fusion system is based wholly on positive mutualistic social relationships. These groups are far more positive in "tone" or social atmosphere than are any of the human and nonhuman primate groups referred to in other chapters of this book. In every case, where behaviours similar or related to those observed by the forementioned scientists are proposed, those shared behaviours or relationships takes a more positive form among the immediate-return foragers, both human and chimpanzee.

To begin with, Chance (1967) proposed the concept of a structure of attention as an important, cohesive element in the social structure of primate groups. His initial observation was that there is a persistent focusing of attention by all subordinate members of rank-ordered primate groups toward the alpha or dominant animal, usually without its taking specific action to attract this constant attention. He refers to this as a "centric" form of attention structure. Later Chance (1976) broadened his concept to also include "acentric" forms of attention structure, in species which lack a rigid dominance order. Using the fissioning and fusing pata monkeys as an example, he hypothesises (Chance, 1976) that in groups characterised by an acentric form of structured attention, rather than the members being strongly focused on one individual, their attention is multi-focused. They watch many members of the group, events going on within the group, and features of the physical environment.

Perhaps influenced by the post-1965 Gombe publications, which were (and still are) considered the most reliable data available, Chance suggests that attention structure takes a centric form among chimpanzees. He argues that at the constantly recurring larger group reunions, through an outstanding demonstration of display behaviour, one individual ultimately stands out and becomes the focus of attention for all. This, in fact, did happen at the Gombe feeding station when the male, Mike, a formerly ostracised, peripheral member of the group, learned to bowl clattering empty oil cans ahead of him as he displayed (Goodall, 1971). But the advantage of this "advanced technology" is not available to wild chimpanzee displayers. More importantly, concentration on one alpha displayer would militate against the adapted pattern of easy division into many small, temporary subgroups with constant realignment of subgroup personnel. In the wild groups, many confident males display impressively and in doing so attract some of the larger social group to follow them, when it once again undergoes fission for more effective foraging.[14]

In the fluid mutual-dependence order, the structure of attention is difficult to distinguish, being multi-focal. But this form of social organisation necessitates an organised network of attention. This network, all interested in and so watching all others, enables control by consensual action in small groups. It adds cohesive strength to what appears to be the fragile social fabric of the constantly fissioning, reuniting and realigning societies of both species of forager. In both centric and acentric forms, the structure of attention is, as Chance suggests, an illuminating statement of the way a group is organised.

From observation of a group of rhesus monkeys in Bristol Zoo, Virgo and Waterhouse (1969) detected two types of structured attention. One was a positive form, in which members occasionally looked at one particular animal in order to facilitate the making of a social contact, or just in order to remain in its vicinity. There was also a negative form of attention, the constant watching of another individual in order to avoid all proximity. Pitcairn (1976) found evidence of these same two kinds of social attention among long-tailed macaque monkeys in the zoo in Basle, Switzerland. He suggests that the positive form indicates a relationship based on friendship, and the negative form a relationship based on dominance rank order. He refers to the animals involved as being "positive" and "negative referents", according to the type of relationship in which they are involved.

Pitcairn's concept of two types of social referent is useful in analysis of both chimpanzee and human foraging societies. In societies organised

[14]An impressive display does attract followers, but we are reminded that leadership in immediate-return foraging groups is charismatic, and charisma is based on personality traits and roles, not vigour, strength or sex, as such. Display behaviour has many highly important ramifications, some of which are suggested in my larger study.

around a system of mutual dependence every temporary charismatic–dependent alliance is a relationship of positive social referents. In both these foraging species, those who are negative social referents are the few aggressive, disruptive individuals whose socially deviant behaviour threatens the peace and cohesion of the group. These are refused important relationships by other members, through simple shunning by individuals, and also through the group-backed sanction of ostracism.

Sanctions are usually thought of as being a human society's reaction to a particular behaviour, either through approval (positive, rewarding sanction) or disapproval (negative, punishing sanction). The most frequently used negative sanction in human foraging groups is ostracism, ranging from simple shunning to exclusion from the group and its privileges by general consensus. There is evidence from which to argue that chimpanzees, too, use this same avoidance behaviour to control or punish socially deviant members whose behaviour threatens the harmony of the group.

Human foragers consciously and deliberately apply the sanction of ostracism to persistent trouble makers. I do not take up the question of whether chimpanzees consciously use this group sanction. However, individual avoidance (shunning) of an animal through simple dislike or fear, when carried out by most members of the group, is functionally ostracism.

The sanction of ostracism is powerfully effective in human societies that are organised around positive social relationships. In work originally published in 1961, Marshall (1976, p. 288) testifies that the desire of the !Kung foragers ". . . to avoid both hostility and rejection leads them to conform in high degree to the unspoken social laws . . . most !Kung cannot bear the sense of rejection that even mild disapproval makes them feel. If they do deviate, they usually yield readily to expressed group opinion and reform their ways". Group opinion is frequently expressed through ostracism.

Marshall also points out that in their harsh (Kalahari Desert) environment the !Kung must belong to a band, they cannot survive alone. They are physically dependent on the food-sharing that is an essential characteristic of contemporary human foragers in difficult environmental conditions. Accordingly, they are not without jealousy and watchful self-interest. At the same time, she states (Marshall, 1976, p. 287), these foragers are ". . . extremely dependent emotionally on the sense of belonging and on companionship". This intense emotional dependence is the factor of particular interest here, as adult chimpanzees in their natural, bountiful, forest environment are not characteristically food-sharers but nutritionally independent.

Through her extensive knowledge of the !Kung foragers, Marshall is convinced that these people find enforced separation from the group emotionally "unendurable". Turnbull (1961) makes much the same observation in regard to the Mbuti pygmy foragers who, in their more food-abundant Ituri forest (north-east Congo) home, are not nearly so dependent on each other in terms of nutrition.

Apparently the !Kung find lack of actual physical contact difficult to endure. Draper (1973) comments on the obvious pleasure that the !Kung get from sitting in groups, in close physical contact with one another. Marshall (1976, Pp. 287–288) also notes this: she believes that, ". . . their wanting to belong and be near is actually visible" in the way that groups and families sit clustered together, ". . . often touching someone, shoulder against shoulder, ankle across ankle". Chimpanzees, too, obviously enjoy sitting in groups, grooming, in close physical contact. Touching is a reassurance gesture of chimpanzees (and humans), so it seems reasonable to assume that the proximity, as well as the grooming, is enjoyed.

The behaviour of the ape, Goliath, who Goodall judged to be the earliest identified dominant male, was socially deviant. This nervous, extremely aggressive chimpanzee was the only male (in the early years) Goodall (1963, p. 297) ". . .ever saw . . . actually attack a female and on one occasion he even drove a young ape from its nest and took over". Goliath was generally avoided and, as mentioned before, only the very confident David Greybeard seemed able to tolerate and calm this animal. Goodall (1971) records that the other apes fled when he (Goliath) leapt into a tree to join them. De Waal (1982, p. 149) also writes of ". . . something lacking" in the leadership of a young adult male, Nikkie, who became dominant through aggressive means; he was acknowledged, but unpopular and avoided, in the caged zoo colony. Reynolds and Luscombe (1976), too, describe an extremely aggressive female who was shunned, not accepted into the "friendly associative exchange" evident among the rest of a captive (New Mexico) group.

The Gombe male, Mike, deposed Goliath from the position of despot through the use of clattering oil cans in display. Yet in the early, pre-1965 years, Mike was cowed and nervous, threatened and attacked by almost every other male (Goodall, 1965). He was not usually included in the male grooming clusters, nor apparently did many females solicit or accept his (or Goliath's) sexual attentions. MacKinnon (1978, p. 79) suggests that neither of these two despotic or negative-referrent apes, ". . . showed much interest in the opposite sex". We cannot know the reason for this; however, wild chimpanzees of both sexes are free to invite, accept or reject signalled invitations to copulate. Mating is multiple, rather than promiscuous, in that it is clearly not indiscriminate; moreover Goodall (1971) reports that Gombe females refused the sexual solicitations of aggressive males in the early years. The outcome of an experiment by Yerkes and Elder (1936) suggests that under the relaxed and congenial conditions of a natural wild group, the female chimpanzee controls mating choice. There is a parallel among human foragers, in that sexual access is the right of women to control and bestow (Woodburn, 1980); !Kung women usually refuse to marry a violent-tempered man (Howell, 1979). As Hinde (1974), in writing on the biological bases of human social behaviour suggests, over-aggression in a male can frighten off prospective mates.

Before his rise to despot power, Mike apparently did not often invite grooming. It is not purely speculative to suggest that perhaps he had learned through past experience that his solicitation would be rebuffed.[15] Perhaps the knowledge that their sexual advances too, would be declined, is the reason underlying the disinterest in copulation shown by these two unpopular males. Whether it is through ostracism or from personal disinterest, this lack of participation is of great social and evolutionary significance.

In social terms, a general refusal of companionship is a powerful sanction among the human foragers. It is reasonable to assume that it is also a powerful social sanction among the gregarious chimpanzees. In terms of adaption, through the refusal to accept the social deviant as a companion, access to the charismatic–dependent alliances—which are survival strategies—is refused. Even more significantly, by denial of sexual access, whether this is self-denial through lack of interest as MacKinnon suggests, or ostracism by the females, (which seems more likely in an egalitarian society based on positive [friendship] referent relationships), the ultimate effect is the same. The negative-referent chimpanzees are eliminated from contributing to the gene pool.

Another consequence of the need for charismatic members to be generally accepting and friendly in order to gain followers is the positive tone of what is, in the more rigidly structured primate rank orders, appeasement behaviour.[16] In order to encompass both negative and positive usages of this same soliciting behaviour, the general term "supplication" is substituted, because what is being requested through the same gesture or behaviour will be very different in emotive tone in a rank-ordered, and an egalitarian, primate society.

In rank-ordered non-human primate societies, supplication is indeed an appeasement behaviour. (In human groups, all of these behaviours are more complex.) Appeasement is a behaviour typical of submissive, lower-ranking members. Characteristically, it is used to mollify higher-ranking

---

[15]In my larger study (see Footnote 3), the argument is developed that grooming by adult chimpanzees is one way of indicating social acceptance of the (usually youthful) solicitor to the adult ranks. Denial of grooming by a charismatic and/or adult ape informs the group at large that the soliciting individual is not accepted socially by that influential member as a fellow adult and peer, irrespective of the solicitor's physical maturity.

[16]I thank Dr. M. R. A. Chance for suggesting this point to me in a personal communication.

animals, in reaction to their threat or actual attack. In the mutual-dependence order of wild chimpanzees, "appeasement" behaviour has a more positive form and use.

Both dependent and charismatic chimpanzees use "appeasement" gestures without the implication of threat and signalled submission. Characteristically, as used in wild chimpanzee groups, supplicatory gestures are made not in order to mollify a superior, but as a request for another member, often but not always a charismatic, to reassure the solicitor by responding to the request for acceptance and support. The request is probably for reassurance. The gesture signals dependence, not submission.

In the egalitarian societies of undisturbed chimpanzees, this kind of social attention, too, is essentially a mutualistic exchange, generalised to all members of the group. Unlike the one-directional appeasement behaviour observed in rank-order societies, the positive form, supplication, requesting reassurance, is both sought by, and obtained from, both charismatic and dependent chimpanzees.

Montagner and his associates (see Chapter 10) are interested in the social behaviour and organisation of contemporary Western pre-school age children in day-care and kindergarten settings. Children at this very early age (12 weeks to 6 years) are, of course, the least socialised members of our society. Montagner et al. (1978; 1979; and Chapter 10) report that among these young children there are individuals with virtually the same charismatic and dependent personality-type, in the same fluid charismatic leader/dependent follower relationship, with free movement from leader to leader and group to group, that is characteristic of human foragers and wild chimpanzees.[17] The phylogenetic implications of this are enormously important, and should be explored. We need to know much more of the potential for positive social behaviour in primates, most urgently in our own stressed and self-endangered human species.

In their chapter (10), Montagner et al. describe supplication gestures and behaviour used within groups in similar positive fashion by very young, leader children towards their cohorts, who are attracted to follow. (Montagner and his team, of course, use the established term "appeasement" for this reassurance-soliciting behaviour.) The result of the use of the supplicatory gestures is, Montagner suggests, "a strong probability of activities together and often in co-operation". It has the same results in chimpanzee groups.

In his chapter (5), primatologist Franz de Waal considers the mechanisms of social cohesion in a colony of chimpanzees in Arnhem zoo, which he has studied since 1976. This group is held in superior zoo conditions, in

[17]Dr. Donald Omark (1980) observed the same fluid movement among American school boys in kindergarten and Grades one and two. He reports that by Grade three the pattern was becoming less fluid in form.

a large "semi-natural" enclosure during the summer months, and indoors in more restricted quarters during the cold months of winter. Even in the outdoor enclosure, the conditions of their lives are less natural than those of the wild groups on which I base my hypotheses. The Arnhem apes are not able to follow a pattern of fission and fusion, which is a major homeostatic mechanism of wild chimpanzee societies. They can never completely isolate themselves from the group (de Waal, 1982). The inability of captive chimpanzees to follow this tension-relieving pattern is undoubtedly one reason why captive apes engage in many more aggressive exchanges than do wild chimpanzees.

De Waal is impressed by the strong tendency of combatant chimpanzees to reconcile after an aggressive encounter. So strong is this tendency that he is led to wonder whether alternating friendly and aggressive interactions may not act to stimulate the formation of social bonds, more so than do continuing smooth affiliative interactions.

De Waal's hypothesis and mine illustrate the same theoretical point, the importance of parting and reunion as a way of reaffirming the social bond. It is possible that under the special (zoo) circumstances, reconciliation has replaced the carnival reunions of wild chimpanzees as a socially cohesive mechanism. Both act as "maintenance rites", but whereas reunion celebrates the social bond, reconciliation must re-establish it.

A constantly repeated pattern of freely parting (without cost) and reuniting, rather than costly aggression and reconciliation, is surely the optimally adaptive, cohesive mechanism. But either reunion or reconciliation revitalises the social bond between the individuals involved. De Waal's recognition of the reconciliatory behaviour of chimpanzees is further testament to the importance of stimulation and reaffirmation of social bonds. As Goffman (1971) suggests, the strength of a cohesive bond will deteriorate if nothing is done to keep it vital.

REFERENCES

Albrecht, H. & Dunnett, S. C. (1971). *Chimpanzees in western Africa*. Munich: Piper.

Baldwin, P. J. (1979). *The natural history of the chimpanzee (Pan troglodytes versus), at Mt. Assirik, Senegal*. Doctoral dissertation, Stirling, Scotland: University of Stirling.

Bowlby, J. (1973). Self-reliance and some conditions that promote it. In R. Gosling (Ed.), *Support, innovation and autonomy* (Pp. 23–48). London: Tavistock.

Chance, M. R. A. (1967). Attention structure as the basis of primate rank order. *Man, 2*, 503–518.

Chance, M. R. A. (1976). Attention, advertance and social control. In M. R. A. Chance & R. R. Larsen (Eds.), *The social structure of attention*. London: John Wiley & Sons Ltd.

de Waal, F. (1982). *Chimpanzee politics: Power and sex among apes*. New York: Harper & Row.

Draper, P. (1973). Crowding among Hunter–Gatherers: The !Kung bushmen. *Science, 182*, 301–303.

Durkheim, E. (1926). *Elementary forms of religious life*. New York: Macmillan.

Festinger, L., Pepitone, A., & Newcomb, T. (1952). Some consequences of de-individualization in a group. *Journal of Abnormal Social Psychology*, *47*, 382–389.
Gardner, P. M. (1972). The Paliyans. In M. G. Bicchieri, (Ed.), *Hunter–Gatherers today* (Pp. 404–407). New York: Holt, Rinehart & Winston.
Garner, R. L. (1896). *Gorillas and Chimpanzees*. London: Osgoode, McIlvaine & Co.
Ghiglieri, M. P. (1979). *The socioecology of chimpanzees in Kibale Forest, Uganda*. Doctoral dissertation, University of California: Davis, California.
Goffman, E. (1971). *Relations in public*. New York: Harper & Row.
Goodall, J. (1963). My life among wild chimpanzees. *National Geographic, 124(2)*, 272–308.
Goodall, J. (1965). Chimpanzees of the Gombe Stream Reserve. In I. DeVore (Ed.), *Primate behavior* (Pp. 425–473). New York: Holt, Rinehart & Winston.
Goodall, van Lawick-, J. (1971). *In the shadow of man*. Glasgow: William Collins, Sons & Co. Ltd.
Hebb, D. O. (1946). Emotion in man and animal: An analysis of the intuitive processes of recognition. *Psychological Review*, *53*, 88–106.
Hinde, R. A. (1974). *Biological bases of human social behavior*. New York: McGraw-Hill.
Hinde, R. A. & Stevenson-Hinde, J. (1976). Towards understanding relationships: Dynamic stability. In P. P. G. Bateson & R. A. Hinde (Eds.), *Growing points in ethology* (Pp. 451–479). Cambridge, England: University of Cambridge Press.
Homans, G. C. (1961). *Social behaviour: Its elementary forms*. London: Routledge & Kegan Paul.
Howell, N. (1979). *Demography of the Dobe !Kung*. New York: Academic Press Inc.
Itani, J. & Suzuki, A. (1967). The social unit of chimpanzees. *Primates*, *8*, 355–381.
Izawa, K. & Itani, J. (1966). Chimpanzees in Kasakati Basin, Tanganyika. (I) Ecological study in the rainy season 1963–1964. *Kyoto University African Studies*, *1*, 73–156.
Kortlandt, A. (1962). Chimpanzees in the wild. *Scientific American*, 206(5), 128–138.
Kummer, H. (1971). *Primate societies*. Chicago: Aldine.
Lee, R. B. (1968). What hunters do for a living. In R. B. Lee & I. DeVore (Eds.). *Man the hunter* (Pp. 30–48). New York: Aldine.
Lee. R. B. (1979). *The !Kung San*. Cambridge, England: University of Chicago Press.
Lee, R. B. & DeVore, I. (1968). *Man the hunter*. New York: Aldine.
Lee, R. B. & DeVore, I. (1976). *Kalahari hunter–gatherers: Studies of the !Kung San and their neighbours*. Cambridge, Mass.: Harvard University Press.
MacKinnon, J. (1978). *The ape within us*. New York: Holt, Rinehart & Winston.
Marshall, L. (1965). The !Kung bushmen of the Kalahari Desert. In Gibbs, J. (Ed.), *Peoples of Africa* (Pp. 243–278). New York: Holt, Rinehart & Winston.
Marshall, L. (1976). *The !Kung of Nyae Nyae*. Cambridge, Mass.: Harvard University Press.
Mason, W. A. (1965). Determinants of social behaviour in young chimpanzees. In A. M. Schrier, H. F. Harlow, & F. Stollnitz (Eds.) *Behaviour of nonhuman primates*. (Pp. 335–362). New York: Academic Press Inc.
Maynard Smith, J. (1978). The evolution of behavior. *Scientific American*, *239*, 176–192.
Mayr, E. (1970). *Populations, species and evolution*. Cambridge, Mass: The Belknap Press.
Montagner, H., Henry, J., Lombardot, M., Restoin, A., Benedini, M., Godard, D., Boillot, F., Pretot, M., Bolzoni, D., Burnod, J., & Nicolas, R. (1979). The ontogeny of communication behaviour and adrenal physiology in the young child. *Child Abuse and Neglect*, *3*, 19–30.
Montagner, H., Henry, J., Lombardot, M., Restoin, A., Bolzoni, D., Durand, M., Humbert, Y., & Myose, A. (1978). Behavioural profiles and corticosteroid excretion rhythms in young children. Part I: Non-verbal communication and setting up of behavioural profiles in children from 1 to 6 years. In N. Blurton-Jones & V. Reynolds (Eds.), *Human behaviour and adaptation* (Pp. 207–228). London: Taylor & Francis Ltd.

Murdock, G. P. (1968). The current status of the world's hunting and gathering peoples. In R. B. Lee & I. DeVore (Eds), *Man the hunter*. (Pp. 13–20). New York: Aldine.
Nisbet, R. & Perrin, R. G. (1977). *The social bond* (2nd. Ed). New York: Knopf Inc.
Omark, D. R. (1980). The umwelt and cognitive development. In D. R. Omark, F. F. Strayer, & D. G. Freedman (Eds.), *Dominance relations*. (Pp. 231–258). New York: Garland STPM Press.
Pitcairn, T. K. (1976). Attention and social structure in *M. fascicularis*. In M. R. A. Chance & R. R. Larsen (Eds), *The Social Structure of attention*. London: John Wiley & Sons Ltd.
Power, M. (1986). The foraging adaptation of chimpanzees, and the recent behaviour of the provisioned apes in Gombe and Mahale National Parks, Tanzania. *Human Evolution*, *3*, 251–266.
Reynolds, V. (1965). *Budongo: An African forest and its chimpanzees*. New York: The Natural History Press.
Reynolds, V. (1966). Open groups in hominid evolution. *Man*, *1*, 441–452.
Reynolds, V. & Luscombe, G. (1976). Greeting behaviour, displays and rank order in a group of free-ranging chimpanzees. In M. R. A. Chance & R. R. Larsen (Eds), *The social structure of attention* (Pp. 105–115). London: John Wiley & Sons Ltd.
Reynolds, V. & Reynolds, F. (1965). Chimpanzees in the Budongo Forest. In I. DeVore (Ed.), *Primate behaviour* (Pp. 368–428). New York: Holt, Rinehart & Winston.
Rowell, T. (1972). *Social behaviour of monkeys*. London: Cox & Wyman Ltd.
Schachter, S. (1959). *The psychology of affiliation*. Stanford, Calif.: Stanford University Press.
Service, E. R. (1966). *The hunters* (Foundations of modern anthropology series). New Jersey: Prentice-Hall.
Shafton, A. (1976). *Conditions of awareness: Subjective factors in the social adaptations of man and other primates*. Portland, Oregon: Riverstone.
Shils, E. (1968). Charisma. *International Encyclopedia of the Social Sciences*, *2*, 386–390.
Sugiyama, Y. (1972). Social characteristics and socialization of wild chimpanzees. In F. E. Poirier (Ed.), *Primate socialization* (Pp. 145–163). New York: Random House.
Teleki, G., Hunt, E. E., & Pfifferling, J. H. (1976). Demographic observations (1963–1973) of the chimpanzees of Gombe National Park, Tanzania. *Journal of Human Evolution*, *5*, 559–598.
Turnbull, C. (1961). *The forest people*. New York: Touchstone Books, Simon & Schuster.
Turnbull, C. (1968a). Hunting and gathering. Part III. Contemporary societies. *International Encyclopedia of the Social Sciences*, 7 (Pp. 21–16).
Turnbull, C. (1968b). The importance of flux in two hunting societies. In R. B. Lee and I. DeVore (Eds.), *Man the hunter* (Pp. 132–137). Chicago: Aldine.
Turnbull, C. (1972). Demography in small-scale societies. In G. A. Harrison & A. J. Boyce (Eds.), *The structure of human populations* (Pp. 283–312). Oxford: Clarendon.
Virgo, H. B. & Waterhouse, M. J. (1969). The emergence of attention structure amongst rhesus macaques. *Man*, *4*, 85–93.
Weber, M. (1947). *The theory of social and economic organization*. New York: Oxford Press.
Woodburn, J. (1980). Hunters and gatherers today and reconstruction of the past. In E. Gellner (Ed.), *Soviet and Western anthropology* (Pp. 95–117). London: Duckworth & Co.
Woodburn, J. (1982). Egalitarian societies. *Man*, *17*, 431–451.
Wrangham, R. W. (1974). Artificial feeding of chimpanzees and baboons in their natural habitat. *Animal Behaviour*, *22*, 83–93.
Wrangham, R. W. (1980). Sociobiology: modification with dissent. *Biological Journal of the Linnean Society*, *13*, 171–177.
Yerkes, R. M. (1925). *Almost human*. London: Cape.
Yerkes, R. M. & Elder, J. (1936). Oestrus, receptivity and mating in chimpanzees. *Comparative Psychology Monographs*, *13(5)*, 1–39.

5 The Reconciled Hierarchy

Frans B. M. de Waal
Regional Primate Research Center, University of Wisconsin, Madison, U.S.A.

INTRODUCTION

The absence of an obvious negative relation between the level of intra-group aggression in a species and the social cohesiveness of the groups in which it lives is one of the major paradoxes in the study of social organisation. The view of aggression as a dispersive, antisocial behaviour must be based on observations of territorial species, because in gregarious species aggression serves a multitude of functions, many of which have very little to do with the regulation of inter-individual distances. Thus, in three different macaque species, high levels of aggression have been reported among individuals who often associate. Weigel (1980, p. 316) has spoken of a "failure to demonstrate a dispersive effect of aggression", and Kurland (1977, p. 78), showing that the highest threat rates occur between close relatives, noted that these threats do not have the same effect as those between non-kin: "Although immature animals run from their mother's threats, within 10–15 s they are back at her side."

As yet, few studies have addressed the nature of mechanisms that social animals use to cope with and buffer antagonism, or how such mechanisms affect affiliative relationships. It is logical to assume that there is great variability in the effects of aggression because of variability in overall social relationships and in the specific contexts in which aggression occurs. It is even possible that aggression, and subsequent reassurance, contribute to bond formation. Kaufman (1974), in laboratory studies of mother–infant relationships in macaques, found that maternal punishment and rejection increase an infant's dependency and attachment. The same mechanisms of mother–infant bonding are likely to be found on the adult level. One well-known example is the way in which male hamadryas baboons (*Papio hamadryas*) herd their females. The male's threats and neckbites result in a "reflected escape", that is, the female does not flee from but rather toward her attacker (Kummer, 1957). The use of aggression to hold a harem together is rather straightforward, but in a more subtle way aggression may serve a similar binding function in a great variety of other relationships. Chance (1961) has already pointed out that threats do not always repel but also may attract.

The view that social bonding is based on a compromise between hostility and attraction, rather than on attraction alone, has been most strongly expressed by Lorenz (1966). His suggestions were based on observations of pair-bonding rituals in ducks and geese. Reynolds (1980, p. 7) objected to the extrapolation of these "bird-and-fish" insights to humans, and complained, "One wishes that Lorenz's knowledge about primates were greater". Lorenz definitely carried the paradox too far, but it is not so obvious that our present knowledge of primates undermines the idea that aggressive tendencies sometimes play a constructive role in bond formation. Although I agree with Hinde (1970) that aggression per se can hardly be regarded as something "good" for society, the central point of this chapter is to demonstrate that it is not fruitful to consider aggression in isolation from the total context of behaviour. Aggressive behaviour is embedded in social mechanisms, such as dominance, which evolved to mitigate its effects. In many species these mechanisms are so powerful that they allow for the use of restrained aggression to define and structure relationships without disrupting them (Bernstein & Gordon, 1974).

In the succeeding parts of this chapter the mechanisms of which aggression is part will be categorised and illustrated by examples from the literature. Then I will present quantitative data on the interrelationship of aggressive and affiliative behaviour in the unique chimpanzee (*Pan troglodytes*) colony of the Arnhem Zoo (Netherlands). Throughout, my emphasis will fall on *proximate* explanations—those based on immediate causes and goals—and on the social histories, experience, and intelligence of the animals involved. This treatment means that the perspective from which social dominance will be discussed will differ from the currently predominant genetic/evolutionary perspective. It will stress the impact of behaviour on the dynamics of social relationships and on the supra-individual system, irrespective of the reproductive interests of individuals.

Quite in contrast to Symons (1978), I believe that the current concentration of theorists on individual reproductive strategies and the taboo on interpretation at the level of the social system have severely restricted the scope of research on group-living animals. It has replaced what Koehler (1933; cited and explained by Lorenz, 1982) called *die Ganzheitsbetrachtung*—the perception of the entirety of a biological system—by an essentially atomistic view, one that regards genes as fundamental and societies as a mere abstraction. From the observer's standpoint, the world has been turned upside down. Actually, animal societies are observable realities, almost like a transparent organism (Kummer, 1984), whereas the influence of genes on behaviour is as yet mainly a theoretical matter. This is not to say that these approaches exclude each other; rather, they look at social animals from different angles. To restore the balance between evolutionary and mechanistic approaches, while maintaining a sharpened awareness of the distinction between them, is a major task for students in this field.

In order to answer evolutionary questions, we need to measure how each individual benefits from living in and exploiting a particular social system. Another type of functional question arises, however—perhaps not a "functional" question in the strictest biological sense, but a question regarding the working of the system as a whole. What would happen to the general social organisation if a particular behavioural mechanism were lacking, or if it were to operate differently? To what extent does the organisation depend on it? This effect we may call the "role" or "contribution" of a behavioural mechanism at the group level. How this higher-order function relates to the adaptive value of the mechanism is still unclear. (Ecological systems can serve here to illustrate the use of multiple functional levels. Hundreds of different organisms together build an intricate system, within which each plays its own role. These are often not roles for which they have been selected—deer are not designed to be food for tigers. Nevertheless, we cannot even begin to understand complex ecosystems without viewing all cause–effect chains as integrated into a well-balanced whole. At this higher level of functional analysis, deer do serve as food for tigers.)

When applied to the dominance concept, the foregoing approach takes the entire social relationship as its object of investigation, rather than confining itself to examining the priority of one individual over another and the reproductive benefits that result from that priority, which is the primary issue in evolutionary approaches to dominance (e.g., Fedigan, 1983; Shively, 1985). Current evolutionary models only explain why individuals strive to become dominant over others, not how the whole set of status-related behaviours (including submissive and appeasement behaviours) came into existence. It is not only the outcome, but also the mode and regulation of competition that demand attention.

Dominance–subordinance relationships will be regarded here as sorts of peace agreements. The determination of priorities is only one consequence of these agreements, and in primates there is remarkable flexibility about this. This variance is due partly to competitive success by individuals that have more influence and better connections in the group than their dominance ranks might suggest (de Waal, 1982; Strum, 1982), and partly to restraints and social tolerance in dominants. To indicate that dominants may refrain from claiming resources, various authors have used terms such as "concessionary behaviour" (Noë, de Waal, & van Hooff, 1980), "respect for possession" (Kummer, Götz, & Angst, 1974; Sigg & Falett, 1985), "control over the desire for exclusive possession" (Kawanaka, 1982), "special tolerance" (Yamada, 1963), or "favoritism" (Dittus, 1979).

One should not be surprised to see a female chimpanzee bow to the alpha male and grunt submissively while quietly taking the leaves he was eating out of his hands, even when he tries to avoid this action by turning away. As long as the female is respectful, such incidents will not disturb her relationship with the male. It is this combination of status communication, mutual trust, and tolerance that makes the psychology of dominance relationships so fascinating. There is more to dominance than simply the question of who eats the leaves.

AGGRESSION–AFFECTION SEQUENCES

Reconciliation Behaviour

There are numerous indications in the literature on primate behaviour that body contact, especially grooming, plays a role in the regulation of aggression (e.g., Blurton-Jones & Trollope, 1968; Ehrlich & Musicant, 1977; Ellefson, 1968; Lindburg, 1973; Poirier, 1968; Seyfarth, 1976). The functions of these contacts are commonly described as "reassurance", "appeasement", or "arousal reduction". Note that this terminology emphasises the effect on an individual's internal state rather than on the individual's social relationships. Three recent studies, however, have drawn attention to the social impact of this behaviour. McKenna (1978) showed that one out of three grooming bouts in a captive group of hanuman langurs (*Presbytis entellus*) occurs in tense situations. Aggressive interactions significantly increase the chance that grooming will occur. Speculating about the multiple meanings of grooming, McKenna (1978, p. 506) stated, "In one context the recipient of aggression grooms to appease the aggressor and to prevent the aggression from continuing, while in another context the aggressor grooms the victim, thereby reassuring the recipient that aggression has ceased and that non-aggressive behaviors can resume between them."

The second study concerns chimpanzees. Goodall's books and articles provide ample qualitative evidence of the outstanding need of wild chimpanzees to reassure and appease one another by means of a variety of contact behaviours (e.g., van Lawick-Goodall, 1968a,b). At Arnhem Zoo, which keeps a large colony of chimpanzees on a one-hectare island, participants in agonistic interactions were followed for three-quarters of an hour after each encounter (de Waal & van Roosmalen, 1979). Our quantitative data indicate that after an aggressive incident the participants have a high tendency to make body contact, especially with the opponent. Contact with outsiders is characterised by embracing, whereas kissing is characteristic of contact with the opponent. Both kinds of interaction may be preceded by inviting gestures, such as holding out a hand toward the other. Although grooming is common, it is not characteristic of post-conflict interactions (i.e., grooming often occurs in other contexts as well). To suggest a socially homeostatic function, we labelled contacts between former opponents "reconciliations". A comparison of inter-individual distances before and after agonistic encounters demonstrated that in our colony the effect of aggression was generally to *decrease* such distances.

The third investigation concerned a reputedly hostile and intolerant primate species: the rhesus monkey (*Macaca mulatta*). De Waal and Yoshihara (1983) compared the post-conflict behaviour of participants in agonistic interactions in a large captive troop with the behaviour of the same individuals during matched control periods. The dispersal hypothesis, predicting that former adversaries will avoid one another, could be rejected. The data, on the contrary, supported the reconciliation hypothesis, i.e., there was selective attraction between former adversaries. Apparently conflicts were resolved rather than forgotten. For another study on macaques that emphasises the role of calming body contact in the agonistic context, see Thierry (1984, 1986).

Besides body contact, reconciliation behaviour may involve gift-giving, at least among humans. This act occurs at an early age. Montagner (1978) has reported that after a fight children often repair their relationship with a peer by spontaneously bringing to their fellow some object or by simulating such an act of giving. To use material exchanges in this context seems typically human. It is doubtful, though, whether it is as effective as body contact in reducing tension. Maybe it is used mainly as an introductory gesture to establish contact. Montagner (1978, p. 90) suggested this function when he wrote "L'offrande (ou la simulation de l'offrande) est un comportement qui permet d'établir et de renforcer le contact, puis de développer des échanges non aggressifs", (translation: "Gift giving, or its simulation, is a behaviour allowing for the establishment and reinforcement of contact, hence of developing non-aggressive exchanges"). A systematic study by Schropp (1985, Note 2) lends further support: object exchanges facilitate interactions among children who are at great social distances from one another.

Other ways in which humans break tensions and de-escalate conflicts have been investigated in children by Gottman and Parkhurst (1978, Note 1), and by Sackin and Thelen (1984). The solutions they observed include asking a question about the other's feelings, making a joke, making co-operative propositions, offering apologies, and talking in fantasy roles in which the children give each other emotional support.

Contrast Value

The investigations summarised here show that aggressive and friendly interactions alternate, and very probably affect each other. One of the unanswered questions in research on social cohesion is whether this alternation in fact stimulates bond-formation. In other words, if we represent conflicts by a minus sign, and grooming, embracing, and other affiliative interactions by a plus sign, does a (+ − + − + − + −) sequence give a stronger bond than a (+ + + + + + + +) sequence? According to Lorenz the second sequence, a purely positive relationship, is hardly possible. He stated (Lorenz, 1966, p. 209), "Thus intraspecific aggression can certainly exist without its counterpart, love, but conversely there is no love without aggression". He meant that love depends, at least partly, on aggression. The arguments are far from convincing, however. How can we distinguish this possibility from the alternative hypothesis that individuals who are often together cannot avoid being aggressive now and then? The big question is: do social bonds exist thanks to a certain dose of aggression, or in spite of it?

Empirical investigation of this intriguing problem requires that agreement be reached about what to call a social bond. In order to avoid circularity and confusion between the intensity of reconciliations and the intensity of the bond itself, the definition should not be framed in terms of those particular interactions. Individuals may be said to have a strong social bond if they: (1), react by distress when separated; (2), are prepared to spend energy to get together again; and (3), show positive contact behaviour upon reunion. Testing of bonds should preferably take place in a normal group context. If monkeys A and B are first removed from their group and then tested to see how they react to being separated, the result may be different from that of a test in which only A is removed from the group and B has to "work" to get A back into it. Only in the second situation can B's response not be explained by mere social deprivation; it reflects a specific need for the presence of A. In addition, only the second test measures all bonds, including those that depend on the group context.

Experimental studies of how bonding is affected by post-conflict contacts should compare these particular interactions with normal, socially positive behaviour. A distinct characteristic of post-conflict contact between former opponents is that it occurs in situations in which one individual is usually full of doubt about the other's intentions. It is conceivable that friendly gestures that put an end to such incertitude have a stronger impact on a relationship than do friendly gestures in a more relaxed situation. If that is true, we might speak of a "contrast value" of conciliatory gestures (analogous to contrast effects in perception and conditioning). In other words, affliative contacts following conflicts are not to be counted as mere (+) interactions; they are of a different order. A regular occurrence of reconciliations, together with an inhibition of aggression, may create mutual trust to a degree that no smooth series of normal (+) interactions can bring about. If so, the prediction is that experiments in which two animals can contact each other as much as they want *except* during a limited period after aggression, are less likely to result in social bonds than experiments in which the contact opportunity is severely limited, except after aggression.

Such experiments, which have yet to be done, are necessary to distinguish the possibility that aggression and subsequent reconciliation promote bond-formation from the alternative possibility—that bonding depends on quite separate mechanisms and that reconciliations merely serve as a "lubricant", rescuing the bond from the undermining effects of aggression. It is well to realise, however, that these two hypotheses need not completely exclude each other. It is conceivable that, *given* some attraction or mutual dependency between two partners, their bond can actually be manipulated and even strengthened by aggression. An example is to be found in human initiation ceremonies, known from many cultures. Hazing and hostilities are unlikely to have any bonding effects on newcomers who lack a desire to join the club. It is only in combination with such a desire that harsh treatment has the potential of strengthening attraction (Gerard & Mathewson, 1966). The ingredients of many initiation rituals are: (1), attraction; (2), aggression; (3), subordination to senior members, or to a Master; and finally (4), reconciliation and the euphoria of social acceptance. Similar ingredients are recognisable in a widespread animal bonding pattern.

Conditional Reassurance

In a series of experiments with young chimpanzees, Mason (1965) showed that they responded to frightening situations by seeking body contact. The amount of distress vocalisations made by the apes in reaction to painful stimuli could be reduced by grooming them or by holding them in a ventro-ventral position. This observation strongly suggests that one function of body contact is to reduce arousal. However, not all contacts serve this function. Social play does not. Mason's reunion experiments rather indicate that play occurs *after* the reduction of arousal through other forms of contact.

It is interesting to compare these results with those of studies on the effects of spatial crowding on an entire group of chimpanzees

(Nieuwenhuijsen & de Waal, 1982). The Arnhem colony is kept indoors during the winter, on a surface one-twentieth the size of their outside enclosure. Under these crowded conditions the members of the colony groom significantly more, and play less. Observed was also an increased level of aggression, but this effect was far less dramatic than one might expect from the results of crowding studies on rodents or even monkeys; the frequency of aggression increased by a factor of less than two, and the intensity of aggression did not change. Our speculation was that primates, and in particular chimpanzees, are better able to cope with stressful situations than are rodents, and that grooming is one of their tools for coping.

The fact that aroused individuals actively seek a calming contact gives an aggressor an effective extra weapon, provided it is with the aggressor that the other tries to make contact. The aggressor can make conditions of the form: if you want contact with me, then you have to show behaviour X, or to stop behaviour Y. The clearest illustration of this mechanism is the way mothers treat their offspring during weaning, at least in species in which the mother–child bond extends beyond the weaning period. A chimpanzee mother, after having pushed or threatened her infant away from her breast, allows it to return on condition that it not suckle any more. The infant often throws a temper tantrum in such situations and urgently wants to be embraced or groomed by its mother. When the mother finally does provide reassurance the infant immediately stops screaming, but as soon as it attempts to re-establish nipple contact the whole noisy interaction may start again.

The weaning procedure may serve as a blueprint for other conditioning interactions. Altmann (1980, p. 167), regarding weaning as a process by which mothers teach infants the "proper" time for contact and nursing, noted: "This is probably one of the earliest forms of socialisation for an infant baboon and may provide the social sensitivity on which later social integration depends." Before discussing the function of this sensitivity in dominance relationships among conspecifics, we note a telling human–chimp example described by Terrace (1979). Nim, a young chimpanzee used for a language experiment, did not let himself be dominated by his caretakers all of the time. An effective technique for disciplining him was to walk away, preferably while signing "You bad" or "I not love you." Nim responded to this threat of separation by what the author calls the "sorry–hug routine". In the course of time, however, this procedure became less and less effective, unless Nim's caretakers postponed the reassuring contact. Sometimes his response to that was a temper tantrum. "After a tantrum Nim's behavior almost always changed dramatically for the better. The exceptions were instances in which a teacher reassured Nim too quickly" (p. 108).

Thus we arrive at a hypothesis that there are situations in which reconciliation occurs on condition that the recipient understand the reason for the previous aggression and adapt its behaviour accordingly. If aggression is the stick, reconciliation often is the carrot. Weaning is one example of such an "arrangement"; establishment of a dominance relationship seems to be another.

THE AFFECTIVE COMPONENT OF THE HIERARCHY

Formal Dominance

Most definitions of dominance are in terms of winning or losing, pecking or being pecked, taking or relinquishing possession, approach or retreat, and so on. The definitions concern the outcome of competitive encounters rather than the exchange of communication signals between dominants and subordinates. Their communication, however, is usually more stable over time and more predictable than are the outcomes. Rowell (1966) has already demonstrated that subordinate behaviours are remarkably consistent in their direction as compared to most dominant or aggressive behaviours. A further distinction within the category of subordinate behaviour was reported for Java monkeys (*Macaca fascicularis*) by Angst (1975) and de Waal (1977). These macaques signal subordinance by means of a stereotypical facial expression in which the teeth are bared. This facial display is completely unidirectional between individuals, i.e., if during a given period A shows the signal to B, B will never show it to A. Non-ritualised expressions of fear, such as avoidance and flight, show exceptions to this rule in up to 5% of the cases.

To test further this remarkable consistency in the direction of the teeth-baring behaviour, de Waal and Luttrell (1985) undertook an exhaustive study of a large captive group of rhesus monkeys. The study involved nearly 2,000 hours of observation over a 30-month period. The facial display was observed to be completely unidirectional in 244 of the 245 dyadic relations in which it was observed. In the one exception, mutual teeth-baring was limited to an unstable period of a few weeks during which two sisters reversed their ranks. We also found that the hierarchy based on teeth-baring approached perfect linearity, which is not the case for other dominance measures, such as the direction of aggressive encounters (Bernstein & Williams, 1983; de Waal, 1977).

The teeth-baring display, however, is not unidirectional among chimpanzees, but that species has as a substitute another common signal that is consistently given by subordinates. This signal is a vocalisation known under such various names as a "rapid ohoh", "pant grunting", "bobbing pants", or "submissive greeting" (Bygott, 1974; Noë, de Waal, & van

Hooff, 1980). This vocalisation is part of greeting rituals in which subordinates utter the sound while making bowing or bobbing movements toward dominant individuals. Compared to those of macaques, the hierarchical relationships of chimpanzees are best characterised as flexible and plastic (de Waal, 1982; Maslow, 1940), with a heavy reliance on changing coalitions (Riss & Goodall, 1977; Nishida, 1983; de Waal, 1982; 1984a). This distinction makes the presence of a highly predictable status indicator in chimpanzees all the more remarkable.

What does it signify that certain signals have a completely fixed direction? Surely, it implies that the direction of that signal is not affected by factors that produce inconsistencies in the outcomes of conflict. The most important factor is the momentary social context (e.g., the presence or absence of supporters). This conclusion is why teeth-baring in the macaque has been called an indicator of "context-free" dominance relationships (de Waal, 1977). The only way to explain this phenomenon is to assume *status awareness*. That is, primates know in which relations they are dominant and in which relations they are subordinate, and they communicate that awareness. Such perceptions do not need to be disturbed by fluctuations in the outcome of conflicts, so long as a predominating outcome remains recognisable. The dependence of ritualised status signals on evaluation processes makes these signals relatively immune to transient variations in context.

Popp and DeVore (1979) are presumably right in assuming that the outcomes of competition are resource-specific and in denying the existence of a *general* hierarchy, but their opinion that there is no status irrespective of social context is untenable in the light of the above evidence. What can be seen is a continuum of dominance-related behaviours that range from signals expressing fixed positions in the hierarchy to the actual outcomes of conflicts, which are more variable and do not automatically correspond with the communicated positions. Priority of access to incentives seems to be at the other end of the scale, as being easily affected by motivation and context.

When Altmann (1962) characterised status indicators as *meta*communication he illustrated his point by quoting Itani's unpublished observation of a male macaque who did not dare to take a tangerine thrown between him and a dominant animal until the latter had reacted to the subordinate's presentation by mounting him. Such interactions illustrate the paradox that confirmation of the fixed social relationship allows flexibility in other domains of behaviour. A good example of the greater variability in competitive situations is seen in a study by Weisbard and Goy (1976) of the changes in rank that follow parturition in female stump-tail macaques (*Macaca arctoides*). The investigators manipulated the social context and found that the presence of top-ranking individuals affected the drinking order among the rest of the group, but not the direction of agonistic behaviour. In this study, competition over access to water was created by water-deprivation for 24 hours and a testing procedure with a single drinking fountain. A shorter deprivation period and a shareable resource (e.g., a water basin) lead to even greater flexibility in the outcome of competition, due to an increased influence of special tolerance relationships, whereas the direction of approach–retreat interactions and submissive teeth-baring remains unaffected (de Waal, 1986a).

The discussion that follows will be restricted to ritualised dominance criteria, because they reflect well-established relationships. Macaques do not achieve these relatively stable positions before they reach puberty. Young monkeys are still in the midst of determining their ultimate ranks. They usually flee from threatening peers without baring their teeth, and they direct an increasing amount of appeal–aggression to older group members that rank below their mothers. Appeal–aggression consists of vocalised threats together with recruitment behaviour shown to third parties. Adult monkeys, on the other hand, have already attained well-recognised rankings. Their agonistic encounters are of a predominantly communicative nature; the threats of the dominant animal are usually silent and look more self-confident, since there is no need to appeal to third parties (de Waal, 1977). See Walters (1980) for similar observations on dominance processes in young baboons (*Papio cynocephalus*).

In chimpanzees the distinction between agonistic and ritualised dominance is even more pronounced, because they communicate their relative positions in rank by means of signals that are not, like most status signals of baboons or macaques, of an agonistic nature. The panting grunts and bowing movements of the subordinate chimpanzee and the dominant's way of moving an arm over the other during display are only rarely seen in the course of escalated aggressive encounters. These status rituals occur mostly sometime after social excitement in the group, or as a greeting behaviour. Social organisation, then, can be viewed as consisting of two layers: a stable *formal* layer and a flexible layer of *real* dominance (de Waal, 1982). After parturition or during oestrus, a female monkey may temporarily assume a more advantageous position in the social network and thus gain in terms of real dominance, but her formal position in the group hierarchy remains the same. In the same manner, high-ranking anoestrous chimpanzee females may win 81% of their competitions and 20% of their aggressive confrontations against adult males, but formally they rank below the males, i.e., they frequently greet males submissively and the reverse behaviour is almost unthinkable (Noë, de Waal, & van Hooff, 1980). Simply to adopt a new terminology cannot explain these phenomena, but may help to formulate more constructive questions than one such as: "What is the value of the dominance concept?", so often heard whenever low correlations between different dominance criteria are found.

Although formal dominance and success in aggressive encounters are not identical, they do overlap. If the discrepancy between the two rank orders exceeds a certain limit, the formal hierarchy is likely to become "adjusted". In addition, it is to be expected that the formal hierarchy will regulate certain forms of competition, although not necessarily in the simplest possible way (dominant animals may control the situation without taking priority).

The correlation between different measures of dominance should not obscure the distinct nature of formal relationships, however. Thus Bygott (1974) noted that the chimpanzees in Gombe Stream, Tanzania, show their whole extensive repertoire of social and agonistic behaviour patterns to baboons, with the notable exception of the pant–grunting signal, a vocalisation that is termed a submissive greeting throughout this chapter, and is to be regarded as the chimpanzee's indicator of formal status. Adult male baboons may be individually known and feared by some chimpanzees (i.e., the baboons have real dominance), but none of them ever reaches an acknowledged position in the chimpanzee hierarchy. Because, as will be argued later, the formalisation of status facilitates social bonding, Bygott's observation reflects the integrity of the chimpanzee community: baboons lack membership therein.

The Key to Reconciliation

Formalisation of status communication permits peaceful and subtle relationships to endure, even to the extent that highly reputable observers have failed to recognise dominance in them. Thus, Heinroth and Heinroth (1928) and Lorenz (1935) have suggested that within goose families there is a kind of rankless tolerance or, at least, that if there is something like a rank order the geese do not really make use of it. Kalas (1977), however, followed the ontogeny of relationships among goose siblings (*Anser anser* L.) and observed fights among them during the first days after hatching. After that period, overt aggression became extremely rare, and positions in the rank order were expressed almost exclusively in greeting behaviour. Kalas did not want to call this a "Hackordnung" (pecking order), and saw it as a fixation of reactions that reduces both aggressive and fearful tendencies. She considered the maintenance of a close and long-lasting family cohesion to be the main function of that process.

The suggestion that the existence of hierarchies helps to reduce tension is widespread and old (e.g., see Collias, 1953). Ideas about their cohesive function are not so general, however, although Sparks (1967) has observed that the amount of grooming in primate species seems to be positively related to the strictness of their hierarchies. More recently, Seyfarth (1977) presented an elaborate model of rank-related grooming, a model now in the process of being evaluated. An experimental study by Maxim (1976), using pairs of rhesus monkeys, illuminates the close connection between the establishment of dominance and the development of a positive attitude. After initial fighting, there was a gradual decrease in the aggressive tendency of one monkey and of fear responses of the other, until in the end they showed expressions of dominance and submission that, according to Maxim's analysis, fell within the "friendly zone".

A preliminary study of female ontogeny in a large captive group of rhesus monkeys (de Waal & Luttrell, 1985) indicated that the establishment of dominance and social integration are one and the same. The age at which a young female began to receive ritualised submission from adult females whose ranks were below that of the young female's mother was negatively correlated with her frequencies of contact and grooming with those same females. This result translates into a positive link between social bonding with adult females and the speed with which formal dominance over them is achieved. The correlation persisted after controlling for other variables, such as rank distance and frequency of aggression. Further research is needed to determine whether bond formation facilitates the establishment of dominance, or vice versa.

In patas monkeys (*Erythrocebus patas*) we can see the consequences of a weakly developed hierarchical system. Kaplan and Zucker (1980) noticed that fear grimaces between females of this species are uncommon, that their competitive interactions in captivity involve an astonishing amount of physical aggression, and that affiliative contacts are relatively rare. These females seem to show a combination of weak formalisation and weak cohesion. The authors suggested that these monkeys have a "strong dependence on space, rather than body gestures, as the social buffer for individuals within the group" (p. 211). Similarly, Rowell and Olsen (1983) noted that social organisation is maintained by behaviours that regulate distance, and that this is of special relevance in species, such as the patas monkey, that show relatively little display behaviour.

Perhaps patas males are even more extreme in this respect than the females: they seem to lack the social technique of ritualised submission. When spatially restricted, they are unable to put an end to their fights; according to Kummer (1971, p. 111), ". . . no matter how far the weaker male now withdrew, the winner would relentlessly seek him out and chase him every few minutes". This is a very different situation from the fights observed among male gelada baboons (*Theropithecus gelada*), which Kummer (1975) also analysed in detail. In the gelada, after a decisive outcome was reached, the loser would nervously avoid the winner, who tried to approach him with appeasing gestures such a lip-smacking and presenting. They then proceeded to mount and groom each other, and finally become relaxed. Kummer (1971, p. 112) concluded from his comparison of patas monkeys and gelada baboons that the "behavioral design that permits troop life in geladas thus is submission".

Bernstein (1969) observed an old alpha male pigtail monkey (*Macaca nemestrina*) who, instead of showing submission, tried to ignore the male who had badly injured him. The alpha male died after a few weeks of continued hostility. In the next "round", the same aggressor thoroughly defeated another male, but this second adversary did show submissive responses. The winner then ceased further attacks upon him, groomed him, and even cleaned the wounds he had inflicted. In other words, acceptance of the relationship by the subordinate monkey was the key to reconciliation with the challenger. It is easy to see the mechanism of conditional reassurance at work in this situation.

Contact Refusals among Male Chimpanzees

In the Arnhem chimpanzee colony I have made numerous, although initially mainly unsystematic, observations of conditional reassurance. Behaviour was observed in five dominance contests among adult males. Three of these contests resulted in a reversal of the previous formal relationship, and the other two in a re-establishment of it. Each contest took two to three months to complete. During this period the two males did not exchange any submissive greeting, showed many mutual bluff displays, and had a high frequency of conflict. They also often contacted one another in a friendly manner, except at the end of this unstable period. In the last few weeks of the process, the losing party's attempts to establish grooming contact with the winner became less and less successful, especially after conflicts or bluff displays. The winner would avoid reconciliation *unless* the loser would utter some submissive grunts. For detailed descriptions, see de Waal (1982).

Fortunately, in one of the contests, non-agonistic contacts between rivals could be systematically recorded from the very beginning. The social instability that arose was predictable, as the alpha male had been temporarily removed from the group. Within a week, two of the other adult males started a fierce competition over the vacant position. Their contest lasted 90 days, and during that time I spent 218 hours of observation spread over 50 days, with the express intention of measuring the effect of submission on affiliative tendencies. During the first nine weeks of the contest both males initiated grooming and other affiliative contacts with one another: 53% of the contacts were initiated by the eventual loser, 47% by the winner (n = 180). There was an average of 1.3 contacts per hour (with weekly averages ranging between 0.9 and 2.4 contacts per hour). The frequency of contact dropped almost to zero in the two weeks before the first submissive greeting occurred. The peak in the frequency of agonistic conflict occurred during this same period. The absence of affiliative contacts resulted from avoidance by the future dominant male. He refused contact up to six times a day. His first refusal occurred four weeks before the loser started showing

TABLE 5.1
Progressive Establishment of a Dominance-Submission Relationship in Chimpanzees

Type of Behaviour	*Frequency per hour* Before First Submission	After First Submission	*Statistics** U	P<0.05
Bluff Displays				
by Yeroen	2.5	1.1	9	
by Luit	7.0	6.1	17	
Agonistic Conflicts	0.9	0.5	3	*
Friendly Contacts	0.2	1.2	0	*
Grooming (in minutes)	0.5	10.1	2	*

*Mann-Whitney U, two-tailed ($n_1 = n_2 = 7$)
NOTE: During the winter of 1977–1978, adult male Yeroen unsuccessfully challenged Luit. On March 16, Yeroen uttered the first submissive grunts to Luit. The Table compares the seven observation days before first submission with the seven observation days after first submission.

submissive greeting, and the last refusal was observed one day after the first signs of submission. The few affiliative contacts observed during the weeks of increased tension were all initiated by the loser (n = 9). Table 5.1 shows that after this period the frequency of contact returned to its previous level and that there was a dramatic increase in the amount of time spent in grooming between the two males. Significantly, the longest grooming bout of the whole period occurred a few hours after the first observed submissive greeting ritual. This analysis confirms that the reestablishment of a peaceful relationship is conditional upon the formal recognition of dominance.

In addition, it was observed that adult male chimpanzees may show pathetic temper tantrums when they lose a conflict or are refused reconciliation afterward. This remarkable response suggests that the possible link with weaning remarked upon earlier is more than an analogy. The psychological mechanisms involved in both weaning and the acceptance of a subordinate position may be very similar.

Linking Dominance to Sharing and Trading

"One of the most difficult problems is to reconcile the finding that dominance is based on fear–aggression with the observation that the dominant animal is often sought or followed by the other members of the group," wrote Mason (1964, p. 292). It is a general shortcoming of the classical hierarchy or pecking-order model that the affective component has been left out. Whereas pure fear would induce an animal to run away and never come back, social animals usually stay and show submission. Submission results from ambivalence between fear and attraction. Schenkel (1967, p. 319) most clearly expressed this view when writing about wolves and dogs: "Submission is the effort of the inferior to attain friendly or harmonic social integration."

In macaques this aspect of submission is visible in a rich continuum of facial and vocal displays that blend fearful and friendly tendencies: screaming, silent teeth-baring, teeth-chattering, and lipsmacking (van Hooff, 1967). In the chimpanzee submissive pant–grunts are usually uttered as a greeting, i.e., with an outspoken approach tendency. The spontaneous element is illustrated by Riss and Goodall (1977, p. 142) in their description of Evered's submissive attitude after his defeat by Figan (both were adult male chimpanzees): "One morning Evered climbed from his nest at dawn and travelled for 5 min., until he came to the tree where Figan was sleeping. He then sat below Figan's nest and waited for another 25 min. until Figan awoke. When Figan looked down towards him, Evered made extremely submissive pant–grunt vocalizations and bobbing gestures."

In their analysis of bureaucratic organisation, Williams, Sjoberg, and Sjoberg (1980) have described bonding mechanisms within the framework of formalised dominance relationships among humans. Subordinates will say and do things to make the boss feel important, and dominant individuals will occasionally make concessions and show that they care about their subordinates (e.g., by remembering their birthdays). According to the authors, these gestures serve to strengthen bonds of loyalty. The same link with bonding mechanisms is present in the discussion by Friedman (1973, p. 123) of the distinction between *de jure* and *de facto* authority, power, and prestige, in which he explains the usefulness of these concepts in defining "the nature of the cohesion or unity characteristic of human societies". Hinde and Datta (1981) have drawn attention to the possible value of applying similar concepts to primatology. In chimpanzees, for example, there may exist a relation between an individual's role as protector and arbitrator in disputes, on the one hand, and its frequent receipt of greeting ceremonies and massive support against rivals, on the other. As this complex of characteristics is not necessarily present in the formal alpha male, it needs to be distinguished from dominance, and has been termed "authority" (de Waal, 1982).

Social attraction and dependency turn fear into submission and respect. At the same time they make dominant individuals conciliatory and "forgiving", rather than despotic. These two effects are interdependent: no submission, no peace. Once the dominance relationship is clearly established the dominant's increasingly tolerant attitude may create room for a certain degree of assertiveness, protest and, in some species, teasing on the part of subordinates. Theoretically, it is even possible that in the course of time the formal dominance–submission relationship becomes a mere convention, necessary to avoid tensions, behind which a relationship on almost equal terms develops (although dominants may in a critical situation "fall back" on the original relationship).

The highest stage in this development toward attenuated status differences is reached when competitive tendencies are kept so well under control that peaceful sharing and trading among adults become possible. In hominoids, resource distribution may depend as much on these mechanisms as on aggression and dominance. With respect to food, alternative modes of distribution have been experimentally demonstrated in captive chimpanzees by Nissen and Crawford (1936) and by Yerkes (1941), and have been observed in the field by Goodall (1963), Nishida (1970), Teleki (1973), and others.

Access to females in oestrus also may be determined by non-competitive mechanisms, described as *sexual bargaining* by de Waal (1982). In the presence of sexually receptive females, adult male chimpanzees in the Arnhem colony showed a dramatic increase in grooming activity, directed at other males who were attracted to the same female. This activity seemed to relax the atmosphere to such an extent that sharing the female became possible. Male–male grooming thus appeared to be the "price" of an undisturbed mating session. Additional data on this phenomenon are now being collected in Arnhem and, Otto Adang tells me, these seem to corroborate the trading model.

If (1), the shift from the principle of aggressive dominance to that of sharing and trading is contingent, among other things, on advanced techniques for regulating tension; and (2), tension regulation is part and parcel of dominance relationships, as I have already argued, it follows that there may exist some degree of evolutionary continuity between the two principles. Non-competitive modes of distributing resources, according to this view, have been achieved through an extension of the status-related mechanisms of inhibition of aggression, of reassurance, and of appeasement. One intriguing indication of such a link is that chimpanzees beg for food with the same hand gesture that they use to invite former adversaries to a reconciliation (de Waal & van Roosmalen, 1979; de Waal, 1982; Mori, 1982). Another finding consistent with this hypothesis is that generosity during feeding is greatest in adult male chimpanzees (McGrew, 1975; Silk, 1979; Teleki, 1973); this is also the age and sex class with the most clear-cut formal status hierarchy.

CHIMPANZEE SOCIAL ORGANISATION

General Data

The best examples of relationships that defy the traditional dichotomy between affiliative and antagonistic tendencies are those between adult male chimpanzees of the same community. They are serious rivals, even to the extent that they may kill one another in captivity (de Waal, 1986b), yet at the same time they can be called friends. Data supporting my statement were collected from the Arnhem chimpanzee colony, which contained four adult males and nine adult females (together with a growing number of immature individuals) during the course of our studies. The data stem from 334 hours of protocol taken by a changing team of co-workers and the author in the course of the five summer periods from 1976 through 1980. On the average, two to three observation hours per week were spent on the long-term project. There were always at least two simultaneous observers, recording detailed reports on tape or video recorder. Besides recording agonistic encounters, greetings, intimidation displays, sexual encounters, and other relatively short-lasting interactions, the observers took five-

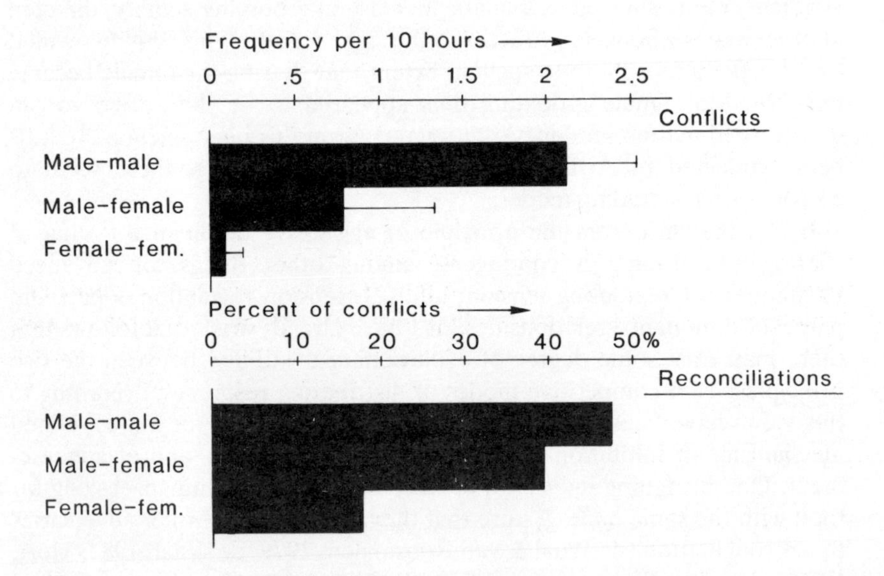

FIG. 5.1 Agonistic interactions and reconciliations in dyadic relationships.
Top: mean and standard deviation of agonistic conflicts (n = 1337) averaged per dyadic relationship among adult chimpanzees. The male–female class comprises behaviour in both directions. Bottom: proportion of agonistic conflicts, observed in a separate study (see text), followed within a half-hour by non-agonistic contact between the former adversaries.

minute time-samples to record play behaviour, grooming, and association patterns (see de Waal & van Hooff, 1981; Nieuwenhuijsen & de Waal, 1982, for details of definitions, methods, and analysis).

Figure 5.1 shows that the frequency of aggression among adult females was about one-twentieth that among adult males. The frequencies given are averages per pair of individuals (6 male–male, 36 male–female, and 36 female–female pairings), irrespective of the direction between them. For example, between any two particular females an average of 1 agonistic conflict per 104 hours was observed.

The average frequency of both agonistic conflicts (Fig. 5.1) and submissive greeting rituals (Fig. 5.2) is significantly higher among male dyads than among female dyads (two-tailed *t*-tests, $P < 0.001$). Interactions between the sexes show intermediate frequencies: higher than among females ($P < 0.001$) and lower than among males ($P < 0.001$ for conflict frequency; $P < 0.05$ for greeting frequency). It should be noted that greeting rituals and aggressive encounters usually occur separately in time.

Associative behaviour and grooming show a different pattern (Fig. 5.3). The amount of time spent within arm's reach (sample categories "contact-sit", "groom", and "proximity" taken together) is on the average around

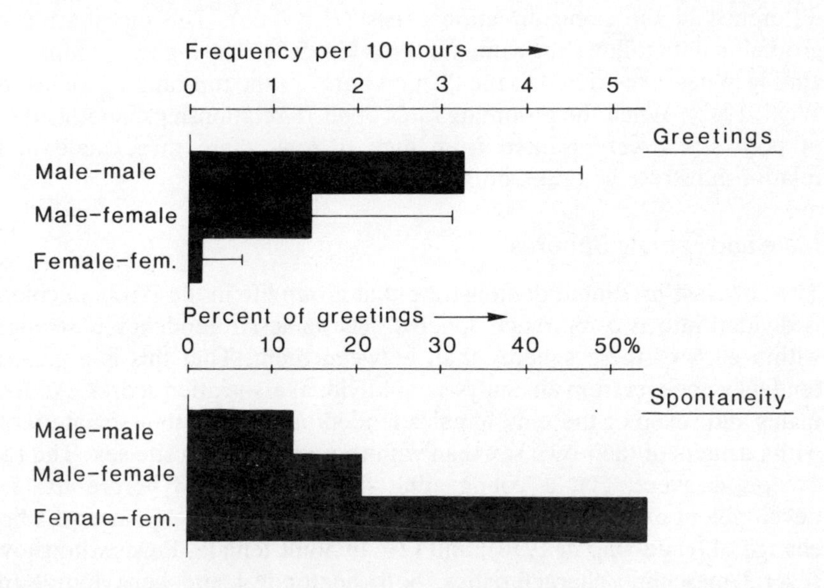

FIG. 5.2 Submissive greeting rituals in dyadic relationships
Top: mean and standard deviation of pant–grunt greetings (n = 2406) averaged per dyadic relationship among adult chimpanzees. Greeting between sexes occurred in 100% of interactions in the female-to-male direction. Bottom: proportion of greetings in the same sample that were of a spontaneous nature. For definitions, see the text.

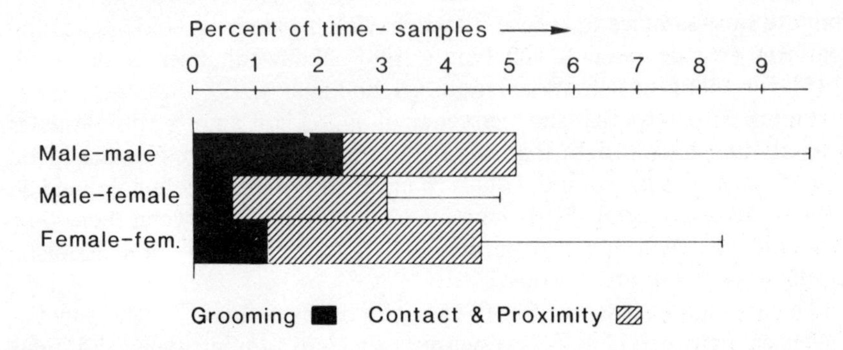

FIG. 5.3 Percentage of time-samples in which individuals associated in dyadic relationships Mean and standard deviation of contact–sit, grooming and proximity (i.e., sitting within arm's reach), combined and averaged per dyadic relationship among adult chimpanzees. Within the bars, average grooming rates are given separately in black. The graph is based on 1868 scan samples.

5% per relationship between individuals of the same sex. Between the sexes, on the other hand, there is less association. For females, the amount of association with members of the opposite sex is significantly lower than with members of their own sex ($P < 0.05$); for males a nearly significant difference in the same direction exists ($P < 0.06$). The distribution of grooming data follows the same pattern (Fig. 5.3). The high grooming rate among males is restricted to the "triumvirate" of the top-ranking males (de Waal, 1982). When the grooming rates of all 78 relationships among adults of both sexes were ranked from high to low, these three male–male relationships received rank numbers 2, 4, and 10.

Male and Female Spheres

The data just presented demonstrate that group life in the Arnhem colony is divided into two unisexual spheres. The cohesive tendency is stronger within each of these spheres than between them. That this is a general tendency appears from an analysis of individual association scores. All four males and seven of the nine females tended on average to associate more with partners of their own sex than with those of the opposite sex. The two exceptions were: (1), a young adult female who died before the full development of the two spheres in the colony (which started after the first change of leadership in 1976); and (2), an adult female, Puist, who shows several masculine characteristics, both anatomical and behavioural (de Waal, 1982).

The segregation of the sexes can be explained partly by hostility. As females meet much more aggression from males than from other females, they may avoid the male sphere. It should be clear by now, however, that a high risk of aggression never suffices as an explanation. Otherwise, it would be hard to explain why males stay together.

Males must have powerful mechanisms to overcome the disruptive effects of aggression. Support for this supposition includes the relatively high, mutual grooming rate among the three top males. A more direct way of investigating the relation between aggression and association is to observe the occurrence of reconciliations. This could not be included in the long-term procedures because it was too time-consuming. Instead, Fig. 5.1 (bottom) includes data gathered by a student, Tine Griede, in the summer of 1980. She followed the participants for a half-hour after any conflicts between them (n = 510 agonistic conflicts). Among adult male opponents there were relatively more non-agonistic post-conflict contacts within this interval of time than among female opponents ($\chi^2 = 6.4$, $P < 0.05$). Interestingly, male–female conflicts were also followed by more contacts of the participants than were conflicts among females themselves ($\chi^2 = 4.0$, $P < 0.05$), and over half of these reunions with males were initiated by females (non-significant). This means that adult females, although they seem to avoid the rough male sphere, do not avoid reconciliations with males.

It is important at this point to emphasise both the Arnhem colony's similarities with and its differences from the lives of chimpanzees in their natural habitat. The patterns of males in the Arnhem Zoo seem similar to those of wild chimpanzees. Also, in the wild the male hierarchy is much more pronounced than is the female hierarchy (Bygott, 1974). Complex coalition strategies of the kind reported for the Arnhem males (de Waal, 1982; 1984b) are known from feral males as well (Nishida, 1983; Riss & Goodall, 1977). On the other hand, for females the situation is quite different. Females in the Arnhem colony frequently associate with one another (Fig. 5.3) and form strong, stable coalitions, which serve them in agonistic contexts and even influence the male hierarchy (de Waal, 1982; 1984b). In their natural habitat, on the contrary, female chimpanzees are noticeably less sociable than males (Halperin, 1979; Nishida, 1979; Wrangham, 1979).

Of two characterisations of intra-sexual relationships in feral chimpanzees made by Nishida (1979), one concerning males agrees with the pattern of group life in the Arnhem colony, whereas the one concerning females does not. Nishida (p. 93) wrote of the males, completely in line with the hypothesis adopted here: "It is likely that complex sequences of threat–submission–reassurance interaction may strengthen the male bond among chimpanzees." Of females he wrote (p. 102), "Generally, social bonds among females may, if anything, be passive homogeneous aggregations based on physiological homogeneity in activity, occasioned by similar age, maternity and sexual receptivity."

Figure 5.2 (bottom) shows clearly that the female hierarchy in Arnhem was of a more spontaneous nature than the male's. A submissive greeting (i.e., the utterance of pant–grunts) was designated spontaneous if it occurred while the subordinate individual approached the dominant one in a completely relaxed context. Cases in which the dominant individual made the approach, or in which one of the chimpanzees (not necessarily the individual being greeted) was displaying or had very recently displayed, were regarded as non-spontaneous greetings. All differences with respect to the spontaneity of greeting in Fig. 5.2 are significant (χ^2-tests, $P < 0.001$). These data show that although ritualised submission among females is rare, the few expressions of subordinance that do occur seem to be relatively free of tension.

The notion (Rowell, 1974) that a subordinance hierarchy exists, rather than a dominance hierarchy, clearly does not have general applicability (for critical comments, see Angst, 1980; Clutton-Brock & Harvey, 1976; Deag, 1977), but her emphasis on the subordinate's behaviour remains valuable. It might prove useful to view relationships among female chimpanzees in this light. Wild female gorillas also seem to lack a pronounced hierarchy (Harcourt, 1979), and captive female orang-utans have been observed to establish stable dominance patterns without any overt threat or intimidation (Nadler & Tilford, 1977). Female anthropoids seem to have weakly formalised status relations that may depend more on expressions of subordinance than on the assertion of dominance.

It can be concluded that the most aggressive age and sex class is not necessarily the least sociable. Male-bonding in chimpanzees seems to have been achieved by a strong emphasis on formalised dominance and by the regulation of tension (as reflected in the high frequencies of ritualised submission, reconciliations, and grooming). These mechanisms fully integrate their rivalries and make the characteristic team-like relationships of male chimpanzees possible. Although this development is definitely more advanced in chimpanzees than in many other Old World primates, it would seem that not all of its elements are unique. For example, male rhesus monkeys, too, are more active in reducing tension than are females of the same species (de Waal, 1984a).

Female chimpanzees appear virtually to lack the set of dominance-related mechanisms required to maintain social bonds in the face of serious competition. Bonding and solidarity are exceptionally strong among the Arnhem females as compared to their free-living counterparts. This is because competition over food is not a life or death matter in a zoo. In the wild, female chimpanzees avoid competition. They live dispersed throughout the forest, each accompanied by her immature offspring (Nishida, 1979; Wrangham, 1979). This rather solitary lifestyle explains why the social mechanisms that allow males to keep tensions under control are less well developed among females. Adult male chimpanzees do need to cope with competition because they often travel in bands. In addition to each individual's "political" motives for staying in touch with the male core (de Waal, 1982; Nishida, 1983), it is vital that all the males in a community close ranks in the face of violent aggression from similar bands in neighbouring territories (Goodall et al., 1979; Nishida, Hiraiwa-Hasegawa, Hasegawa, & Takahata, 1985).

In her recent book, (Goodall (1986) confirms the picture sketched above of greater tolerance and frequent reassurance among adult male chimpanzees. New data show that reassurance behaviour followed agonistic conflicts between males nearly four times more often than it followed conflicts between females. Goodall (1986, p. 364) notes: "These differences between the sexes reflect the lack of a clear-cut hierarchy among the Gombe females." Although this statement is not followed by an explanation of the exact mechanisms involved, it hints at similar social mechanisms as observed in the Arnhem colony.

Striving and Buffering

The idea that animals seek to improve their status by aggressive means (Maslow, 1937) has been ignored for a long time, because it implies intentional striving caused by an appetence for dominance. Recently, growing numbers of primatologists are seriously making suggestions of this sort (e.g., Angst, 1980; Chance, Emory, & Payne, 1977; de Waal, 1977; 1982; Walker, 1979). De Waal (1982) has depicted male chimpanzees as highly ambitious creatures primarily because male bluff displays and aggression are of a relatively spontaneous nature (de Waal & Hoekstra, 1980) and decrease in frequency and intensity after submission by the targeted individual. In other words, male aggression seems to be a part of a negative feedback system.

One of the Arnhem males, Nikkie, showed an enormous amount of aggression when he rose in dominance over the adult females. A comparison between the summer period during his rise and the one just after (i.e., after females began to defer to him) showed a significant reduction in both the number of conflicts he provoked and the intensity of his aggression against females. This was not a non-specific change in Nikkie's behaviour, since his aggression toward males increased during the same period. Another, more pronounced difference between Nikkie's relationships during the two summers was a decreased tendency of females to counter-attack him. The interpretation of these data seems straightforward: the females had given up and hoisted the white flag by greeting Nikkie and, now that he was satisfied, he no longer needed to provoke them.

Dominance struggles were also clearly recognisable among the adult males themselves, and will be used here to investigate two questions. First, does aggression increase in the absence of clear-cut formal dominance relations? Second, if it does, how do males cope with the tense situation that arises?

Instability in the male rank order occurred in 1976, 1977, and 1978. The logbook of these years records 40 damaging fights among the three senior males. In the case of unobserved fights we reasoned that if two males showed wounds they had fought together. Three of the unobserved fights have been excluded from the analysis because only one male was injured and his opponent could not be identified. For each relationship, the three years were divided into periods with or without submissive greetings, which is taken to be the indicator of formalised status relations. We have calculated that 9.1 of the 37 damaging fights would have occurred during periods without greeting if each male dyad had distributed its fights at random. The observed number of fights during such periods was, however, as high as 22, and all three male–male relationships contributed to this difference ($\chi^2 = 24.3$, $P < 0.001$). This means that, in the absence of formalised status, the probability of violence among adult males increased by a factor of 4.6.

Figure 5.4 illustrates frequencies of grooming between male rivals during each of three contests that resulted in a reversal of dominance, either during the period before the subordinate male ceased to greet the other submissively (period I), or during the period without any greeting (II), or during the period after there had been a reversal of the direction of

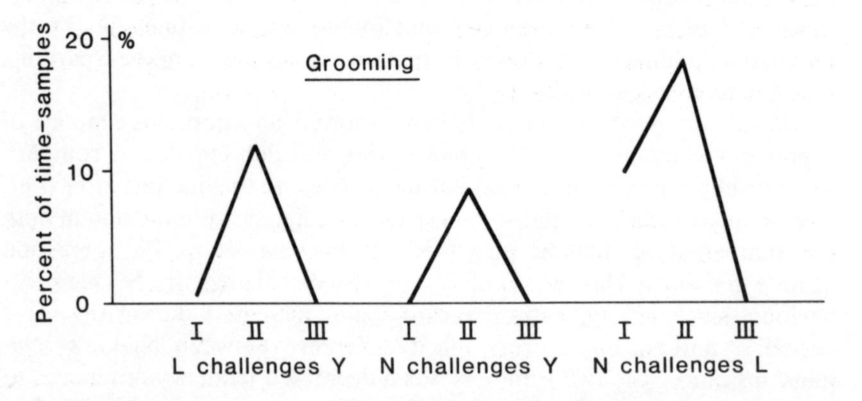

FIG. 5.4 Percentages of time-samples in which grooming activity occurred during three dominance reversals among male chimpanzees Y, L, and N
Period I is before, and period III after the reversal of formal dominance status. Period II is the unstable contest period, during which a formalised status relationship was lacking.

greeting (III). The lengths of periods I and III have been made equal to the duration of the unstable situation II, which in each case lasted about two months. All periods occurred during the outdoor condition (the dominance reversals occurred at the end of the summers of 1976 and 1977; for period III, data were collected at the beginning of the next summer). According to χ-square tests, period II, in which the rivals did not greet each other, scored higher than either the preceding or the following period in all three contests. Except for the comparison of periods I to II in the third contest, for which $P < 0.05$, the differences were all significant at the 0.01 level. Our conclusion is that although the absence of formalised status relationships among males did increase the risk of serious fighting, the males also groomed each other much more during those tense periods. Coe and Levine (1980) found a similar grooming peak occurring during another dominance reversal among adult male chimpanzees in captivity.

In fact, the relation between grooming and struggles for dominance may be more complex. As Table 5.1. (which concerns a process of re-establishment of dominance between two males rather than a process of reversal) shows, the grooming level was not constant during the unstable period, and reached a peak immediately thereafter. Our long-term method, which provided the data for Fig. 5.4, was likely, however, to miss shortlasting peaks. On the basis of qualitative observations I would suggest that grooming between male rivals occurs mainly during the first half of a status contest period and immediately following the contest. In between, there is a phase in which an absence of grooming may contribute to the emotional pressure to which the future dominant male subjects his adversary. It is during this period that he applies the mechanism of conditional reassurance. Further research is needed to verify whether grooming among male chimpanzees, rather than being a general calming behaviour, is really used to manipulate relationships through hills and valleys of tension toward the goal of a stable, formalised status relationship.

CONCLUSION

The view that dominance relationships have an affective component is conspicuously lacking from many recent theoretical discussions. For example, Popp and DeVore (1979) expressed wonder why dominant individuals do not simply settle disputes by killing their opponents. They considered the main disadvantage of such behaviour to be the risk that the weaker party will fight for its life. Then they observed: "By contrast, the only benefit for the potential assassin would be the elimination of just one of many competitors" (p. 329). Where is the notion of the *social* primate in this model? Our theories are really losing touch with reality if the elimination of competitors, which most of the time are group mates, is counted not as a loss, but as a possible benefit. See Vehrencamp (1983) for evolutionary modelling based on a more balanced set of assumptions.

The current one-sided view of social animals as selfish competitors and fighters reminds me of the impressionistic characterisation of piranhas by Meyers (1949, p. 42): "The fishes swam slowly about, each keeping well away from its fellows and showing a plain desire not to have another directly behind it, where the neighbor could attack unseen. Like a set of ruthless gunmen, each with a pistol in his pocket and each one quite mindful that all the rest were ready to use them at any moment." Primates are very different indeed. Members of the same group maintain a cohesive network of social bonds and mutual dependencies. There is a general interest among them in keeping the costs of competition low.

My exploration of the mechanisms through which primates achieve this end brings me to the conclusion that the study of dominance demands distinctions between: (1), formalised relationships, as expressed in ritualised status signals; (2), competitive ability, as expressed in successful claims to access to resources by means of force or the threat of force; and (3), competitive tendency, as expressed in the degree to which dominant individuals make use of their competitive ability, show social tolerance toward subordinates, or share with them.

Formalisation, conditional reassurance, and status-striving presumably exist in a great variety of group-living mammals and birds. Chimpanzees may show these mechanisms in their most recognisable form, but there are as yet no indications for differences in principle. In dwarf mongooses (*Helogale undulata*), aggressive interactions are broken off when the loser initiates a grooming session (Rasa, 1977), and after fights between male mouflons (*Ovis ammon*) the loser submits and is invited to lick the dominant animal's neck and ears as "marks of respect" (Pfeffer, 1967). Species that do not show reassuring post-conflict behaviour may have at least a form of conditional tolerance, that is, some link between submission and acceptance within the group's territory. Thus, although the reconciled hierarchy (Table 5.2) is not the sole principle of social organisation, it is presumed to be a very general one.

We should note that the reconciled hierarchy model as presented here limits itself to processes within dyadic relationships and to their possible impact on group dynamics. The situation, at least in primates, is really more complex than this. The balance within a dyad depends partly on the triadic configurations of which it is part. For example, individual A may need B's support to dominate C. This implies that A has to be careful both not to estrange B and not to provoke C in B's absence. These inhibitions result in an attenuation of status differences to the extent that it is sometimes more appropriate to speak of a balance of power than a rank order. The distinction between formal and real dominance is particularly

TABLE 5.2
The Reconciled Hierarchy Model and the Three Social Mechanisms on Which it Rests

Inter-individual Mechanism	*Description*	*Effects at the Group Level*	*Description*
Conditional reassurance	The winner offers reconciliation and social tolerance in return for the loser's submission	Cohesion	A well-recognised hierarchy promotes social bonds and reduces violence
Formalisation	Ritualised status communication independent of social context	Harmony and stability	Mutual trust between dominants and subordinates, even during incidental reversals of the outcomes of competition
Status-striving	An individual attempts to become the dominant party in as many relationships as possible	Dynamism and tensions	Dominance processes involve much aggression and may transform both the hierarchy and the association pattern

NOTE: Three social mechanisms operate at the level of inter-individual relationships and affect those relationships in such a manner that certain overall patterns emerge at the group level. The first two mechanisms, conditional reassurance and formalisation, go hand in hand; the last two, formalisation and status-striving, exclude each other and alternate with one another.

useful at this triadic level, because the dependence of dominant individuals on subordinate ones can reach a point where, behind the scenes, certain formally subordinate individuals seem to pull the strings (de Waal, 1982).

The integration of dyads at these higher levels (including their transitive arrangement into hierarchies and the effect of the magnitude of status differences within those hierarchies) is beyond the scope of this chapter: The pioneering ideas of Chance (1967) and the work by Seyfarth (e.g. 1977) and by Strayer (e.g. Strayer & Trudel, 1984) strongly suggest, however, that at this level, too, dominance and social bonding interact closely; the hierarchical structure seems to co-ordinate the patterns of cohesion. Study of the interface between these two main dimensions of social life will undoubtedly lead to a less "piranha-like" view of primate social relationships.

ACKNOWLEDGEMENTS

The research on chimpanzees was supported by the Research Pool of the University of Utrecht (1975–1979) and the Dutch Organization for the Advancement of Pure Research, Z.W.O. (1979–1981). The study occurred under the auspices of the

Laboratorium voor Vergelijkende Fysiologie of the University of Utrecht. I am indebted to Dr. J. van Hoof for his encouragement, and to the many students who helped collect data, especially Tine Griede. I also thank the directorate of the Arnhem Zoo for its co-operation. The text of this chapter was first published as a paper in the *Quarterly Review of Biology* (1986) *61*, 459–479, and is reproduced (under a different title, and with minor changes) by kind permission of the editor. That paper was written at the Wisconsin Regional Primate Research Center (Publication No. 21-023), supported by NIH Grant RR-00167. I am grateful for critical feedback from my colleagues D. Goldfoot, R. Goy, and C. Snowdon, and from two anonymous reviewers. The manuscript was prepared with the assistance of Jackie Kinney and Mary Schatz; drawings were made by Linda Endlich.

REFERENCE NOTES

1. Gottman, J. & Parkhurst, J. (1978, October). *A developmental theory of friendship and acquaintanceship processes*. Paper presented at Minnesota Symposium on Child Psychology, Minneapolis, U.S.A.
2. Schropp, R. (1985). *Children's use of objects: Competitive or interactive?* Paper presented at the 19th International Ethology Conference, Toulouse, France.

REFERENCES

Altmann, J. (1980). *Baboon mothers and infants*. Cambridge, Mass.: Harvard University Press.

Altmann, S. (1962). A field study of the sociobiology of rhesus monkeys, *Macaca mulatta, Annals of the New York Academy of Sciences*, *102*, 338–435.

Angst, W. (1975). Basic data and concepts in the social organization of *Macaca fascicularis*. In L. Rosenblum (Ed.), *Primate behavior*, vol. 4 (Pp. 325–388). New York: Academic Press Inc.

Angst, W. (1980). *Aggression bei Affen und Menschen*. Berlin: Springer.

Bernstein, I. (1969). Spontaneous reorganization of a pigtail monkey group. In *Proceedings of the 2nd Congress of the International Primatological Society*, Atlanta, 1968, vol. 1 (Pp. 48–51). Basel: Karger.

Bernstein, I. & Gordon, T. (1974). The function of aggression in primate societies. *American Scientist*, *62*, 304–311.

Bernstein, I. & Williams, L. (1983). Ontogenetic changes and the stability of rhesus monkey dominance relationships. *Behavioral Processes*, *8*, 379–392.

Blurton-Jones, N. & Trollope, J. (1968). Social behaviour of stump-tailed macaques in captivity. *Primates*, *9*, 365–394.

Bygott, D. (1974). *Agonistic behaviour and dominance in wild chimpanzees*. Ph.D. Thesis, University of Cambridge, Cambridge, England.

Chance, M. (1961). The nature and special features of the instinctive social bond of primates. In S. Washburn (Ed.), *Social life of early man. Viking Fund Publications in Anthropology*, *31*, 17–33.

Chance, M. (1967). Attention structure as the basis of primate rank orders. *Man*, *2*, 503–518.

Chance, M., Emory, G., & Payne, R. (1977). Status referents in long-tailed macaques (*Macaca fascicularis*): precursors and effects of a female rebellion. *Primates*, *18*, 611–632.

Clutton-Brock, T. and Harvey, P. (1976). Evolutionary rules and primate societies. In P.

Bateson & R. Hinde (Eds.), *Growing points in ethology* (Pp. 195–237). Cambridge, England: Cambridge University Press.

Coe, C. & Levine, R. (1980). Dominance assertion in male chimpanzees (*Pan troglodytes*). *Aggressive Behavior*, *6*, 161–174.

Collias, N. (1953). Social behavior in animals. *Ecology*, *34*, 810–811.

Deag, J. (1977). Aggression and submission in monkey societies. *Animal Behaviour*, 25:465–474.

De Waal, F. (1977). The organization of agonistic relations within two captive groups of Java monkeys (*Macaca fascicularis*). *Zeitschrift für Tierpsychologie, 44*, 225–282.

de Waal, F. (1982). *Chimpanzee politics*. London: Jonathan Cape.

de Waal, F. (1984a). Coping with social tension: sex differences in the effect of food provision to small rhesus monkey groups. *Animal Behaviour*, *32*, 765–773.

de Waal, F. (1984b). Sex differences in the formation of coalitions among chimpanzees. *Ethology and Sociobiology*, *5*, 239–255.

de Waal, F. (1986a). Class structure in a rhesus monkey group: the interplay between dominance and tolerance. *Animal Behaviour*, *34*, 1033–1040.

de Waal, F. (1986b). The brutal elimination of a rival among captive male chimpanzees. *Ethology and Sociobiology*, *7*, 237–251.

de Waal, F. & Hoekstra, J. (1980). Contexts and predictability of aggression in chimpanzees. *Animal Behaviour*, *28*, 929–937.

de Waal, F. & Luttrell, L. (1985). The formal hierarchy of rhesus monkeys: an investigation of the bared-teeth display. *American Journal of Primatology*, *9*, 73–85.

de Waal, F. & van Hooff, J. (1981). Side-directed communication and agonistic interactions in chimpanzees. *Behaviour*, *77*, 164–198.

de Waal, F. & van Roosmalen, A. (1979). Reconciliation and consolation among chimpanzees. *Behavioral Ecology and Sociobiology*, *5*, 55–66.

de Waal, F. & Yoshihara, D. (1983). Reconciliation and redirected affection in rhesus monkeys. *Behaviour*, *85*, 224–241.

Dittus, W. (1979). The evolution of behaviors regulating density and age-specific sex ratios in a primate population. *Behaviour*, *69*, 265–302.

Ehrlich, A. & Musicant, A. (1977). Social and individual behaviors in captive slow lorises. *Behaviour*, *60*, 195–220.

Ellefson, J. (1968). Territorial behavior in the common white-handed gibbon. *Thylobatus lar*. In P. Jay (Ed.), *Primates: Studies in adaptation and variability*. (Pp. 180–199). New York: Holt.

Fedigan, L. (1983). Dominance and reproductive success in primates. *Yearbook of Physical Anthropology*, *26*, 91–129.

Friedman, R. (1973). On the concept of authority in political philosophy. In R. Flathman (Ed.), *Concepts in Social and Political Philosophy* (Pp. 121–145). New York: Macmillan.

Gerard, H. & Mathewson, G. (1966). The effects of severity of initiation on liking for a group: a replication. *Journal of Experimental Social Psychology*, *2*, 278–287.

Goodall, J. (1963). My life among wild chimpanzees. *National Geographic Magazine*, *124*, 272–308.

Goodall, J. van Lawick. (1968a). A preliminary report on expressive movements and communication in the Gombe Stream chimpanzees. In P. Jay (Ed.), *Primate: Studies in adaptation and variability* (Pp. 313–374). New York: Holt.

Goodall, J. van Lawick. (1968b). Behaviour of free-living chimpanzees of the Gombe Stream area. *Animal Behaviour Monographs*, *3*, 161–311.

Goodall, J. van Lawick. (1986). *The Chimpanzees of Gombe: Patterns of behavior*. Cambridge, Mass.: Belknap (Harvard University Press).

Goodall, J., Bandora, A., Bergman, E., Busse, C., Matama, H., Mpongo, E., Pierce, A., & Riss, D. (1979). Intercommunity interactions in the chimpanzee population of the

Gombe National Park. In D. Hamburg and E. McCown (Eds.), *The great apes* (Pp. 13–53). Menlo Park, Calif.: Benjamin Cummings.

Halperin, S. (1979). Temporary association patterns in free ranging chimpanzees. In D. Hamburg and E. McCown (Eds.), *The great apes* (Pp. 491–499). Menlo Park, Calif.: Benjamin Cummings.

Harcourt, A. (1979). Social relationships among adult female mountain gorillas. *Animal Behaviour*, *27*, 251–264.

Heinroth, O. & Heinroth, M. (1928). *Die Vögel Mitteleuropas*. Berlin: Behrmüller.

Hinde, R. (1970). Aggression. In J. Pringle (Ed.), *Biology and the human sciences* (Pp. 1–23). Oxford: Clarendon Press.

Hinde, R. & Datta, S. (1981). Dominance: an intervening variable. *Behavioral and Brain Sciences*, *4*, 442.

Kalas, S. (1977). Ontogenie und Funktion der Rangordnung innerhalb einer Geschwisterschar von Graugansen (*Anser anser* L.). *Zeitschrift für Tierpsychologie*, *45*, 174–198.

Kaplan, J. & Zucker, E. (1980). Social organization in a group of free-ranging patas monkeys. *Folia Primatologica*, *34*, 196–213.

Kaufman, I. (1974). Mother/infant relations in monkeys and humans: a reply to Professor Hinde. In N. White (Ed.), *Ethology and psychiatry* (Pp. 46–68). Toronto: University of Toronto Press.

Kawanaka, K. (1982). Further studies on predation by chimpanzees of the Mahale Mountains. *Primates*, *23*, 364–384.

Kummer, H. (1957). *Soziales Verhalten einer Mantelpavian Gruppe*. Bern: Huber.

Kummer, H. (1971). *Primate Societies*. Chicago: Aldine.

Kummer, H. (1975). Rules of dyad and group formation among captive gelada baboons (*Theropithecus gelada*). In *Symposium of the 5th Congress of the International Primatological Society (1974)*, (Pp. 129–159). Tokyo: Japan Science Press.

Kummer, H. (1984). From laboratory to desert and back: a social system of hamadryas baboons. *Animal Behaviour*, *32*, 965–971.

Kummer, H., Götz, W., & Angst, W. (1974). Triadic differentiation: an inhibitory process protecting pair bonds in baboons. *Behaviour*, *49*, 62–87.

Kurland, J. (1977). Kin selection in the Japanese monkey. In F. S. Szalay (Ed.), *Contributions to primatology*, vol. 12. Basel: Karger.

Lindburg, D. (1973). Grooming behavior as a regulator of social interactions in rhesus monkeys. In C. Carpenter (Ed.), *Behavioral Regulators of Behavior in Primates* (Pp. 85–105). Lewisburg, Penn.: Bucknell University.

Lorenz, K. (1935).Der Kumpan in der Umwelt des Vogels. *Journal of Ornithology*, *83*, 137–213.

Lorenz, K. (1966). *On aggression*. New York: Bantam Books.

Lorenz, K. (1982). *The foundations of ethology*. New York: Simon & Schuster.

Maslow, A. (1937). The role of dominance in social and sexual behavior of infra-human primates. IV. The determination of a hierarchy in pairs and in a group. *Journal of Genetics and Psychology*, *49*, 161–198.

Maslow, A. (1940). Dominance-quality and social behavior in infra-human primates. *Journal of Social Psychology*, *11*, 313–324.

Mason, W. (1964). Sociability and social organization in monkeys and apes. In L. Berkowitz (Ed.), *Advances in experimental social psychology* (Pp. 277–305). New York: Academic Press Inc.

Mason, W. (1965). Determinants of social behavior in young chimpanzees. In A. Schrier, H. Harlow and F. Stollnitz (Eds.), *Behavior of nonhuman primates* (Pp. 335–364). New York: Academic Press Inc.

Maxim, P. (1976). An interval scale for studying and quantifying social relations in pairs of rhesus monkeys. *Journal of Experimental Psychology*, *105*, 123–147.

McGrew, W. (1975). Patterns of plant food sharing by wild chimpanzees. *Proceedings of the 5th Congress of the International Primatological Society*, (Nagoya, Japan), (Pp. 304–309). Basel: Karger.

McKenna, J. (1978). Biosocial function of grooming behavior among the common langur monkey (*Presbytis entellus*). *American Journal of Physical Anthropology*, *48*, 503–510.

Meyers, G. (1949). A monograph on the piranha. In G. Meyers (Ed.), *The piranha book* [1972 reprint.] Neptune City: T.F.H. Publications.

Montagner, H. (1978). *L'enfant et la communication*. Paris: Stock.

Mori, A. (1982). An ethological study of chimpanzees at the artificial feeding place in the Mahale Mountains, Tanzania—with special reference to the booming situation. *Primates*, *23*, 45–65.

Nadler, R. & Tilford, B. (1977). Agonistic interactions of captive female orangutans with infants. *Folia Primatologica*, *28*, 298–305.

Nieuwenhuijsen, K. & de Waal, F. (1982). Effects of spatial crowding on social behavior in a chimpanzee colony. *Zoo Biology*, *1*, 5–28.

Nishida, T. (1970). Social behavior and relationships among wild chimpanzees of the Mahale Mountains, *Primates*, *11*, 47–87.

Nishida, T. (1979). The social structure of chimpanzees of the Mahale Mountains. In D. Hamburg and E. McCown (Eds.), *The great apes* (Pp. 73–121). Menlo Park, Calif.: Benjamin Cummings.

Nishida, T. (1983). Alpha status and agonistic alliance in wild chimpanzees. *Primates*, *24*, 318–336.

Nishida, T., Hiraiwa-Hasegawa, M., Hasegawa, T., & Takahata, Y. (1985). Group extinction and female transfer in wild chimpanzees in the Mahale National Park, Tanzania. *Zeitschrift für Tierpsychologie*, *67*, 284–301.

Nissen, H. & Crawford, M. (1936). A preliminary study of food-sharing behavior in young chimpanzees. *Journal of Comparative Psychology*, *22*, 383–419.

Noë, R., de Waal, F., & van Hooff, J. (1980). Types of dominance in a chimpanzee colony. *Folia Primatologica*, *34*, 90–110.

Pfeffer, P. (1967). Le mouflon de Corse (*Ovis ammon musimom*): position systématique, écologie et éthologie comparées. *Mammalia*, *31 (Suppl.)*.

Poirier, F. (1968). Dominance structure of the Nilgiri langur (*Presbytis johnii*) of South India. *Folia Primatologica*, *12*, 161–186.

Popp, J. & De Vore, I. (1979). Aggressive competition and social dominance theory: synopsis. In D. Hamburg and E. McCown (Eds.), *The great apes* (Pp. 317–338). Menlo Park, Calif.: Benjamin Cummings.

Rasa, O. A. (1977). The ethology and sociology of the Dwarf Mongoose (*Helogale undulata rufula*). *Zeitschrift für Tierpsychologie*, *43*, 377–406.

Reynolds, V. (1980). *The biology of human action* (2nd ed.). Oxford: Freeman.

Riss, D. & Goodall, J. (1977). The recent rise to the alpha-rank in a population of free-living chimpanzees. *Folia Primatologica*, *27*, 134–151.

Rowell, T. (1966). Hierarchy and the organization of a captive baboon group. *Animal Behaviour*, *14*, 430–443.

Rowell, T. (1974). The concept of social dominance. *Behavioral Biology*, *11*, 131–154.

Rowell, T. & Olson, D. (1983). Alternative mechanisms of social organization in monkeys. *Behaviour*, *86*, 31–54.

Sackin, S. & Thelen, E. (1984). An ethological study of peaceful associative outcomes to conflict in preschool children. *Child Development*, *55*, 1098–1102.

Schenkel, R. (1967). Submission: its features and function in the wolf and dog. *American Zoologist*, *7*, 319–329.

Seyfarth, R. (1976). Social relationships among adult female baboons. *Animal Behaviour*, *24*, 917–938.

Seyfarth, R. (1977). A model of social grooming among adult female monkeys. *Journal of Theoretical Biology*, *65*, 671–698.
Shively, C. (1985). The evolution of dominance hierarchies in nonhuman primate society. In S. Ellyson and I. Dovidio (Eds.), *Power, dominance, and nonverbal behavior* (Pp. 67–87). Berlin: Springer.
Sigg, H. & Falett, J. (1985). Experiments on respect of possession and property in hamadryas baboons (*Papio hamadryas*). *Animal Behaviour*, *33*, 978–984.
Silk, J. (1979). Feeding, foraging, and food sharing behavior of immature chimpanzees. *Folia Primatologica*, *31*, 123–142.
Sparks, J. (1967). Allogrooming in primates: a review. In D. Morris (Ed.), *Primate ethology* (Pp. 190–225). [Reprint, 1969]. New York: Anchor Books.
Strayer, F. & Trudel, M. (1984). Developmental changes in the nature and function of social dominance among young children. *Ethology and Sociobiology*, *5*, 279–295.
Strum, S. (1982). Agonistic dominance in male baboons: an alternative view. *International Journal of Primatology*, *3*, 175–202.
Symons, D. (1978). The question of function: dominance and play. In E. O. Smith (Ed.), *Social play in primates* (Pp. 193–230). New York: Academic Press Inc.
Teleki, G. (1973). *The predatory behavior of wild chimpanzees*. Lewisburg, Penn.: Bucknell University Press.
Terrace, H. (1979). *Nim: A chimpanzee who learned sign language*. New York: Washington Square Press.
Thierry, B. (1984). Clasping behavior in *Macaca tonkeana. Behaviour*, *89*, 1–28.
Thierry, B. (1986). A comparative study of aggression and response to aggression in three species of macaque. In I. Else and P. Lee (Eds.), *Primate ontogeny, cognition and social behaviour* (Pp. 307–313). Cambridge, England: Cambridge University Press.
van Hooff, J. (1967). The facial displays of the catarrhine monkeys and apes. In D. Morris (Ed.) *Primate ethology* (Pp. 3–88). [1969 reprint.] New York: Anchor Books.
Vehrencamp, S. (1983). A model for the evolution of despotic versus egalitarian societies. *Animal Behaviour*, *31*, 667–682.
Walker, L. J. (1979). A strategy approach to the study of primate dominance behaviour. *Behavioral Processes*, *4*, 155–172.
Walters, J. (1980). Interventions and the development of dominance relationships in female baboons. *Folia Primatologica*, *34*, 61–89.
Weigel, R. (1980). Dyadic spatial relationships in pigtail and stumptail macaques: a multiple regression analysis. *International Journal of Primatology*, *1*, 287–321.
Weisbard, C. & Goy, R. (1976). Effect of parturition and group composition on competitive drinking order in stumptail macaques (*Macaca arctoides*). *Folia Primatologica*, *25*, 95–121.
Wrangham, R. (1979). Sex differences in chimpanzee dispersion. In D. Hamburg and E. McCown (Eds.), *The great apes* (Pp. 481–490). Menlo Park, Calif.: Benjamin Cummings.
Williams, N., Sjoberg, G. & Sjoberg, A. (1980). The bureaucratic personality: an alternate view. *Journal of Applied Behavioral Analysis*, *16*, 389–405.
Yamada, M. (1963). A study of blood relationships in the natural society of the Japanese Macaque. *Primates*, *4*, 43–65.
Yerkes, R. (1941). Conjugal contrasts among chimpanzees. *Journal of Abnormal Psychology*, *36*, 75–199.

6 The Origin of Human Equality

Junichiro Itani
Centre for African Area Studies, Kyoto University, Yoshida Sakyo, Kyoto, Japan

INTRODUCTION

Looked at in retrospect, over my 30-year career as an anthropologist, my subjects have changed from the Japanese monkey to the great apes, and from the apes to the nature-dependent peoples—hunter–gatherers, slash-and-burn agriculturalists, and pastoralists. I have often been asked: "How are your primatology and ecological anthropology interrelated?" The answer has been: "For the investigation of the evolution of human society, these two fields are prerequisite. Non-human primates in the wild and the nature-dependent peoples are indispensable subjects." However, I have not had any firm theory to connect these fields. Rather, with an intuitive belief that: "Evolution must link the two", I have continued my research because of the charms of the subjects themselves.

Here I do not intend to retrace my career. My intention is to review the research of my colleagues and myself from a certain standpoint, or from within a certain integrative framework. The framework is related to inequality and equality, in which terms I have already attempted to compare the Japanese monkey and two chimpanzee species in a paper entitled "Inequality versus equality for coexistence in primate societies" (Itani, 1984).

In his *Discourse on the origins and foundations of inequality among men*, Jean-Jacque Rousseau (1755), in English translation, p. 78) stated: "The philosophers who have examined the foundations of society have all felt it necessary to go back to the state of nature, but none of them has succeeded in getting there."

The discrepancy, noted by Rousseau, between the natural state and the civil state, has undoubtedly increased over the past 230 years. Although I am not sure whether I can discover the state of nature in Rousseau's sense, given the data on non-human primates and African nature-dependent peoples, I believe that a reconsideration of inequality and equality is anthropologically worthwhile.

Rouseau (1755, same translation, pp. 104–105) wrote about man's state of nature as follows:

> Savage man, wandering in the forests, without work, without speech, without a home, without war, and without relationships, was equally without any need of his fellow men and without any desire to hurt them, perhaps not even recognising any one of them individually. Being subject to so few passions, and sufficient unto himself, he had only such feelings and such knowledge as suited his condition; he felt only his true needs, saw only what he believed it was necessary to see, and his intelligence made no more progress than his vanity. If by chance he made some discovery, he was all the least able to communicate it to others because he did not even recognise his own children. Every art would perish with the inventor. There was neither education nor progress; the generations multiplied uselessly, and as each began afresh from the same starting-point, centuries rolled on as underdeveloped as the first ages; the species was already old, and man remained eternally a child.

Our present knowledge suggests that, however primitive, the first "man" must have graduated from the stage assumed by Rousseau. Even though described in contrast with the civil state, Rousseau's image of "natural man", in particular their lack of sociality, is far from being acceptable. He assumed that the species members, without any interreliance, had only anonymous groups, that they could not recognise their offspring, and that they could not transmit their discoveries or traditions from one generation to the next.

The chimpanzee and even the Japanese monkey are less primitive than such a savage. To a certain degree, both chimpanzee and monkey are highly curious and intelligent, handle symbols, transmit discoveries and innovations through generations, are often subject to passions, and sometimes even so vicious as to hurt their conspecifics.

Incidentally, Rousseau (1755, same translation, p. 155) quoted travellers' reports in one of his footnotes, which refer to great apes: "In the Kingdom of Congo . . . one finds many of those large animals which are called *orang-outangs* in the East Indies . . . in the forests of Mayomba in the Kingdom of Loango, one sees two sorts of monsters, of which the larger are called *pongos* and the smaller *enjocos*." The latter two kinds of animals apparently indicated gorillas and chimpanzees; however, they were confused with orang-utans. He also quoted the information on the physical features and the folk-tales about the apes that the travellers had heard from the natives, but information on the behaviour and society of these animals was virtually unavailable.

Thomas Huxley (1863) published *Man's place in nature* 108 years later. His work was based on the zoological knowledge of the day. The orang-utan, the chimpanzee, and the gorilla were already recognised as three distinct species. In the first chapter of this book, he summarised the reports by Battell and other travellers, which had also been quoted by Rousseau (1755). Knowledge of the apes' ecology and society was then no greater than in Rousseau's day.

Comparison of these two books reveals that 100 years have changed monsters into zoological species. If Rousseau had been born one century later, or if he had had today's knowledge of ape society, how would he have described his "natural man"? Even so, the civil state described by Rousseau (1755) began not at the stage of the first "man" but much earlier. The conditions of that natural state he described, if closely examined, may be found only among some nocturnal prosimians, which lead solitary lives and whose societies I call "elemental" (Itani, 1977).

Nevertheless, the elemental society is not a true egalitarian society. It is a society in which neither inequality nor equality exists, or one in which—and here it is difficult to choose exact terms—the species members are biosocially equiponderous or equipotent. Rousseau's assumptions about the egalitarian society were too strict. Admitting more sociality—the existence of cohorts, inter-individual alliances, and of mutual recognition among them and between mothers and their offspring—we can form a model of such an egalitarian society. That is the pair-type (monogamous) society, which is found in many families or subfamilies of non-human primates.

When Rousseau published his *Discourse* (1755), Lamarck had already been born (1744); it was just before the age of the evolutionary theories. Although knowledge of non-human primates and primitive human society was scarce in his time, Rousseau planned to consider human society through examination of its original state. The great philosopher's insight should be reconsidered in the light of scientific knowledge today.

INEQUALITY VERSUS EQUALITY

Many nocturnal prosimians that have elemental societies have been studied well enough to reveal the unique features that differentiate them one from another (e.g., Charles-Dominique, 1977). However, without any exception, these species have no stable social unit in their social organisations. In such a society, each individual is intolerant of any other. Each behaves so as to avoid others within its own territory, which functions to distance it from them. Each has no identity other than being a member of its species; the social rule is intolerance between member individuals. Therefore, such a society requires equipotency of each individual, but neither inequality nor equality, which can be formed only through inter-individual interactions.

Even in the elemental societies of these nocturnal prosimians, a budding outgrowth toward monogamy or polygamy can be found in the territoriality and sexual interactions that can be observed in the short breeding seasons. Probably the stable basic social unit (BSU; see Itani, 1977; 1985) first attained by primates was that of monogamy, in which a particular female was tied to a particular male. Each tolerated one of the opposite sex, thus forming a stable BSU. This structure appeared in the stage of nocturnal life, and was carried on into the stage of diurnal life. It is found in almost all taxonomic families of social primates that have BSUs. In such monogamous BSU, there seems to be a sort of equality between the male and female; their sexual dimorphism remains minimal as if to guarantee their equality.

Higher primate social structures have diverged from this type in various directions by differentiating modes of inter-individual tolerance. We can trace this macro-evolution only through inferences drawn from comparisons between various structures in each family of primates. It is likely that there was a trend toward large-sized BSUs and the elaboration of kinship-related structures. Each individual has become not only a member of a particular BSU but also a member of a particular kin group, age group, etc. Each has come to have a multiple identity, and to show complex behaviour according to this multiplicity. The structure is no longer the same as with the equal relationship between a monogamous pair, which is based on the tolerance only between the paired male and female. In more complex social structures, some other principles for co-existence must be functioning. My aim is to extract them and trace the pathways of primate social evolution.

I have before described (Itani, 1984) two contrasting principles, the inequality principle and the equality principle, citing representative examples from primate behaviour. Here I should like to review them again in more detail.

In the vocalisation of the Japanese monkey, there are two types of quiet sounds emitted face-to-face between two individuals. The sounds themselves resemble each other, but their contexts are entirely different (Itani, 1963). The first sound, <ngu>, which I coded A–5, is nasal as well as guttural. The other sound, <ngu ngu> or <nguru nguru>, coded A–6, is shorter than A–5 and often emitted in a series. A–6 differs from A–5 in that the emitter produces it by puffing out the upper lip while lowering the head. A–5 is a sort of salutation exchanged between two individuals, but does not expose the dominant/subordinate relationship between them. On the other hand, A–6 is emitted by the subordinate individual toward the dominant one in contexts such as trying to win the superior's favour, to please them, or to beg their pardon. It should be noted here that either situation assumes peaceful co-existence between the two, although A–6 is based on the inequality principle and A–5 on the equality principle. Japanese monkeys form an order of stable dominant/subordinate relationships within a troop. A–6 is emitted in accordance with this rank order, whereas A–5 is emitted as if the order were non-existent. Except for special relationships, such as mother–offspring or consort relationships, the organisation of social co-existence among individuals is based on either of these two principles.

The co-existence of individuals within a Japanese monkey troop is thus founded on two layers of different principles. Apparently, the inequality principle is the basic one, and the equality principle functions to adjust or to complement it. We might have a better view of the origin of the inequality principle, if there were more detailed studies of primitive societies having BSUs other than the monogamous type. Without such studies, we can only speculate that the social co-existence and integration based on the inequality principle was born as soon as large-sized BSUs began to be formed; in other words, the formation of these was concurrent with the establishment of the inequality principle. Had there not been BSUs there would not have been the inequality principle, and vice versa.

The inequality principle is a rule that exists *a priori* in a society that has BSUs more elaborate than the monogamous type. This is a rule based on the recognition of strong–weak or superior–inferior. The inequality principle functions provided that the two are not equal and their dominant/subordinate relationship is stable over time. The principle can be extended, as known empirically among a large number of individuals to form a linear rank order among them. In other words, this principle is a reliable one among those who belong to a particular BSU. The mechanism of its formation in a Japanese monkey troop will be given closer examination later.

The equality principle does not have as firm a foundation as the inequality principle; it is founded only on the negation of the inequality principle, and it functions in a society ruled by the inequality principle as if that principle were non-existent. Therefore, the equality assumed here is not identical to the original equality assumed in a monogamous BSU but of a different dimension; I shall call it the "conditional equality". The conditional equality is supported by an implicit agreement for achieving peaceful co-existence among the troop members.

The agreement for the negation of the inequality principle is exchanged between both interactors. Its context is the same as the "false" or "borrowed" world described by Kitamura (1985) in his study of interactions among pygmy chimpanzees. This must be the origin of "social contract". Without a foundation as firm as that of the inequality principle the equality principle is amorphous, but because of this flexibility it is more creative. In peaceful co-existence the interactors can share "joy" (Kitamura, 1985).

In his *Discourse*, Rousseau (1755, English translation, p. 88), remarked on one human ability which distinguishes man from beasts:

> It would be sad for us to be forced to admit that this distinguishing and almost unlimited faculty of man is the source of all his misfortunes; that it is this faculty which, by the action of time, drags man out of that original condition in which he would pass peaceful and innocent days; that it is this faculty, which, bringing to fruition over the centuries his insights and his errors, his vices and his virtues, makes man in the end a tyrant over himself and over nature.

This ability of man, "*perfectibilité*", must have originated from the same source.

Rousseau traced the origins of the escalation of inequality in human society, and criticised this situation. He contrasted the equality in the natural state with the inequality in the civil state. Differing from him, I should like to trace the evolutionary stages from the original equality to the fundamental inequality, and from the fundamental inequality to the conditional equality. These stages were unknown to Rousseau. In particular, I should like to put emphasis on the conditional equality. The title of my chapter may be thus justified. The inequality among men or social inequality, as discussed by Rousseau, followed these stages, but must be deeply related to the conditional equality I wil discuss here. As already mentioned, the origin of the fundamental inequality cannot be traced; however, it is possible to describe its developmental process. For the conditional equality a more detailed analysis can be made.

THE ORIGINAL INEQUALITY

The egalitarianism assumed in the early stage of human society is equivalent to the conditional equality discussed in this chapter. At the beginning of his *Discourse*, Rousseau (1755, English translation, p. 77) also distinguished two kinds of inequality: "I discern two sorts of inequality in the human species: the first I call natural or physical because it is established by nature, and consists of differences in age, health, strength of the body and qualities of the mind or soul." He did not give any further account on this natural inequality. This inequality was considered inerasable. His entire *Discourse* was on the social inequality, which developed out of the equality which is commonly assumed in the early stages of human society. This is the second inequality:

> The second we might call moral or political inequality because it derives from a sort of convention, and is established, or at least authorized, by the consent of men. This latter inequality consists of the different privileges which some enjoy to the prejudice of others—such as their being richer, more honoured, more powerful than others, and even getting themselves obeyed by others.

This will be called the "civil inequality" in my terminology.

The fundamental inequality I am going to discuss is both natural and social. The long-term and detailed observations and analyses of the dominant/subordinate relationships among Japanese monkeys provide us with an insight into how this inequality is actually formed within a troop. The structure that supports the inequality among primates is rooted in the matrilineal nature of their society.

The co-existence based on the inequality principle is supported by the self-restraint of the subordinate. As the subordinate individual restrains itself, it avoids unnecessary fights with its dominant; thus the co-existence between the two is assured. At the beginning of this research, food items were often thrown between two individuals to judge their dominant/subordinate relationship. Because of the foods used in the test, economic aspects of the rank order were often emphasised. Or correlation between rank and reproductive success was noted in many of the analyses. These were attempts to find the biological bases of dominance rank, which is a sociological entity, so the attempts were largely unsuccessful. On the other hand, there were some who saw the dominant/subordinate relationship only in the context of aggression. It is easy to understand that these approaches were far from explaining the dominant/subordinate relationship if we look at the process of its formation within a troop.

A troop of Japanese monkeys comprises several matrilineal groups. The monkeys seem to recognise their kin at least to the third degree of consanguinity (Takahata, 1982). A matrilineal group (lineage) consists of three to four generations (Koyama, 1970). Males older than four or five years move away from their natal lineage and in the end they leave their natal troop. Therefore, the lineage is carried on by its female members. The lineage may be split by the death of an old mother who has been providing a link between the sisters. Thus the size of a coherent lineage fluctuates between 4–5 and 13–14. The members of a troop move together when foraging, follow particular males, and are tied by the same troop identity. Within the troop, the component lineages do not dissolve; each lineage remains to form a group of individuals who are tolerant of one another and interdependent.

Between different lineages, there is an apparent rank order. The rank between lineages reflects the rank between individuals. Every member of a particular lineage is higher in rank than any member of another lineage; this fact reveals the coherence within a lineage. Koyama (1967) recognised 16 lineages in the Arashiyama troop in 1965, and found a linear rank order among them. He described the process of the troop fission in 1966; the higher-ranking seven lineages formed the A troop, while the lower-ranking nine lineages formed the B troop. Probably, the lineages adjacent in rank are close in kinship also.

The kin relationships recorded by the researchers, and the lineages revealed from the records, were to give insights into the developmental mechanism of the rank order. In a lineage of Japanese monkeys, the mother is always dominant over her daughters (Koyama, 1970). However, between the sisters, the rank system is not directly ruled by age but by the youngest ascendancy principle, which was found by Kawamura (1958), and which marks one of the most important discoveries in the early days of Japanese monkey research. This principle rules the dominant/subordinate relationship between sisters born from the same mother and, accordingly, the younger individual always becomes dominant over the older because the younger can enjoy more protection from the mother. Kawai (1958) distinguished the basic rank, which is between individuals, and the dependent rank, which may be temporarily formed in the presence of a third individual. The youngest ascendancy principle clearly indicates that dependence on the mother much affects the basic rank between the sisters. Koyama (1970) reported that this principle, first discovered in the Minoo–B troop, was consistent in 28 dyads of sisters over the age of four in the Arashiyama troop, and that in 60 mother–daughter dyads the mother was always dominant. The dominant/subordinate relationships within a lineage are determined by these rules.

Furuichi (1983a; 1983b) reported that the youngest ascendancy principle was not applicable to the Yakushima M troop; he could not explain the reason for this exception. This troop, an average-sized one on the Yakushima Island, consisted of 27 members within 3 lineages, each comprising 2–3 adult females and their offspring. Although we must await future studies to find the reason for this exception, one possibility may be its relation to the frequent allo-mothering behaviour observed among the sisters and nulliparous young females (Okayasu, personal communication). Allo-mothering is not an uncommon behaviour: its high frequency may affect the youngest ascendancy principle, which functions on the basis of the young's dependence on the mother. When an infant is taken care of by an allo-mother, and if the allo-mother is that infant's sister, then the infant will become subordinate to that sister, i.e., the allo-mother. It may further be speculated that his mechanism may also affect the bonds between sisters, so as to prevent a kin group from becoming large; in such a situation, troop fission will be frequent, resulting in the formation of small troops.

Males also go through the same early developmental stages in such a way as to be incorporated into the matrilineal kin groups. However, they increase the distance, as they grow, from their natal lineages, and in the end they leave their natal troop and move into other troops. Therefore, as a rule, the adult males in a troop were born in other troops. It should be noted that the clear linear rank order among the troop males (Itani, 1954) is formed separately from the matrilineal kin relationships. The factors determining rank are age, date of entry into the troop, experience, physical strength, and probably also the peer relationships between those from the same natal troop.

For no species with bilateral BSUs, do we have such detailed data. Probably a bilateral BSU has a less rigid structure; we may assume a structure in which both sexes have a rank system similar to that of Japanese monkey males, but also that structure may retain some remnants of the more primitive, monogamous BSU stage. In other words, the rank system was established when the females began to remain in the BSU to maintain the matrilineality. It should be noted that fundamental inequality, which is natural as well as social, is also a product of primate social evolution, and that it is deeply rooted in the matrilineal structure.

Another point to note is that this equality is not the one involving possession of goods, which was determined by Rousseau to be the source of all evils, but this is an inequality only in social status. Certainly a dominant individual can take food in front of a subordinate one. However, as Rousseau (1755, English translation, p. 106) assumed, in the natural state the subordinate can go to another tree to get food: "What sort of chains of dependence could exist among men who possess nothing? I am chased from one tree, I am free to go to the next; if I am tormented in one place, who will prevent my moving somewhere else?" The fundamental inequality is nothing but this social code ruling the world before the advent of the possession of goods.

THE ORIGIN AND EVOLUTION OF EQUALITY

Before proceeding to the conditional equality, I should like to add a few paragraphs on the characteristics of the original inequality, which preceded the conditional equality in the drama of social evolution.

The Japanese language has a word *kashakunai* (merciless), of which the literal meaning is "with neither falsehood nor loan". This signifies an attitude in which one exerts pressure on another without exception or evasion. The word means to force somebody to solve some problems without any outside help or without falsification of the problem. The co-existence of individuals based on the inequality principle is supported by such *kashakunai* intuitive rules.

In my previous publication on the problem of inequality and equality (Itani, 1984, p. 174), I did not distinguish original and conditional equality:

> The relationships between kin of being depended on and dependent, such as between the mother and offspring, are thought to be original forms of the bonds based on the equality principle. They are relationships in which the two individuals are really identical rather than equal. Even after becoming two individuals by parturition, the mother and offspring are psychologically and socially unseparated for a certain period of time. The importance of this period for the offspring's development is evident from the cases of orphaned chimpanzees under the age of 4.5 years, who are reported inactive, playless, and eventually die (Goodall, 1983), and the cases of orphaned Japanese monkeys (Hasegawa & Hiraiwa, 1980).

This equality is apparently neither the conditional equality nor its origin, but is rather akin to the original equality.

Now I should like to proceed to the world of the conditional equality. Even in the society of Japanese monkeys, which is basically ruled by the fundamental inequality principle, there are various social interactions based on the conditional equality principle. From the studies concerned with topics, I should like to cite a few excellent ones.

Play is a social interaction widely found among mammals. There have been many studies on play, and it is well known that this topic is not an easy one to deal with. One factor that makes it difficult is that play appears most prominently in the developmental stage at which the behaviour itself is, though active, indeterminate or amorphous. Recent studies of primate play have succeeded in dealing with this topic, revealing that play interactions are manifestation of the conditional equality.

According to Norikoshi (1974), in a cohort of infants born in the same year the rank order appears within only 12 weeks of age, and in such a way as to reflect that of their mothers. As play becomes active after this stage, the existence of rank order among the infants cannot be ignored as one of the preconditions for play interactions. In choosing play partners, as I discussed elsewhere (Itani, 1984), those with little difference in rank and age, and with close kin relationships, are chosen. Even though play partners are thus chosen, still more sensitive artificiality is required to achieve the condition for play in which no dominant/subordinate relationship shall appear. Hayaki (1983a; 1983b) revealed in a study of play among juvenile Japanese monkeys: Players restrain themselves so as not to force their partners into play. Behaviour that may hurt the partner is inhibited. When one stops moving, the partner will also withdraw into a pause. Even in a rough tumbling play similar to wrestling, which may appear to be an aggressive interaction, the dominant individual may be topped by the subordinate partner, or the subordinate one may chase the dominant one (Hayaki, 1983a; 1983b). Here a fictitious world is unfolded.

Furthermore, in a study of play among juvenile chimpanzees, Hayaki (1985) grasped the concept of the minimum unit of play as a "play bout", and called interactions that did not successfully turn into play "para-play". He analysed the mechanism of play continuation, the significance of pause and restart, and the termination process of play sessions. The desire to play may be exposed in a special expression, such as "play face". More frequently it is expressed in an exaggeration of behaviour, e.g., by differentiating the timing or rhythm of an ordinary behaviour. Play is induced by the partner's positive reaction to such enticement. Hayaki pointed out that one of the most interesting phenomena involved in play is self-handicapping, in which the dominant individual restrains its strength so as to balance with that of the subordinate. He also referred to the imaginary play in which a juvenile plays with an object such as a stick. Hayaki concluded that each segment of behaviour itself should not be called a play, but play itself is the context that enables such a behaviour to be interpreted as play. In a foregoing paragraph, I phrased this a "fictitious world". Indeed the actors perform in the play context while enjoying it is as a real entity.

In this line of argument there are many important problems to consider in relation to the evolution of mind and society. First, to open the fictitious world an agreement is necessary. The agreement is to assume that the fundamental inequality is non-existent. Then, efforts should be made in accordance with this agreement. One self-handicaps, while the other outstretches. Thus a world disconnected from reality unfolds; there the actors escalate the unreality. A sudden pause in a play session is a pause in fear of breaking the agreement, and a pause to ensure it again.

Second, such interactions require an ability to communicate that is much higher than required for daily life; one must correctly convey one's message to the partner, and correctly receive messages from them. The play participants apparently seek reliable media for signalling their messages. In an advanced stage, these media may convey concepts or values; however, at the play stage, the media are necessary to buy the ticket to the "false" world. The play participants attempt to form the media even out of their daily behaviour by changing its ordinary tempo or rhythm.

Third, this fictitious world is not limited to play. Social grooming mutually exchanged in peace, "paternal care" of a particular one-year-old infant by a particular male (Itani, 1959), allo-mothering behaviour, "peculiar proximate relationship", which is a stable, non-sexual, affinitive relationship between a particular male–female pair (Takahata, 1982), etc., all of which are seen in the society of Japanese monkeys, are not linked to the co-existence based on the fundamental inequality. They are linked to a "borrowed" co-existence founded on an agreement in accordance with each separate context.

Some of these behaviours and interactions must be operative in forming the social structure. Probably, a true understanding of the multi-layered social structure of hamadryas baboons (Kummer, 1968; Sigg, Stolba, Abegglen & Dasser, 1982) can be reached only from this viewpoint. It is hardly possible to think that there is a species with a society entirely dependent on conditional equality. The society of Japanese monkeys, though containing many buds of equality as we have seen, has not made conditional equality a major social principle ruling all members of the species.

However, some groups of primates did not proceed toward the perfection of the fundamental inequality. These have "non-matrilineal" BSUs (Itani, 1980), in particular the Pongidae. In species with such BSUs, females leave their natal BSUs and therefore do not form matrilineal groups. No rigid rank system, which is only inherent in a matrilineal BSU, can be formed. The common chimpanzee and the pygmy chimpanzee, in the Pongidae, have elaborated their societies into BSUs with patrilineal structures. The female–female relationships found among them differ from those of the Japanese monkey. Even without matrilineality functioning as the growth medium for the formation of rank order, dominant/subordinate relationships have been found among the females of these species, though unstable (Kuroda, 1980; Nishida, 1970; 1979).

These two chimpanzee species, despite their similarities, show a great difference in the social interactions between females, prominently in their grooming association patterns, which represent their patterns of affinitive interactions in general. The frequency of grooming between females is only one-twentieth of that between males among the common chimpanzees (Nishida, 1970); in the pygmy chimpanzee that between females is twice as high as between males (Kuroda, 1980). These two species of chimpanzees may have taken opposite pathways to an accommodation over the odd social relationships between females from different natal BSUs. The pygmy chimpanzee females attempt to compensate for the initial lack of common group identity by over-frequent positive contacts, whereas the common chimpanzee females attempt to maintain the fragile social relationships by keeping negative attitudes.

The societies of the chimpanzees, without cohesion founded on fundamental inequality, must have formulated another system for coexistence; for they are living in peace in multimale–multifemale unit groups, each of which has a sizeable population of about 40–100.

This must be related to the rise of conditional equality. Kano (1980), Kuroda (1980), and Kitamura (1985) have described various aspects of the genito-genital contacts exchanged among pygmy chimpanzees; this unique behaviour is undoubtedly related to the conditional equality principle. Furthermore, the food-sharing behaviour observed in both species of chimpanzees poses an important question about the evolution of mentality and society (Itani, 1984, p. 178):

> First, their favourite food items move between individuals. This never occurs in Japanese monkeys. Direct consumption of food from hand to mouth is delayed through transfers between individuals, and the food is consumed also by some individuals who have not originally obtained it. Although it is an exaggeration to say that this is the origin of economic flow of goods, it is true that without this, the economic system in the human society would not work. Moreover, the objects flow from those who have to those who do not have; this flow is opposite to that of exploitation found in an inegalitarian society. The objects that move become the medium that assures the growth of equality in their society.
> . . . Second, begging individuals are psychologically free from the inhibitions that maintain the coexistence of subordinate and dominant individuals in the Japanese monkey society. Therefore, begging is a social interaction not inherent in the inequality principle. Furthermore, food sharing indicates their recognition that one's favorite food is another's favorite food too, which is a mentality related to sympathy and objectification. Only on the basis of sympathy does food sharing become possible. As opposed to the inequality principle, this bud of mentality is oriented toward some social system based on the equality principle.

I believe that the egalitarianism seen among the hunter–gatherer and nature-dependent peoples of today is nothing but a product of the evolutionary elaboration of its counterpart found among the chimpanzees. The demand for equality among the group members must have permeated into every sector of life, sharing must have intensified itself, and the whole process must have been irreversible.

Studies of non-human primate societies have been puzzling us with some fascinating data. The desire for pleasure shown by the two species of chimpanzees, which even appears superfluous, and their ability to share foods with others, cannot be evils. However, in the common chimpanzee, which has such *humanity*, 14 cases of infanticide have been reported (Goodall, 1977; Kawanaka, 1981; Itani, 1984; Nishida & Kawanaka, 1985; Takahata, 1985). Although this cannot be determined an outright evil, some evaluation is necessary when considering the evolution of primate societies.

As known from many other simians, infanticide probably appeared in the stage of higher primates with advanced forms of BSUs, and it occurs in more complicated episodes among the gorilla and the common chimpanzee (Itani, 1982). In my earlier paper, after pointing out the possibility that male infants are selectively victimised, I concluded that (Itani, 1984, p. 183):

> In effect, primates themselves modify their societies, and that this is more prominent in higher primate taxa and more highly developed *specia* [species society; Imanishi, 1950] (Itani, 1982). Continuing sociological studies of wild primates, we have been observing societies that are "artificially modified" by themselves. The implications of negated coexistence and its seemingly paradoxical links with human characteristics are important topics in future studies of human evolution.

Note that this infanticide is not a phenomenon found in a society ruled by fundamental inequality, but one observed in the chimpanzee society, which appears to be searching for conditional equality.

FEAR OF CIVIL INEQUALITY

Many researchers have unfailingly mentioned the egalitarianism in hunter–gatherer societies. They have stated that this is seen not only in their daily food consumption but also in the inhibition of private acquisitions or accumulations of profit or wealth, in the elimination of the differentiation of social status, and in the prevention of concentration of individual fame and honours (Ichikawa, 1982; Tanaka, 1971; 1980; Turnbull, 1983; Woodburn, 1981).

Among such descriptions, the most impressive to me was the narrative by Ichikawa (1982, p. 93) of an episode of elephant hunting in a band of Mbuti pygmies:

> Saranbongo came back. He held his spear on the shoulder; this special way of holding a spear is allowed only when the hunter has caught a game animal. The people there, freed from the tension, began expressing their joy. I had thought that Saranbongo would come back triumphantly in front of the people praising him with applause. But to the contrary, his behavior did not show any sign of display. When I praised him by waving my hands, he shyly turned his face down. If he had not held his spear in the special way, and had I not known his personality as I knew, I would have failed in seeing that it was he who felled the elephant. Much praise of any particular individual is not a usual practice among the Mbuti who live in an egalitarian society.

For the maintenance of such a small-band society, self-restraint, elimination of tenacity for goods, and enforced sharing are required, and these are the egalitarianism in its realisation. Turnbull (1983, p. 11) writes: "This is what they *have to be* in order to survive." Nobody, however, has attempted to consider its origin.

It is obvious that this shows the perfect form of a society based on the conditional equality. A hunter–gatherer band has families within itself. The family apparently is a partial structure that originates in the agreement based on the conditional equality. It also is the social unit that Rousseau (1762; English translation, p. 50) wrote of in the *Social contract* as: "The oldest of all societies, and the only natural one, is that of the family." Hunter–gatherers balance the family with their more primitive unit group, i.e., the band, by their ever-strong social code of egalitarianism.

Here I should like to comment on the original form of the band structure. There has been much dispute over this since Service (1962; 1966) concluded that a composite band is deformed by pressure from the outside. His statement did not conflict with the conclusions reached by the authors of classic texts on the Mbuti—Schebesta, Putnam, and Gusinde. Turnbull (1961; 1965), however, denied it and claimed that its structure is not lineal. Many papers in *Man the hunter*, written and edited by Lee and DeVore (1968), have concluded, through citing new information including Turnbull's report, that the Bushmen, the Hadza, and other hunter–gatherer societies have non-lineal social units. The book everywhere contained a critical response to Service's position. Tanaka (1971; 1980), who studied the Central Kalahari at the same period as Lee and others, also reached the same conclusion.

The succeeding Japanese researchers of the Mbuti, however, unanimously claimed a patrilineal or patrilocal band as the standard. They were Harako (1976), who studied archers and net-hunters, Tanno (1976) and Ichikawa (1978; 1982), who studied net-hunters, and Terashima (1984), who studied archers. Turnbull, (1983), though apparently aware of these Japanese researchers' results, has not changed his opinion at all in the *Mbuti pygmies: Change and adaptation*.

I will not go too far in detailing this discussion, because Ichikawa (1985) gives a good review of it. In his second paper on the Mbuti, Tereshima (1985) withdrew from his previous conclusion and claimed that the Mbuti band was originally bilateral and highly flexible, and he attempted to apply this to all hunter–gatherer societies in Africa. Ichikawa (1985), agreeing with Terashima in regard to the key role of band flexibility, attempts to end the discussion. This seems only to increase the confusion. Perhaps, some important point has been missed or, if not, the discussion itself may be unimportant. To assume that a flexible, bilateral structure was the original one so as to finish the discussion does nothing but throw the results of research into confusion because the flexibility is explained only in terms of adaptation.

Although Ichikawa includes the possibility of matrilineality in the flexible range of the original human society, I do not. From the evolutionary viewpoint, whether it was patrilineal, or non-lineal, or bilateral, does not make much difference. As I proposed in the "pre-band" hypothesis (Itani, 1966), the original human society may have reached the stage of patrilineality as found in the chimpanzees, or it may have only reached the stage of a less rigid but more flexible, bilateral-like structure. Whatever, the problem remains unsolved. However, it is important to note that the pathway chosen by the ancestors of man was not the matrilineal line chosen by some other non-human primates. The original human society was not in the matrilineal line in which the fundamental inequality left little possibility for the conditional equality.

Societies that fear civil inequality are not only found among the hunter–gatherers. One such example can be found in the society of the Tongwe, who are slash-and-burn agriculturalists living in the savanna woodland. In their life we can clearly see an indication of such fear. Kakeya (1974; 1976; 1977b), in his studies of the Tongwe's livelihood, society, and their supernatural world, pointed out the tendency toward minimum effort and equalisation in their food production. No one produces more than one needs in the year, and no one thinks of surplus or stock. Moreover, the people living in the scattered small hamlets often visit one another to consume their food stock by following their tradition of extravagant hospitality. As it is calculated that the amount of the food served to guests may reach 40% of the total food production, one can survive only by visiting others in turn as their guest. This is nothing but fear of a surplus, fear confirmed by their belief in the gods, fairies, ancestor spirits, witchcraft, and other supernatural beings, who are to cause various illnesses and disasters (Kakeya, 1977a; 1977b; 1984). What we can see in his detailed analyses is the image of an egalitarian society, although different from that of the hunter–gatherer, which fears the civil inequality.

As our research on nature-dependent peoples progressed, we began to appreciate that the concept of property differs greatly from one society to another. It is a reflection of the difference in the value system, and of the difference in the meaning of possessing goods. Turnbull (1983, p. 11) stated that, ". . . One of the most dramatic things, for me, that emerges from any study of the Mbuti is their preoccupation with values, and their lack of concern with the material world, or, the world of material well-being." Therefore, before concluding this chapter, I should like to mention another society, the Turkana, who are pastoralists living in northwestern Kenya, in which the human relationship to goods greatly differs or at least appears to be different from that of other societies.

Those who know the Turkana will never forget their persistent begging. Ohta (1985) presented an interesting analysis of the meaning of this *kashakunai* (merciless) behaviour. They force their opponent to solve the problem of being begged. The beggar's attitude is that unless there is any negative element in the begged's relationship with them, the begged shall give to the beggar. The begged, however, will never be thanked. Ohta points out that this is a sort of reconfirmation of their intimate relationship.

The society of the Turkana cannot be called egalitarian. All the same, their condition of possessing livestock is distant from the state of possessing goods in our sense. Their livestock provide them with milk and blood, which constitute their staple foods. The livestock are not to be sold, or to be slaughtered other than for rituals. A few hundred cattle may go from one man to another, as a bride price, for example; but that man may receive more cattle from another. Be repeated exchanges of raids with the neighbouring tribes, their livestock may disappear or may multiply overnight. Their livestock, in other words, resemble a pile of chips on a roulette table.

Their persistent begging among themselves leads us to believe that they should have quite a stock of belongings. In reality, however, their stock of goods differs little from the arrows and necklaces passed around among the band members of the Hadza, as filmed by Woodburn and Hudson (1966). Something given to a particular person today will be found to be somebody else's tomorrow.

The scene of departure among the Turkana, who constantly change their camps, tells everything. When neighbouring homesteads were heard to have been raided by the Jie, the family I lived with decided to move. It was not until two hours before the departure, however, that I noticed their decision. The women began to gather the donkeys in the enclosure. It took less than 40 minutes for all the women to load all of their belongings onto the saddle of each donkey. The load for a donkey seemed to be at most 60kg.

On this occasion, I was impressed to find how few belongings they had for their minimal life, and that having more than that meant an offence against their basic lifestyle. Although each of the women could have used two donkeys to carry many things, none of them did so. The number of the loaded donkeys and that of the women was exactly the same. I should like to add a few words to Ohta's conclusion. The Turkana's persistent begging might be an outlet or drive of helpless desire in a people who cannot step over the border into the world ruled by the civil inequality principle.

If the flow of social evolution has no stagnation point, and if every society follows as the flow directs, every society would change as the flow determines. However, when will or can the Mbuti discard their "disposal culture" (Tanno, 1976; 1981), which is so distant from the possession of property? How can the Tongwe break the shell of their supernatural cosmos, which thickly binds their self-restraint culture? Or can the Turkana share the same concept of private possession as us? To none of these questions, can we give any definite answer, nor can we predict the path each people will take regarding each question.

At the beginning of the *Social contract*, Rousseau (1762, English translation, p. 49) stated that: "My purpose is to consider if, in political society, there can be any legitimate and sure principle of government, taking men as they are and laws as they might be."

In Africa also, kingdoms were born and prospered in fertile lands. In such cultures, the transition from the world based on the conditional equality to that based on the civil inequality may have occurred naturally. In the confused world of today, however, the transition from the natural state to the civil state seems to occur only by acculturation or invasion from the outside civilisations. There seems to be little room for Rousseau's theory of a "general will" to work.

ACKNOWLEDGEMENTS

My thanks are due to: Dr. M. R. A. Chance, who invited me to contribute this chapter, and Mr. P. Scarabaeus, who drafted the English translation of my manuscript. This work was supported in part by a Grant-in-Aid for Special Project Research *Biological Aspects of Optimal Strategy and Social Structure* from the Japanese Ministry of Education, Science, and Culture.

REFERENCES

Charles-Dominique, P. (1977). *Ecology and behaviour of nocturnal primates*. New York: Columbia University Press.

Furuichi, T. (1983a). Dominant/subordinate relationships in the social life of Japanese macaques. *Iden, 37(4)*, 3–9, (in Japanese).

Furuichi, T. (1983b). Interindividual distance and influence of dominance on feeding in a natural Japanese macaque troop. *Primates, 24*, 445–455.

Goodall, J. (1977). Infant killing and cannibalism in free-living chimpanzees. *Folia primatologica, 28*, 259–282.

Goodall, J. (1983). Population dynamics during a 15 year period in one community of free-living chimpanzees in the Gombe National Park, Tanzania. *Zeitschrift für Tierpsychologie, 61*, 1–60.

Harako, R. (1976). The Mbuti as hunters: A study of ecological anthropology of the Mbuti pygmies (1). *Kyoto University African Studies, 10*, 37–99.

Hasegawa, T., & Hiraiwa. M. (1980). Social interactions of orphans observed in a free-ranging troop of Japanese monkeys. *Folia Primatolologica, 33*, 129–158.

Hayaki, H. (1983a). Social interactions of juvenile Japanese monkeys on Koshima Islet. *Primates, 24*, 139–153.

Hayaki, H. (1983b). Social play in juvenile Japanese monkeys. *Iden, 37(4)*, 44–50 (in Japanese).

Hayaki, H. (1985). Social interactions of juvenile and adolescent chimpanzees. *Primates, 26(4)*, 343–360.

Huxley, T. H. (1863). *Man's place in nature* (reprinted 1959). New York: University of Michigan Press.

Ichikawa, M. (1978). The residential groups of the Mbuti pygmies. *Senri Ethnological Studies 1; Africa, 1*, 131–188.

Ichikawa, M. (1982). *Hunters in the tropical forest*. Kyoto: Jinbun-Shoin, (in Japanese).

Ichikawa, M. (1985). Nature recognition. In J. Itani & J. Tanaka (Eds.), *Africa—its nature and peoples*. Kyoto: Akademia Shuppankai, (in Japanese).

Imanishi, K. (1950). Social life of semi-wild horse—a proposal for *specia*, *oikia* and *oikion*. *Physiology and Ecology of Japan*, *4*, 28–41, (in Japanese).

Itani, J. (1954). Japanese monkeys in Takasakiyama. In K. Imanishi (Ed.), *Social life of animals in Japan*. (vol. 2). Tokyo: Kobunsha, (in Japanese).

Itani, J. (1959). Paternal care in the wild Japanese monkey, *Macaca fuscata fuscata*. *Primates*, *2*, 61–93.

Itani, J. (1963). Vocal communication of the wild Japanese monkey. *Primates 4*, 11–66.

Itani, J. (1966). The social structure of chimpanzees. *Shizen*, *21(8)*, 17–30, (in Japanese).

Itani, J. (1977). Evolution of primate social structure. *Journal of Human Evolution*, *6*, 235–243.

Itani, J. (1980). Social structures of African great apes. *Journal of Reproduction and Fertility* (*Suppl*), *28*, 33–41.

Itani, J. (1982). Intraspecific killing among non-human primates. *Journal of Social and Biological Structures*, *5*, 361–368.

Itani, J. (1984). Inequality versus equality for coexistence in primate societies. In *Absolute values and the new cultural revolution* (Pp. 16–189). Chicago: I.C.U.S. Books.

Itani, J. (1985). The evolution of primate social structures. *Man*, *20*, 593–611.

Kakeya, M. (1974). The Tongwe's subsistence system—their habitat, livelihood, and food culture. *Kikan Jinruigaku*, *5(3)*, 3–90, (in Japanese).

Kakeya, M. (1976). Subsistence ecology of the Tongwe, Tanzania. *Kyoto University African Studies*, *10*, 143–212.

Kakeya, M. (1977a). Subsistence, society, and supernatural existences—the Tongwe's case. In *Jinruigaku Kôza* (vol. 12) (Pp. 369–385). Tokyo: Yûzankaku. (in Japanese).

Kayeya, M. (1977b). The world of Tongwe witch doctors. In J. Itani & R. Harako (Eds.), *Natural history of man* (Pp. 377–439). Tokyo: Yûzankaku, (in Japanese).

Kakeya, M. (1984). Curing ritual of the Tongwe witch doctors—its process and logic. In J. Itani & T. Yoneyama (Eds.), *Studies of African cultures* (Pp. 729–776). Kyoto: Akademia Shuppankai, (in Japanese).

Kano, T. (1980). Social behavior of wild pygmy chimpanzees (*Pan paniscus*) of Wamba: A preliminary report. *Journal of Human Evolution*, *9*, 243–260.

Kawai, M. (1958). On the rank system in a natural group of Japanese monkey (I) The basic and dependent rank. *Primates*, *1*, 111–130, (in Japanese).

Kawamura, S. (1958). Matriarchal social ranks in the Minoo-B troop: A study of the rank system of Japanese monkeys. *Primates*, *1*, 149–156.

Kawanaka, K. (1981). Infanticide and cannibalism in chimpanzees, with special reference to the newly observed case in the Mahale Mountains. *African Studies Monographs*, *1*, 69–99.

Kitamura, K. (1985). The pygmy chimpanzee—Its social structure, interindividual relationships, and behavior. In J. Itani & J. Tanaka (Eds.), *Africa—its nature and peoples* Kyoto: Akademia Shuppankai, 43–70, (in Japanese).

Koyama, N. (1967). On dominance rank and kinship of a wild Japanese monkey troop in Arashiyama. *Primates*, *8*, 189–216.

Koyama, N. (1970). Changes in dominance rank and division of a wild Japanese monkey troop in Arashiyama. *Primates*, *11(4)*, 335–390.

Kummer, H. (1968). *Social organization of hamadryas baboons: A field study*. Basel: Karger.

Kuroda, S. (1980). Social behavior of the pygmy chimpanzees. *Primates*, *21*, 181–197.

Lee, R. B. & DeVore, I. (1968). *Man the hunter*. Chicago: Aldine.

Nishida, T. (1970). Social behavior and relationship among wild chimpanzees in the Mahale Mountains. *Primates*, *11*, 47–87.

Nishida, T. (1979). The social structure of chimpanzees of the Mahale Mountains. In D. A.

Hamburg & E. R. McCown (Eds.), *The great apes* (Pp. 73–121). Menlo Park, Calif.: Benjamin/Cummings.

Nishida, T. & Kawanaka, K. (1985). Within-group cannibalism by adult male chimpanzees. *Primates*, 26(3), 274–284.

Norikoshi, K. (1974). The development of peer–mate relationships of free–ranging Japanese monkeys in food-getting situations. *Primates*, *12*, 113–124.

Ohta, I. (1985). Reciprocity among the Turkana, northwestern Kenya. (in Japanese)

Rousseau, J-J. (1755). *Discours sur l'origine et les fondement de l'inégalité parmi les hommes*. (English translation, 1984) Harmondsworth, England: Penguin.

Rousseau, J-J. (1762). *Du contrat social; ou, principes du droit politique*. (English translation, 1968) Harmondsworth, England: Penguin.

Service, E. R. (1962). *Primitive social organization: An evolutionary perspective*. New York: Random House, Inc.

Service, E. R. (1966). *The hunters*. (*Foundation of modern anthropology series*). New Jersey: Prentice-Hall.

Sigg, H., Stolba, A., Abegglen, J. J. & Dasser, V. (1982). Life history of hamadryas baboons: physical development, infant mortality, reproductive parameters and family relationships. *Primates*, *23*, 473–487.

Takahata, Y. (1982). The socio-sexual behaviour of Japanese monkeys. *Zeitschrift für Tierpsychologie, 59*, 89–108.

Takahata, Y. (1985). Adult male chimpanzees kill and eat a male newborn infant: newly observed intragroup infanticide and cannibalism in Mahale National Park, Tanzania. *Folia primatologica*, *44*, 161–170.

Tanaka, J. (1971). *Bushman*. Tokyo: Shisakusha, (in Japanese).

Tanaka, J. (1980). *The San: Hunter–gatherers of the Kalahari—a study of ecological anthropology*. Tokyo: University of Tokyo Press.

Tanno, T. (1976). The Mbuti net-hunters in the Ituri Forest, eastern Zaire: Their hunting activities and band composition. *Kyoto University African Study*, *10*, 101–135.

Tanno, T. (1981). Plant utilization of the Mbuti pygmies: with special reference to their material culture and use of wild vegetable foods. *African Studies Monographs*, *1*, 1–53.

Terashima, H. (1984). The structure of Mbuti archers' band. In J. Itani & T. Yoneyama (Eds.), *Studies of African culture* (Pp. 3–41). Kyoto: Akademia Shuppankai, (in Japanese).

Terashima, H. (1985). Variation and composition principles of the residence group (band) of the Mbuti pygmies: beyond a typical/atypical dichotomy. *African Study Monographs Supplementary Issue*, *4*, 103–120.

Turnbull, C. M. (1961). *The forest people*. New York: Simon & Schuster.

Turnbull, C. M. (1965). *Wayward servants: The two worlds of the African pygmies*. New York: Natural History Press.

Turnbull, C. M. (1983). *The Mbuti pygmies: Change and adaptation*. New York: Holt, Rinehart & Winston.

Woodburn, J. C. (1981). Egalitarian societies. *Man*, *17*, 431–451.

Woodburn, J. C. & Hudson, S. (1966). *The Hadza: The food quest of a hunting and gathering tribe of Tanzania* (16mm film). London: London School of Economics.

7 Alternative Channels for Negotiating Asymmetry in Social Relationships

John Price
Department of Psychiatry, The General Hospital, Milton Keynes, U.K

THE YIELDING SUBROUTINE OF RITUAL AGONISTIC BEHAVIOUR

Darwin (1871) was the first to point out that socially mediated asymmetry between members of the same sex of the same species would, if associated with differential reproduction, be an important force in evolution. He described it as the intra-sexual component of sexual selection (the other component being mate choice), and he suggested that sexual selection might have an importance in evolution approaching that of natural selection. The theory of sexual selection has received support over the years (Campbell, 1973); recently, increased reproduction in men of high rank has been demonstrated in more than 100 pre- and non-industrial societies (Betzig, 1986), and reduced reproduction in low-ranking females of a variety of animal species (Kevles, 1986). Male animals are well known to show an association between reproduction and social rank, particularly those that have polygynous or lek forms of mating system; and although there are no data available for human females, social heredity (Schiff & Lewontin, 1986) ensures that high-ranking females will have excess grandchildren because of the increased reproduction of their sons.

Ritual Agonistic Behaviour

It follows from Darwin's theory that mechanisms for generating social asymmetry are likely to be selected for. In fact, only one such mechanism, ritual agonistic behaviour (RAB), appears to be common in vertebrate species. The discovery and description of RAB is one of the main achievements of classical ethology (Lorenz, 1981). RAB is a form of signalling between two individuals, the result of which is the creation, readjustment or reinforcement of asymmetry between them (see Hinde, 1979). RAB converts a symmetrical relationship (characterised by a potential for escalating conflict) into an asymmetrical or complementary relationship with an agreed upon winner and loser, and having the properties of a stable system (Bateson, 1972). The subsequent roles are differentiated in that the winner occupies a role described as dominant, or higher-ranking, or territory owning, whereas the loser occupies a role described as subordinate, or lower-ranking, or non-territory owning. A matrix of asymmetrical relationships forms a social hierarchy (Kaufman, 1983).

It is a remarkable fact that all the main elements of RAB are seen in reptiles. Some species of lizard defend territories; others establish dominance hierarchies based on individual recognition (Carpenter, 1978). The process of RAB can be seen with particular clarity in lizards, partly because states of dominance and subordinacy are reflected by different skin colours, and partly because the agonistic behaviour is not complicated by any form of affiliative behaviour. Subordinate lizards secrete excess corticosteroids, as do subordinate mammals (Greenberg & Crews, 1983).

The Ritual Agonistic Encounter

When human strangers come together, they usually meet to perform a task, or for recreation, and the issue of dominance does not immediately arise. They are in what Chance and Jolly (1970; see also Introduction and Chapter 1) called the hedonic mode. This state of affairs may continue indefinitely, particularly if the strangers are reasonable people and not too many controversial decisions have to be made. However, in many groups, conflicts arise that cannot be reconciled according to the methods of the hedonic mode, and then there is a switch to the agonic or agonistic mode. This is a well recognised phenomenon in group psychotherapy (Kennedy & MacKenzie, 1986), in which the task of the group is increasingly interfered with by some mysterious process that often turns out to be rivalry for leadership. What happens then is something we share with most other vertebrate species.

Many species of bird (Schjelderup-Ebbe, 1935) and primate (Kummer, 1971) cannot live in symmetrical relations with members of the same sex, and so every meeting with a stranger starts with a ritual agonistic encounter. Schjelderup-Ebbe (1935) has described the meeting of two hens with its inevitable outcome that one dominates the other. Although the end is always the same, the method varies according to the aggressiveness of the hens. If both are fearful, they may both retreat, and then it is the one who recovers first who dominates the other. If one is fearful and the other aggressive, the fearful one submits quickly after a brief period of mutual appraisal. If both are aggressive, they start pecking each other and continue to do so until one gives up and flees; the one who gives up subsequently becomes subordinate to the other and does not retaliate to the other's pecks. Even after the asymmetrical dominant/subordinate relationship has been established between two hens, pecks continue to be delivered by the dominant to the subordinate hen, although at a lower intensity and frequency (on what appears to be an unprovoked and random basis), and such pecks appear to have the function of keeping the subordinate bird in her place; this "down-hierarchy" pecking may have a different motivational base than the symmetrical pecking of the agonistic encounter. If the subordinate bird pecks the dominant, retaliatory pecking by the dominant occurs with great intensity but, particularly if the dominant bird is old or sick, the subordinate bird may reverse the rank order and assume the dominant role. This work by Schjelderup-Ebbe has been confirmed (Wood-Gush, 1955), and the same general scheme has been found to occur in reptiles and mammals, usually with some species-specific activity taking the place of pecking.

Of course, human social life contains many other mechanisms for generating and maintaining asymmetry between members of the same sex, such as warfare, social class, primogeniture, elections, racial prejudice, vocational competition, gambling, hero-worship, and externally imposed rank. Therefore we might expect RAB to operate as a primitive mechanism underlying the products of more recent evolution (such as the mammalian affiliative systems of parental care, pair-bonding, and alliances) as suggested by MacLean (1985), and also underlying the immensely complex structure of human cultural development. This is in the tradition of thought that Durant (1981) has called "the beast within", and is in contrast to an opposite view that human behaviour and its variations are entirely due to culture and have not been subjected to the forces of natural selection (see Hodes, 1986). This is not the forum to debate this important controversy, but research that is directed according to *tabula rasa* assumptions may ignore important variables, as I hope to demonstrate later in regard to the expression of hostility in depression. Human life is so complex that certain patterns may not be visible unless first we know what to look for, and one source of this prior knowledge is comparative ethology, the study of behaviour patterns with determinants deeply embedded in the vertebrate genome.

RAB and Psychopathology

It was M. R. A. Chance, the editor of this book, who first drew my attention to RAB as a potential cause of psychopathology, in his lecture on the social behaviour of a colony of long-tailed macaque monkeys. It certainly seemed that the punishment meted out by the high-ranking to the low-ranking monkeys might cause them some mental perturbation, and the cringing behaviour of the low-ranked was consistent with the presence of what we recognise in human patients as depression and anxiety. Moreover, a mental state such as that of my depressed patients, characterised by ideas of inferiority and unworthiness, seemed ideally suited to reconcile the low-ranking monkeys to their inferior situation. From this, it seemed possible that the function of depression, selected for during evolution, might be to subserve low rank and fall in rank, whereas the converse function of elevated mood might be to subserve rise in rank (Price, 1968). This function of depression in terms of social competition seemed more likely than the other alternatives which had been put forward (Price, 1972), and also more likely than the conclusion reached by Klerman (1974) that depression has no adaptive function at all.

I should say that by depression I mean not depressed mood as an emotion, but an episode of altered behaviour lasting weeks or months that, when it appears inappropriately, or with exaggerated intensity or duration, is liable to be labelled depressive illness. Good descriptions of the current state of knowledge about depression are given by Gilbert (1984, and in press).

Recently Leon Sloman and I (Price & Sloman, 1987) re-examined the hypothesis that depression evolved as the yielding component of RAB. We called this yielding component the yielding subroutine (YS) of ritual agonistic behaviour (RAB), or YS/RAB for short. The instructions required of such a yielding subroutine are two-fold: first, they must ensure that the yielder yields and does not try to make a come-back, either by recontesting the issue or escalating to a less ritualised level of conflict; and, secondly, they must ensure that the yielder reassures the winner that yielding is really taking place, so that the winner ends his RAB and ceases to inflict punishment on the yielder. In that re-examination, we devised theoretical instructions for the yielding subroutine that satisfied these requirements, and also were consistent with the clinical features of depressive illness. We showed that two different kinds of yielding subroutine are required, one subserving social change and the other subserving social homeostasis; the first helps the yielder adjust to subordinate rank after being dominant or equal, the other helps the yielder to maintain low rank and to forego the temptation of attempting to rise in rank. Both kinds of yielding require the same signal aimed at the dominant individual, one that reassures that the yielder is really yielding and not planning a come-back. In our paper, we dealt briefly with the signal of the yielding subroutine (YS/RAB), as outlined next.

Previous Formulation

We suggested that in YS/RAB-modelled depression the appearance of the depressed person has the function of reassuring a hypothetical winner that further discouragement is not necessary. There is, admittedly, no positive signal of incapacity, but the general woebegone appearance is such as to reassure the adversary that the depressed person is incapable of making a come-back. In fact the very lack of a signal (which would represent at least signalling capacity) is likely to reassure others that the absence of capacity is genuine and not a deceptive signal, designed to lull them into a false security while, underneath, preparations for a come-back are secretly being made. This signal of depressive incapacity appears to be similar across species, which is presumably the reason why Schjelderup-Ebbe as a human being was able to recognise it in the bird.

In the language of communication theory, the YS/RAB is a communication couched in terms of metacommunication. It appears, on the surface, to be a metacommunication to the effect that "I am too incapacitated to communicate", whereas underneath it is really a communication saying "I am too incapacitated to fight back." To send such a complex message is quite a manifestation of social skill. In fact, we could say that the depressed individual, as yielder, has the social skill to send a message that denies such skill, in a communication that is received *not as a communication but as an observation of fact.* (Italics by request of the Editor.)

I shall expand this discussion of the signalling aspects of the yielding subroutine by asking, first of all, what signals cause the yielding subroutine to occur; and, secondly, what instructions are necessary to ensure that the signals emitted by the yielder reassure the winner, and how do these instructions compare with the signal aspects of depressive illness. I argue that the signal given off by the depressed person is different from normal signals in two main respects.

Firstly, it is a presentation of the total self for evaluation by the other that is apprehended by the other, not in the way that a submission signal is received, but as a form of judgement about the sender made by the receiver. In other words, the receiver does not identify the sender as someone who is signalling. If the depressive signal is given at the same time as a submission signal, it is acting as a paracommunication, given simultaneously along another channel, and conveying the message, "Even if I did not want to submit there is nothing I could do about it." To the extent that the depressive signal is a message about a signal, conveying the message, "I am too incapacitated to signal properly", it comes into the category of metacommunication, being a signal at a higher logical level than the other signal (Watzlawick, Beavin & Jackson, 1967). On looking at a (male) adversary signalling defeat, the winner registers, "He knows he has been beaten"; on looking at an adversary manifesting the yielding subroutine, the winner registers, "That is a broken man (or woman)."

Secondly, the signal is transmitted in the hedonic mode but it relates to the agenda of the agonistic mode, so that it is likely to be dealt with by the receiver along a channel other than the one dealing with ongoing business; and, to an observer, it is likely to appear incomprehensible.

A Psychiatric Look at Ritual Agonistic Behaviour

Giving Up. Looking at ritual agonistic behaviour from the perspective of a psychiatrist, I instinctively look at the behaviour of the loser of the agonistic encounter. The medical services are concerned with losers rather than winners and, whereas the loser of a non-ritual encounter is likely to be taken to the Casualty Department or the Orthopaedic Ward, the loser of a ritual encounter may need help of a psychiatric nature. As Chance (1977, p. 191) put it, it is ". . . escape-motivated forms of agonism that lead to breakdown" such as depression, and ". . . any mental cure must aim for a transfer of the individual from the agonic to the hedonic mode".

Another word for losing and yielding is giving up. And, indeed, losers have a lot to give up: they have to give up whatever the conflict was about; and, regardless of the actual issue at stake, they have to give up the satisfaction of getting their own way. This is consistent with the predominating mood of the depressed patient, which is one of "giving up".

An Internal Referee. Another thing to strike me is that the constraint of giving up is imposed on the yielder from within. The winner does not bother to stand guard over whatever has been won; the winner assumes that the loser will behave like a loser and not either recontest the issue at a symbolic level, or escalate the fight to a level at which death or serious wounds are likely. The instructions of the yielding subroutine are like an internalised referee who says: "You have lost. Behave like a loser." The yielding subroutine is a symbolic or ritual equivalent of death or serious physical incapacity, in just the same way as the elaborate threat signals of RAB are a symbolic equivalent of hitting and biting. This is consistent with the fact that many depressed patients feel incapacitated, some feel "dead", and others feel that they are "losers".

Reverted Escape. The loser who lives in a group has a lot more to give up than the loser who is a member of a solitary species. The solitary is usually fighting about the possession of territory and, if the battle over one territory is lost, goes on to contest the ownership of another. As long as more territories are available to contest, it is not adaptive to have more than a brief "giving up" feeling over territory just lost, before moving on to try somewhere else. Group-living individuals, on the other hand, require a much more prolonged and complex yielding subroutine, for they may have to give up habits of dominance and leadership developed over many years.

Unless these habits and attitudes are modified in some way, they are unlikely to be able to stay in the group.

Chance (1986) has given the name "reverted escape" to the movement of an individual, who is motivated to escape, back towards a dominant member of the group, so that the individual is contained in the group. It is similar to the anthropological concept of circumscription (Betzig, 1986), except that reverted escape depends more on the attraction of the group, whereas circumscription depends on the unattractiveness (or unavailability, in the case of imprisoned humans or caged animals) of prospects outside the group.

Gaylin (1983, p. 173) has summarised the psychoanalytical view of the depressive as someone who cannot win but lacks the option of escape: "The depressive has given up all hope of either fight or flight"; his only remaining option is submission, and so, ". . . hopeless and helpless he gives up the struggle".

Time Scale. The yielding behaviours so far described by ethologists are brief matters occupying a matter of seconds or minutes, such as fleeing, or the emission of submissive signals like the subordinate wolf who, rolling over on its back and exposing its vulnerable neck and underparts, gives a signal thought to be ritualised from the cub's presentation of its perineal area to the mother for cleaning (Schenkel, 1967). But, if two group members have been in a relationship for months or years, and have then engaged in a ritual agonistic encounter that may have lasted several weeks or even months, as a result of which a previously equal or higher ranking individual has become lower ranking, then it is likely that the phase of adjustment to the new relationship is going to take longer than a few minutes, and that the yielding subroutine will need to operate for weeks or months. This is consistent with the time scale of depressive states.

Signalling of Resource-Holding Potential

Behavioural ecologists have for many years been concerned with the mathematics of RAB, or pairwise contests, and the selective forces acting on the strategies used in them (Maynard Smith, 1982). They have introduced an intervening variable, which they call resource-holding potential (RHP), to assist in the mathematical analysis of such contests (Parker, 1974; 1984). RHP is a measure of fighting capacity and, on the input side, it is determined by such factors as age, size, weapons, previous success, and availability of allies. On the output side it determines probability of fighting (rather than submitting or withdrawing) in a contest, and also duration and intensity of fighting once a contest has begun. All the attack components of agonistic behaviour, including dominance display, threat display, challenge, attack, and chasing, are looked on as signals of RHP.

There is, to my knowledge, no concept currently in use in psychology that expresses the equivalent of RHP: it is related to the ideas of self-esteem, self-confidence, and ego strength, but these are poorly defined and in any case refer to confidence in other areas in addition to fighting ability; it is also related to the "dominance feeling", as described by Maslow (1937), and to the sociological concept of structural power (Kemper, 1978). I think RHP is a helpful term in the conceptualisation of RAB and I hope that in extending it somewhat in its psychological meaning I am not distorting the meaning it already has in behavioural ecology.

The Calculation of Relative RHP. In a contest, or ritual agonistic encounter, we are concerned with each contestant's evaluation of their own RHP compared with their assessment of their adversary's RHP—what Parker (1984) has called relative RHP. This is a somewhat complex evaluation, and it might be useful to recognise the following subdivisions of RHP:

1. *Absolute RHP.* Each individual has some general idea of his or her own fighting capacity in relation to other individuals, regardless of who may be his or her specific adversary on any one occasion. This value of RHP determines any undirected dominance display that the individual signals to the world at large, and the directed dominance display (challenge or threat) that each makes to a potential adversary at the beginning of an encounter, before there has been time to assess the other's RHP.

2. *Signal of Absolute RHP.* This is the signal given in the dominance displays mentioned in the preceding paragraph. Such a signal can obviously be faked, but I will assume here that it is an accurate reflection of absolute RHP, in order not to complicate further an already sufficiently complex subject.

3. *Estimate of Adversary's RHP.* I will assume that this is an estimate reflecting the information received from the adversary's "signal of absolute RHP" but, if faking is suspected, the estimate could be revised up or down.

3. *Estimate of Relative RHP.* This is derived from a comparison of (1) and (3) above. We do not know how this comparison is made, but in the simplest case it must give a result that is either favourable or unfavourable, in order to determine the choice between two possible courses of action: escalation and submission. *These actions then become signals in the next round of the conflict.*

There is likely to be both genetic and environmental variation in the degree of superiority of own RHP over adversary's RHP, which an individual requires for a favourable estimate of relative RHP to be made. Individuals requiring more superiority would tend towards the "dove" phenotype described by Maynard Smith (1982); those requiring less superiority would tend towards the "hawk" phenotype.

5. *Signal of Relative RHP.* A contest may last for several "bouts", so that, at each stage, each contestant is estimating the adversary's absolute

RHP, comparing this estimate with his or her own absolute RHP to calculate his or her own relative RHP (which may be either favourable or unfavourable), and signalling this relative RHP by either escalating or submitting. For simplicity I will assume a two-stage contest in which a period of mutual *assessment* is followed either by the submission of one contestant or by a period of *engagement* in which there is mutual attack. During the assessment stage, either contestant may submit by giving a signal of unfavourable relative RHP, and thus leave the encounter in a subordinate role but without loss of RHP—what Sloman and Price (1987) called "voluntary yielding"; or the contestant may enter the engagement stage by giving a signal of favourable relative RHP and thus have a chance of winning, but at the risk of losing and being forced into what we called "agonistic yielding", with associated loss of RHP.

6. *Verbal Reports of the Subjective Experience of RHP.* RHP is an intervening variable that determines signalling in agonistic encounters and, in animals, it is easy to maintain this conceptual level. In human beings, however, there may be another output in the form of verbal reports to a third party, such as a friend, a psychiatrist, or an investigator, giving an account of the subjective experience of either high or low absolute RHP, such reports may differ from the signal given to a potential adversary. These verbal reports convey an equivalent of the usual meaning of the terms self-confidence and lack of confidence. Reports of the subjective aspects of signals of *relative* RHP are less common, because the subject is likely to be too busy fighting or submitting, but the signal of favourable relative RHP is likely to be accompanied by feelings of anger, indignation, or irritation, and the signal of unfavourable relative RHP by feelings of being overwhelmed by helplessness. It is remarkable that patients have a very poor memory for their own signals of relative RHP, and in the case of favourable relative RHP there may be total amnesia, as in episodes of "blind rage".

The Two Components of the RHP Signal. Although it may be convenient to amalgamate them for mathematical purposes (Parker, 1984), from the psychological point of view "signal of absolute RHP" is very different from "signal of relative RHP". The two differ in the following ways:

1. *Stage of Assessment versus Stage of Engagement.* Signal of absolute RHP is saying: "This is what I am like; examine me and assess my power", and it occurs in the assessment phase of the agonistic encounter, when the adversaries are confronting each other, or circling round each other in mutual appraisal. Signal of relative RHP is saying: "I am better than you and I will prove it"; it occurs in the engagement phase of the encounter, and consists of very ritualised and species-specific activity, such as re-

peatedly charging at each other head on, as in the African buffalo (Sinclair, 1977), the American bison (Lott, 1967), and the marine iguana of the Galapagos Islands (Carpenter, 1978).

2. *Semantic versus Shannon Information*. In signalling absolute RHP, the adversaries present themselves for examination of all their aspects, having little control over what aspects are attended to; therefore, the information offered to the adversary is very extensive, even infinite, and is, I think, what Krebs and Dawkins (1984) have called Shannon information, and what Lockard (1980) has called a composite signal. In signalling relative RHP, on the other hand, they offer only one "bit" of information (namely, whether escalating or not), and the nature of the signal insists that the adversary pay attention to it and to no other. If the signal varies, it varies in quantity rather than in quality. This is what Krebs and Dawkins call semantic information, and what Lockard calls a graded signal.

This difference is probably due to the fact that in the process of sender/receiver co-evolution the exchange of signals of *absolute* RHP has been a *co-operative* matter, in that, if there is a real disparity between the RHP of the contestants, it is in both their interests that the one with lower RHP should rapidly and efficiently identify the disparity and submit. In contrast, the exchange of signals of *relative* RHP has been a competitive matter during sender/receiver co-evolution, because it occurs only if there is no great difference in absolute RHP and each contestant has a fair chance of winning; each is interested not only that the winner should be decided quickly but also that they should be that winner.

3. *Species Similarity versus Species Specificity*. Signals of absolute RHP tend to be common across species, such as upright posture, confident gait, display of weapons, and large size. Exceptions, such as the blue colouring of the rainbow lizard (Harris, 1964), are rare. Signals of relative RHP, on the other hand, tend to be highly species-specific in that each species has its own form of "combat"; some forms, like the head charging of the bison, involve bodily contact whereas others, such as the gill erection of the Siamese fighting fish, do not, consisting entirely of an exchange of signals at a distance. Within these categories the signals are similar in general form but highly specific in detail.

4. *Different Effect on Allies*. The signalling of absolute and relative RHP can be further differentiated if we postulate the presence of an ally. Displays of absolute RHP *are* exchanged by allies, and they boost rather than lower their RHP. Signals of favourable relative RHP are *not* usually directed at allies (except in mock fights for practice) and, if they were, they would lower RHP.

Establishment of RHP. In man and other primates, RHP is probably established at adolescence, in fact the agenda of adolescence may be largely to determine RHP and group membership. Savin-Williams (1987) has described the spontaneous formation of social hierarchies in adolescent boys in a situation in which considerable efforts were made to encourage symmetrical relationships. The rough-and-tumble play of children, and the competitive tendency of adolescent boys, ensure a wide variation in RHP before the stage of adult fighting is reached, and it is likely that for each individual a basal level of RHP is determined at this stage, around which there is only a limited fluctuation in later life. We can expect genetic variation between, on the one hand, adult RHP which is independent of outside influences, and, on the other, RHP that depends for its maintenance on continuous boosting by what Fenichel called "narcissistic supplies" (Gaylin, 1983), and I later call anathetic signals.

As allies are an important determinant of RHP, and allies depend partly on popularity, which in turn depends partly on task competence, the result is likely to be a very complex interaction of the various sources of self-esteem, of which power over people (RHP) is one (Coopersmith, 1967). Those who emerge from adolescence as members of a social group may have within-group and between-group components to their RHP.

Catathetic Signals

Because "signal of favourable relative RHP" is a cumbersome phrase, and because there is no exact ethological equivalent, I propose the term catathetic to describe the signals that are exchanged during the engagement phase of the agonistic encounter (and at other times to reinforce dominance). Catathetic comes from the Greek words for "put" and "down", so expressing the function of catathetic signals, which is to put the other individual down, in the sense of making them yield and/or lowering their RHP. (I am aware that "cathairetic" would be more correct from the etymological point of view, but "catathetic" is easier to use).

It is worth noting that catathetic signals can be defined in relation to either the sender or the receiver. For the sender, they are signals of favourable relative RHP. For the receiver, they are signals that lower RHP. It may at first sight seem surprising that a technical term can be defined in two apparently independent ways; but, on reflection, it is of the essence of a signal that it should have a specific meaning for both sender and receiver. An S.O.S. message, for example, is a signal given when the sender is in distress, and it is also a signal that motivates the receiver to go to the rescue. The catathetic signal is given when the sender is confident of winning and, at the same time, it is a signal that lowers the receiver's confidence (RHP).

Are catathetic signals the only signals that lower RHP? In animals, probably yes. In human beings, because of speech, RHP may be lowered by a message bringing bad news, such as the death or desertion of an ally; and this may be the reason why the bearer of bad tidings, although in no way responsible for the content of the message, may receive punishment at the hands of the recipient. As Shakespeare put it "The nature of bad news infects the teller" (*Anthony and Cleopatra*, Act I, Scene 2). Bad news is received, incorrectly, as a catathetic signal from the messenger, and so elicits a catathetic signal in return.

Considering the species-specificity and low information content of catethetic signals, it is likely that some very specific neural structures have co-evolved for the sending and receiving of these signals. Ethologists I have consulted are not happy to accept that catathetic signals are sign stimuli acting on an innate releasing mechanism to release the fixed action pattern of thc yielding subroutine; but something similar to this classical ethological process seems likely.

Catathetic Signals in Humans. Human beings are unique in the animal kingdom in being able to verbalise the signal of favourable relative RHP (catathetic signal). The message of the signal is "I am superior to you", and whereas other species need to indulge in various pushing and pulling contests in order to get the message across, human beings can simply say it. Of course, if both say it, they are in a contest, and they have to keep on saying it until one gives up or escalates to the next stage, which is physical attack. Therefore, human ritual agonistic encounters take the form of slanging matches in which each contestant continues verbally to assert superiority over the other, with varying degrees of imagination and sophistication. This verbal interchange is the human species-specific form of catathetic signalling. As it appears to be generally true that the structures responsible for catathetic signals tend to become hypertrophied through sexual selection, the same argument must give one reason for the development of the richness of human language.

Human catathetic signals may consist of a simple, comparative statement (e.g., "I am cleverer than you"), or, rarely, a statement of the speaker's high RHP (boasting), but more usually it is a statement emphasising or implying the other's low RHP, such as criticism, sarcasm, insult, disparagement (of another and their allies), or even silence, implying "you are not worth speaking to". Escalated catathetic signals involve physical contact, such as hitting, scratching, biting, caning, flogging, etc. Catathetic signals are equivalent to the sociological concept of processual power (Kemper, 1978).

Raush, Barry, Hertel, & Swain, (1974) were able to elicit mutual catathetic signalling in married couples put in a situation of artificial conflict. When told to choose between a baseball match on T.V. and a programme on naming a baby, some couples discussed the matter rationally and came to a decision; some avoided conflict altogether; but a third group generalised the conflict into what was clearly a ritual agonistic encounter. In the third group, the verbal content typically included criticism of the spouse's mother, and complaints about ill-deeds committed many years ago; the content had a stereotyped quality and was reproduced on subsequent occasions.

McLean (1976) has used the term "microstressors" for repeated slight stresses such as the receipt of catathetic signals from one's spouse. He thinks these may be more important in causing depression than are large events. The sender was often unaware of the catathetic nature of the signals sent; for instance, the comment, "you'd feel much better if you didn't cry all the time", was intended as helpful and supportive but was received as criticism.

Down-hierarchy Catathetic Signals. The discussion so far has been concerned with the exchange of catathetic signals between individuals of equal rank. However, catathetic signals are also exchanged between individuals in asymmetrical relationships. They are usually directed from the dominant to the subordinate, and have the function of confirming and reinforcing the dominance. The classic example is the pecking of the domestic chicken, first described by Schjelderup-Ebbe (1935).

The sending of catathetic signals implies the calculation of relative RHP and, indeed, it must be the case that both dominant and subordinate continue to signal absolute RHP to each other. In particular, the dominant monitors the "RHP gap" to ensure that it remains sufficiently wide. When the gap is wide enough the signal of favourable relative RHP is inhibited, this inhibition being the main change that takes place in the winner of an encounter (see equation 4a in Price & Sloman, 1987). It is only when the gap is too narrow for comfort that more catathetic signals are sent to the subordinate to lower their RHP further, and so restore the gap to a satisfactory size. It is worth noting that in a complementary relationship the signal of favourable relative RHP carries the additional meaning of *insufficiently favourable* relative RHP. Thus catathetic signals will be sent down the hierarchy when the dominant feels that their RHP has fallen for any reason (such as getting depressed, or receiving catathetic signals from further up the hierarchy) or if they feel the subordinate's RHP has risen to an unacceptable level.

The usual strategy of the subordinate is to refrain from sending catathetic signals up the hierarchy and to signal low absolute RHP, which ensures that the dominant perceives a sufficient RHP gap. If catathetic signals are received from the dominant, the response is a *reduction* of the recipient's catathetic signals. This is the opposite of the increase that occurs in a symmetrical relationship (see equation 3b in Price & Sloman, 1987). This altered response to catathetic signals is the main change that characterises subordinate behaviour; it creates a negative-feedback loop, which confers homeostatic properties on the relationship, and it would be consistent with an original analysis by Bateson (1972) to use it as the defining criterion of a complementary relationship (in contrast to a relationship that is less specifically asymmetrical). If the dominant can recognise this complementary response in the subordinate, it is a further source of reassurance about the stability of the relationship.

Sometimes the subordinate challenges the dominant and strives either for equality or for a reversal of the dominance, in which case catathetic signals may be directed up the hierarchy. The subordinate may also, without challenging the actual dominance itself, rebel against the degree of dominance exerted, so that an agonistic encounter occurs within the context of the complementary relationship. Schjelderup-Ebbe observed that such attempts at rebellion elicited very vigorous retaliatory pecking by the dominant bird.

Thus the subordinate suffers a baseline bombardment of catathetic signals from the dominant, and may suffer an increase in these signals either because of an attempt to rebel, or because the dominant is insecure for some reason. These down-hierarchy catathetic signals by definition lower the subordinate's RHP. If the drop in RHP of the subordinate exceeds certain limits, a yielding subroutine is triggered, leading to a further fall in RHP (analogous to the effect of a currency devaluation), in addition to changes described elsewhere (Price & Sloman, 1987). We called this a "confirmation" YS/RAB because it confirms an existing rank difference and thus subserves *social homeostasis*, in contrast to the "conversion" YS/RAB, which is associated with the conversion of a symmetrical to a complementary relationship, and the "reversal" YS/RAB, which is associated with reversal of rank—both of which subserve *social change*. The signalling characteristics of all three forms of YS/RAB are the same, except that the reversal YS/RAB also includes signals that disqualify former high rank in statements such as, "It was all a sham" and, "I did not deserve such respect, I was a fraud." Such statements may appear delusional and give rise to a label of psychotic depression; and it may be that these two types of YS/RAB, subserving, respectively, social homeostasis and social change, underly the clinical impression that neurotic and psychotic depression are distinct entities.

The subordinate also has subordinates, and does not signal to them low RHP in the form of inferiority or incapacity. This is why depressed mothers (and schoolteachers) are able to deal competently with children, particularly when no dominant adult is present. Moreover, the RHP gap between adults and children is usually so large that a depressive increase in down-hierarchy catathetic signalling is not likely to occur unless the teacher is in charge of a class of rebellious teenagers, or the mother has an excessively demanding child at home. In such cases, the adult may be "strict" (catathetic) with the children, or there may be a rank reversal in which the depressed adult has a reversal YS/RAB superimposed on their confirmation YS/RAB, and then the depression is likely to snowball to an end-point, such as attempted suicide and/or hospitalisation.

It is an interesting fact that the quality of catathetic signals is similar whether they are directed to an equal, a subordinate or a dominant. It is the quantity that varies and that differentiates symmetry from asymmetry, and dominance from subordinacy. In fact, it is the relative quantity of catethetic signals that is used to *define* dominance in many studies; and it is the consistency over time of this relative quantity, and its correlation with other measures such as supplanting, precedence, and paying attention, that gave rise to the concept of dominance/subordinacy in the first place (Deag, 1977; Kaufmann, 1983; Richards, 1974); a concept that has stood the test of time in spite of a suggestion that it might be an artifact of captivity (Rowell, 1974).

Asymmetrical Catathetic Signals. At least in human beings, however, catathetic signals may differ in quality, depending on whether they are directed up or down a hierarchy. To take an extreme example, a pupil may be cheeky to a teacher, and a teacher may cane a pupil; but we cannot imagine the pupil caning the teacher or the teacher being cheeky to the pupil.

The down-hierarchy catathetic signal contains two messages at different logical levels. First of all it is a straightforward catathetic signal, which lowers RHP in the usual way. But, secondly, it contains the message, "I am in a position to give you a signal which is only given by dominant people to subordinate people." This higher level message is also catathetic and lowers RHP. Thus the pupil suffers loss of RHP both from the caning and from the fact of being caned.

Anathetic Signals. A contestant may give a signal of unfavourable relative RHP after the assessment stage of a ritual agonistic encounter, or at any time during the engagement stage, in an act of what is usually called submission. The signal may take one of three forms:

1. *A Comment on the Sender's Low RHP.* This may be, for example, running away, prostration or self-denigration.
2. *A Comment on Comparative RHP.* This can be said either in speech, as "You are greater than I", or in symbolic form. It is interesting that of the two main social asymmetries (parent/child and male/female), one has been adopted as a submission signal by canids and the other by monkeys.
3. *A Comment on the Receiver's High RHP.* Apart from deferential attention, this is hard to make except with speech. Given the power of speech, however, the possibilities are infinite.

How can these signals be defined with respect to the receiver? It would be tidy, and possibly not unreasonable, to suggest that they can be defined as signals that *raise* the RHP of the receiver. Then we could call them anathetic signals (from the Greek words for "put" and "up"), and note that they are in most ways opposite to catathetic signals (I will deal with an important exception later).

Asymmetrical Anathetic and Neutral Signals. Like catathetic signals, most anathetic signals are similar whether directed up or down-hierarchy; but, likewise, some are not, such as patronising behaviour (e.g., tipping). Equally, some RHP-neutral signals, such as the giving of orders related to the task in hand, may be asymmetrical; for instance, to give orders in a certain tone of voice may imply dominance.

It seems likely that the receipt of such an asymmetrical anathetic or neutral signal has a catathetic effect, in that it is a threat to the recipient's RHP and challenges their underlying dominance (or equality). Thus an asymmetrical anathetic signal boosts the recipient's RHP at one logical level and lowers it at another. The net effect may be to lower RHP and/or trigger a yielding subroutine.

A Case History. Although there has not been space to present case reports to illustrate the phenomena discussed here, it might be helpful to mention briefly a patient who came to me for treatment of depression whilst I was preparing this chapter, and who appeared to be an example of depression triggered by an asymmetrical anathetic signal. The patient had just had a baby, and lived near her mother-in-law, with whom she had a symmetrical relationship. After the birth of the baby, the mother-in-law tried to be helpful and took to coming into the patient's house frequently, without invitation, to help her with the baby. The patient sensed that this was an infringement of her territorial rights, such as a dominant person might claim, but because the mother-in-law was being so helpful, even with menial chores, she felt she could not complain. She became increasingly depressed, and the more incapacitated she became with depression, the more the mother-in-law came in to help, until she actually moved in altogether and slept in her daughter-in-law's house. With family therapy and a course of antidepressant drugs the patient recovered and was able to establish a satisfactory protocol for the mother-in-law's visits, and they ended up on the best of terms.

This case is informative in a number of ways. First of all, it is an example of depression triggered by behaviour that, in the patient's culture, was an asymmetrical anathetic signal; i.e., coming into the house to help without being invited. Secondly, it illustrates a case of "conversion" yielding subroutine-modelled depression, because it depended on the conversion of the symmetrical relationship between the patient and her mother-in-law to a complementary one in which the mother-in-law was dominant. Thirdly, it illustrates the positive-feedback nature of some depressions: the more depressed the patient got, the more her mother-in-law stayed; the more her mother-in-law stayed, the more depressed the patient got, until she nearly reached the end-point of suicide or hospitalisation. Finally, it illustrates the difficulty of assigning cause and effect in depression, because it could perfectly well be explained as an endogenous or puerperal depression, by which the mother-in-law was forced into a dominant role because of the depressive helplessness of the patient.

Defining the Relationship. Dominance is an elusive concept that is liable to infinite regression of the type, "I insist that you make the decisions", "I insist that you decide who makes the decisions", etc. (Hinde, 1979). The ultimate decision-maker is the one who defines the relationship, in the terminology of family therapy (Haley, 1963). This leads to some paradoxical effects. If one member of a pair (for example, the male) insists on equality, by this very insistence he is defining the relationship, making himself dominant, and excluding the possibility of equality. Even if he defines himself as subordinate, he is exercising the defining role, so that his statement is an anathetic signal on the surface but carries a catathetic message at a deeper logical level. This is probably why young people are reluctant to define what they should call older people, so that if, as often happens, the older person fails to give a lead, the younger person has no means of addressing the older directly.

An example of what may happen if a junior person takes the lead in defining the relationship is portrayed in a recent novel *The mission* by Robert Bolt (Penguin, 1986). The hero, being a subordinate member of a gang, defines himself as subordinate to the leader; a short time later the leader tries to kill him. This interaction reveals the author's perception of the catathetic nature of a definition of subordinacy.

The Components of RHP. In the discussion so far I have said that catathetic signals lower RHP. However, they clearly do not lower RHP in a substantive way, as it might be lowered by physical illness or the infliction of wounds. Often a catathetic signal is a comment on low RHP, such as "You are pathetic." Is there a difference between lowering RHP and commenting on low RHP?

Here we must remember that we are dealing with *ritual* agonistic behaviour. The catathetic signal achieves a ritual lowering of RHP, which is the ritual equivalent of the substantive lowering of RHP produced by wounds in non-ritual fighting. We can list the following components of RHP:

1. *Substantive RHP.* This includes size, strength, weapons, allies and other real resources. We can include here RHP secondary to group membership.

2. *Ritual RHP.* This is the RHP attributable to the ritual signals of others, and is increased by anathetic signals and reduced by catathetic signals.

3. *Endogenous or Thymic RHP.* It is part of the thesis argued here that RHP is lowered in YS/RAB-modelled depressive states (and raised in manic states). Manic patients behave as though they have excess of both substantive and ritual RHP, whereas depressed patients behave in the reverse way. Possibly one could speak of expanded RHP in mania and contracted RHP in depression. Contracted RHP would be equivalent to the "self-deceit downwards" of Hartung (in press).

Pain and Depression. The lowering of RHP by ritual catethetic signals appears to have been associated during evolution with a ritualisation of the pain sense (if one can talk of ritualisation in a sensory modality). Physical pain, which presumably first evolved to protect organisms from harmful non-social aspects of the environment, became ritualised in such a way that it is now experienced on the receipt of catathetic signals, such as a blow or a slap in the face; further ritualisation has led to the evolution of mental pain, which is experienced on the receipt of non-contact catathetic signals, such as criticism. The experience of depression is one of mental pain, and it may be that the prolonged and diffuse mental pain of depression evolved out of the brief and localised mental pain experienced on receipt of a non-contact catathetic signal.

THE YIELDING SUBROUTINE AND DEPRESSION

This part of the chapter is mainly about the signals given off in depression, and I hope to show that they are consistent with the instructions of the yielding subroutine model, and also similar to the signals of animals undergoing a yielding subroutine. But for the model to achieve any credibility, it is desirable to demonstrate that depression does in fact occur in situations in which one would anticipate a yielding subroutine (or at least explain why we do not expect it to occur in those situations).

Signals That Cause YS/RAB-Modelled Depression

There is not space here to give a discussion of the relevance of the large amount of work on "life events" (Paykel, 1978) and the social origins of depression (Brown & Harris, 1978), but I would like to say a word about the relation between phylogenetic and ontogenetic causation.

A number of theorists have tried to account for the existence of depression by postulating adaptive value in the form of disengagement from goals and incentives (Klinger, 1975). If a goal is unrealistic or unattainable, failure to achieve it is followed by depression, with loss of interest and associated detachment from the goal. The individual is then free to pursue a more realistic goal. The problem with this theory is the *pervasiveness* of depression, in that interest is withdrawn not only from the goal that is unattainable but also from all other alternative goals. Moreover, the poor concentration and impaired decision-making of depression are hardly likely to help in selecting an alternative goal. Classically as in Shakespeare's *Hamlet*, the problems of life induce a depression that takes away the capacity to deal with the problems.

Let us consider the case of our hunter–gatherer ancestor. If a man's hunting is going badly, he should devote more resources to gathering. What is required is a disillusionment with hunting that does not affect enthusiasm for gathering. But depression as a response to goal-non-attainment is pervasive and leaves him lacking all enthusiasm, so that he is unlikely to switch to gathering.

The solution to this problem lies in focusing up from the individual to the level of the small family group. Depression is pervasive and therefore cannot subserve switching from one goal to another within the same individual, but the pervasiveness does not extend to other group members and, therefore, it is likely to be an effective means by which a group can switch its goal from that espoused by one member to that espoused by another. If the hunter's brother is keen on gathering, depression in the hunter would reduce his status and authority in the family and allow the brother to have his way. The group would devote themselves to gathering and the brother would assume the role of leader. The cause of the hunter's depression might well be seen by the hunter and his family as his failure in hunting, but the adaptive value would be the yielding of authority to another. There would be no adaptive value in the depression if the hunter were acting entirely on his own, or if other contenders for power in the group were equally devoted to hunting.

The two competing goals of hunting and gathering espoused by the two group members would be elements in the "concept pool", which Hill (1984) has proposed as the raw material of socio-cultural selection, and the survival of one goal rather than another is associated with the distribution and redistribution of prestige between group members.

Depression Subserving Social Change. We (Sloman & Price, 1987) reported the case of a farmer who became depressed because his milking herd was doing badly. During his depression he sold his herd and virtually handed over his farm to his son, who was strongly opposed to milking and wanted to raise beef cattle. Treatment was directed at defusing the confrontation with his son by getting him formally to delegate the running of the farm to the son, and by encouraging him to invest his self-esteem in his hobby of showing pedigree dogs. At the time they came for treatment the farmer could have been said to be undergoing a "reversal" yielding subroutine, and he was in danger of becoming permanently subordinate, in the role of a hypochondriac, to his son; with treatment they were able to achieve a symmetrical relationship (thanks to the division of territorial interest). Our farmer was in the position of the hunter whose hunting was going badly, and the son in the position of the brother who was keen on gathering. Although the adaptive value was in the reversal of rank between the two, the apparent cause of the depression was the failure of the milking herd; the father never realised that he was in serious conflict with the son. According to our hypothesis, the critical life event is not the failure to attain the goal, but the failure to attain the goal in the presence of a competing individual who espouses an alternative goal. This hypothesis is testable; for instance, it predicts that if the farmer had had no son he would not have become depressed.

In the previous section I mentioned a case in which depression had been precipitated by an asymmetrical anathetic signal. I can also report a case in which depression appeared to be precipitated in a young lady by an offer of marriage. The young man, who was a very forceful person, realised that his prospective fiancée was a strong feminist, and included in his proposal an *insistence that they should have a totally equal relationship*. Neither of them realised that this proposed definition of the relationship was an asymmetrical signal with catathetic effect, and the lady's depression was complicated by (although not caused by) the double bind of being made equal at one logical level but subordinate at another. I should like to be able to report that with therapy she administered a counter double bind by insisting that, on the contrary, she would prefer to have a subordinate role in the relationship (and thus by redefining the relationship making a bid for the dominant role), but in fact, she broke off the relationship.

Depression Subserving Social Homeostasis. Depression due to being chronically bullied by a dominant person is also difficult to measure exactly, but is probably more familiar both to psychiatrists and general practitioners. Rippere and Williams (1985) give some examples of depression due to aggravation by superiors at work. From the Adlerian school, von Andics (1947) described a series of women who committed suicide because of "mortification at being abused and humiliated". McLean (1976) has described depression occurring as a result of the "microstressors" of married life, and it is common in clinical practice to see a subordinate spouse with chronic depression and low self-esteem that appears to be due to a constant barrage of disparagement by the marital partner. These spouses, usually but not always wives, are in the situation of the low-ranking birds described in the next section, and the epithet "hen-pecked" is applied with good reason. They are in a situation similar to dogs with "learned helplessness" (Seligman, 1975) in that they suffer random noxious stimulation over which they have no control; and they are in what

Gardner (1982, and this volume) has called an in-group omega psalic. Sometimes they appear to undergo an actual depressive episode due to a particularly severe bout of abuse, either because the dominant spouse is being bullied in another relationship and is "redirecting the aggression", or because the patient has made a bid for equality or freedom and is being punished for the rebellion, and then they appear to have one "confirmation" yielding subroutine superimposed on another. The depressed spouse is "stuck" in the unhappy marriage, too lacking in energy to make a bid for independence, too lacking in interest to contemplate another partner. Therapy is difficult because any improvement tends to lead to fighting back or attempts to escape, either of which leads in turn to further punishment by the dominant spouse.

The Depressed Bird. It seems likely that birds and mammals both inherited agonistic behaviour directly from our common reptilian ancestor, whereas affiliative behaviour in birds developed quite independently. Therefore we can detect the agonistic signals of birds and empathise with them, whereas we cannot empathise with bird affiliative behaviour (what little they show between members of the same sex) in the way that we can empathise with the signals given by two chimpanzees being reconciled with each other after a fight.

Although separated by 100 million years of independent evolution in each lineage, we as human beings can be intimidated by the threat stare of a bird of prey, and we can feel a mixture of pity and contempt for the "hen-pecked" barnyard fowl. Schjelderup-Ebbe (1935), who was the first to describe in vertebrates a social hierarchy based on individual recognition, reported a mild, chronic depression in birds of low rank and a severe, self-limiting depression in birds falling in rank. He described the difference in countenance and behaviour between high-ranking and low-ranking birds in the following terms: whereas the face of the superior bird would "radiate with joy of satisfied pecking-lust", the subordinate had "a much less enjoyable and anxious existence", and if it tried to revolt against the despot, the subordinate "fights with less display of energy than usual. It seems as if the spirit of the bird were dulled by a premonition of hopelessness" (p. 955). When the high-ranking bird falls in rank "its behaviour becomes entirely changed. Deeply depressed in spirit, humble, with drooping wings and head in the dust, it is—at any rate, directly upon being vanquished—overcome with paralysis, although one cannot detect any physical injury. The bird's resistance now seems broken, and in some cases the effects of the psychological condition are so strong that the bird sooner or later comes to grief. This is especially true if the bird has been absolute ruler for a long time, and the reaction has been, therefore, most complete. In most cases, however, time heals the disappointment and the bird becomes used to its new position" p. 966.

Language Problems. Subsequent accounts of mood in animals have been less vivid, no doubt due to an aversion to anthropomorphism, and also, I suspect, to a lack of suitable terminology. For this reason I have extended the previous use of the term RHP and introduced the terms catathetic and anathetic. It is unsatisfactory to call the hostile interaction between husband and wife "threat display" or "pecking", and equally unsatisfactory to call the pecking of the subordinate by the dominant bird "nagging", but to call both phenomena catathetic signals enables us to describe them with the same conceptual system. It is more satisfactory to borrow animal terms for use in man, hence the use of RHP: it is easier to think of RHP in man than of ego strength in animals. I have followed the advice of Bandura (1983) in trying to avoid the use of "aggression" altogether.

Of course, there is no point in trying to use the same conceptual system for animals and man if there is merely analogy based on convergent evolution. But it seems quite possible, and is certainly heuristic, to think that RAB and its associated yielding subroutine have been such a fundamental part of vertebrate social life than they have continued relatively unchanged since we divided from present-day reptiles and birds, and that the underlying structures and processes are homologous (see Chapter 8).

In regard to the value of animal data for human psychology, it would have been quite impossible to formulate the present hypothesis relating depression to RAB without the extensive ethological data about RAB. Ritual agonistic behaviour is such an extraordinary and unlikely phenomenon that, without the knowledge that it is an ubiquitous and fundamental vertebrate characteristic, the evidence of RAB in man, overlaid as it is with inherited affiliative systems and culture, would probably have been ignored, or considered a trivial matter of no evolutionary significance. And even if the hypothesis should turn out to be wrong, it would be hard to maintain that it was not worth formulating. Totman (1985) said that there is nothing new in social psychology because there are no investigative tools that have not been available in at least rudimentary form since ancient times; but the data of comparative ethology constitute just such a new, investigative, conceptual tool for human psychology.

Yielding Signals Given by the Depressed Patient

The yielding hypothesis predicts that the following changes in signalling should be observed in depressed patients:

1. A reduction of catathetic signals (hostility) directed towards dominant people and towards those equals who are regarded as competitors.
2. An increase in catathetic signals (hostility) to subordinate individuals. This prediction was derived from the application of the yielding subroutine

instructions to the mutual pecking equations of our bird/mathematical model (Price & Sloman, 1987).

3. An increase in anathetic signals (adulation) to dominant people.

4. A reduction in anathetic signals (morale-boosting) to subordinate people, such as children.

5. Signalling of lowered absolute RHP.

6. Demonstration of altered responsiveness to the receipt of catathetic signals.

7. Depressed patients should share some signalling characteristics with animals who have been defeated and/or occupy low rank.

Catathetic Signals (Hostility) in Depression. Although the association of rage and hatred with depression received comment from early psychoanalytical writers (described by Gaylin, 1983), studies of depressed patients in hospital found that aggression (hostility) is profoundly inhibited. However, closer investigation of less severely depressed patients has led to very conflicting results (Cochrane & Nielson, 1977; Snaith & Taylor, 1985).

In the enormous literature on hostility in depression, two main dimensions of variation are recognised. One concerns the direction of the hostility, whether it is directed to other people (or things) or to the self; the other dimension is concerned with whether or not the hostility is expressed to its object (as opposed to being reported to the psychiatrist or investigator). Catathetic signals are equivalent to hostility expressed outwards. Our hypothesis predicts that expressed hostility to higher ranking persons will be reduced in depression. The same should apply to equal-ranking people who are perceived as competitors. Hostility to lower ranking persons, on the other hand, is predicted to be increased by the onset of a depressive episode (Price & Sloman, 1987).

One study gives some information relevant to this latter prediction. Weissman and Paykel (1974) found that depressed women expressed more hostility to intimates and less hostility to non-intimates than a control group. As the intimates were often children, and children usually rank below their mothers, the result tends to support our hypothesis.

Our own study (Price & Sloman, Note 1) attempted to test the hypothesis on a sample of married couples, one of whom had presented as a depressed patient. We rated them on two variables, husband dominant/wife dominant, and hostility to spouse increased/reduced by the depression. We predicted that when it was the dominant spouse who was the depressed patient the hostility expressed to the partner would be increased and *vice versa*. In spite of relatively small numbers and considerable difficulty in measuring both variables, the results were in line with the hypothesis and nearly reached the 5% level of significance.

It is an interesting sidelight on the lack of ethological contribution to psychiatric research that, of all the many studies which have been conducted in this field, not a single one (apart from our own) has measured whether the hostility is directed at a higher- or lower-ranking person, nor has the desirability or possibility of such a measurement been discussed.

Anathetic Signals (Adulation) in Depression. The depressed patient does not produce the elaborate signal of submission that is sometimes directed at a specific high-ranking individual, such as the flattering speech of eulogy associated with avowals of allegiance, and this is at first sight surprising as we should expect depressed people to do all they can to reassure those in power of their desire to serve and not to challenge. This is probably because the depressive signal is using a much more primitive channel of communication, one which cannot be falsified as can the flowery speech, and therefore likely to be more fundamentally reassuring. Even more to the point, the flowery speech requires social skill, a component of RHP, and therefore the presentation of a flowery speech indicates the presence of the resources needed for a come-back. But the absence of a flowery speech when one might be expected, or some nervous mumblings of a self-absorbed and depressive nature, would serve to demonstrate lack of social skill. This presentation of the incapacitated self would not be received as a signal but as a signal about a signal, a metacommunication to the effect that the signaller lacks the capacity to communicate appropriately.

The same would apply to the impairment of attention in the depressed patient. The good subordinate should always be attending to the dominant, giving what Callan (1976) has called "advertence". The depressed patient is signalling, "I am too incapacitated to give you the normal politeness of my attention."

In general, we may say that by failing to give elaborate and skilled signals of submission the depressed patient is submitting at a different logical level (Watzlawick et al., 1967); saying, "I am so incapacitated by my submissive state that I am not even able to submit properly."

Being signals of unfavourable *relative* RHP, anathetic signals also include self-denigration, and this form of anathetic signal *is* given in depression. As predicted, it is given to dominant people and not to equal or subordinate people, and this may explain the common, clinical observation of in-patients who express strongly depressive ideas to the consultant in the ward round but give no sign of depression to fellow patients on the ward or to nursing staff (and whom nursing staff tend to think are hoodwinking the consultant in order to escape being discharged). Self-denigration is not a skilled task and does not betray the general impression of incapacity.

The various forms of down-hierarchy anathetic signalling, such as morale-boosting and encouragement of children and other subordinates, are reduced in depression; the depressive has no energy to spare for making others feel better.

Signal of Low RHP in the Co-operative Mode. Chance (1986) distinguishes between social behaviour in which ritual agonistic behaviour (RAB) is occurring and the agonic mode in which group members are prepared for RAB even though no actual RAB is taking place. For convenience I will call both these modes the competitive mode, and contrast it with the co-operative mode in which competitive within-group differences are forgotten and the pair or group is able to concentate on other things. In the co-operative mode are included Chance's hedonic mode, in which the group is egalitarian and oriented towards affectionate interaction, and other states in which the group is engaged in performing a task, in recreation, or in competing against another group. In these latter groups there may be an underlying hierarchy which has been established in the competitive mode, or some other form of underlying social asymmetry; but, by definition, any such asymmetry is not being contested in the co-operative mode.

If an asymmetry such as leadership is contested, the group has switched to the competitive mode—to the agonistic mode if the contest is open, to the agonic mode if the contest is unspoken. After a bout of RAB there may be a rapid switch back to the co-operative mode, particularly if the outcome of the RAB is accepted by all parties; this process is well illustrated by de Waal's chimpanzee group, in which there is active reconciliation after a contest, albeit the reconciliation may be conditional on submissive behaviour by the loser. If the skirmish has been light, the submission may be entirely voluntary, but if it has been intense and prolonged, and particularly if there has been a change of leadership, the loser may undergo a yielding subroutine (YS) and suffer a prolonged period of incapacity. The YS of the deposed leader (a male, for the sake of argument) allows the RAB to be concluded, and allows the other group members to return to the co-operative mode. The deposed leader himself, however, cannot return to the co-operative mode because the YS is a component of RAB. He therefore continues his submissive form of RAB, which may well appear inappropriate in the task or recreational setting of the co-operative mode, and he is not likely to concentrate on the task or enjoy the recreation with the other group members. The function of his behaviour is to reassure the new leader that he is not likely to make a come-back, thus permitting the group to remain in the co-operative mode. This is achieved by the symptoms of depression, which make the deposed leader appear "not his old self" and to have lost former leadership qualities such as initiative, decisiveness and attractiveness. The capacity for obeying orders and carrying out routine tasks is less impaired in depression.

A YS may occur in the leader while the group is in the co-operative mode, for instance after a bereavement or other loss of RHP. In that case the leader is likely to stand down from the leadership voluntarily, if there is any competition for the role, and the group as a whole is thereby saved from a switch into the competitive mode. It is unlikely that any of the group members will have insight into the reason for the change of leadership, which is likely to be attributed to some incidental cause like old age or ill health.

Altered Response to Catathetic Signals. What is likely to give great reassurance to a dominant person is to recognise the altered dynamic whereby their catathetic signals no longer elicit catathetic signals from the other, but rather a reduction in catathetic signals or the equivalent, in the form of an increase in anathetic signals; so that their assertions are no longer met with "Yes, but . . .", or other forms of answering back, but, with agreement. "A soft answer turneth away wrath" (Proverbs 15: 1).

Similarity to Defeated or Low-ranking Animals. There are certain characteristics of posture, gait and gaze that occur generally in low-ranking birds and mammals, and these features are found in depressed patients (Adams, 1980). The posture of the depressed patient tends to be bowed or hunched, appropriate to submission. The gait lacks "jauntiness" and resembles that seen in low-ranking animals. Eye contact is reduced both in depressed patients and those in whom depressed mood is experimentally induced (Kleinke, 1986), particularly if the interlocutor is seen as hostile.

Previous Theories of the Depressive Signal. The most common function suggested for the depressive signal is recruitment of support or a "cry for help" such as Salvador Rado's "great despairing cry for love" (Gaylin, 1983). However, this hypothesis is confusing because of the different meaning of "support" in the two modes. In the competitive mode, recruitment of support means the enlisting of allies or subordinates to challenge a dominant individual or group, and this kind of activity is profoundly inhibited in depression. In the co-operative mode, support usually means nurturance or care given to a weaker or needy person, and the elicitation of this kind of support by "care-eliciting behaviour" (Henderson, 1974) would not conflict with the yielding function of depression, although Klerman (1974) found that depression, if at all prolonged, tended to alienate rather than recruit the support of husbands and other family members for a series of depressed women. McLean (1976) expressed this in behavioural terms: "Depressed people are usually not reinforcing to be with and consequently are often tactfully avoided." (p. 313)

If, as I suggest, the depressive signal evolved as a yielding signal, its functional relevance is to the competitive mode, and when it is received in the hedonic mode its meaning is likely to be obscure. This obscurity is likely to be the usual state of affairs, as human beings spend the great majority of thier time in the co-operative mode, either engaged in a task or enjoying recreation; and this is even more the case in the medical setting, when they are receiving help. It is not surprising, therefore, that various psychiatric observers have disagreed about the meaning of the depressive signal, and some have concluded that it has no meaning at all (Klerman, 1974).

Some Problems for the Theory

The Use of Depressive Symptoms to Get One's Own Way. According to the yielding model, depression is a primitive means of not getting one's own way. And yet sometimes depressed patients seem to get their own way by the very fact of being depressed. For instance, a wife may complain to her husband that she is too depressed to visit his parents. Hooper, Vaughan, Hinchcliffe, & Roberts (1978) even go as far as to say that "it is possible to see the whole depressive stance as a massive attempt to exercise control over the marital relationship".

I think here the depressed patient is slipping into the role of support seeker (Heard and Lake, 1986), and taking advantage of the parental, nurturing instincts of the other, much as children manipulate their parents by crying. Like any other symptoms, depressive symptoms can be manipulated for "secondary gain".

In the case of relations between patient and therapist, patients are expected to let the therapist get his own way, and it is a matter for comment if they do not. Freud said (Gaylin, 1983) about depressed patients, "They are far from evincing towards those around them the attitude of humility and submission that alone would befit such worthless persons; on the contrary, they give a great deal of trouble, perpetually taking offence and behaving as if they had been treated with great injustice." It is not impossible that these troublesome patients of Freud's were high-ranking Viennese who saw themselves as socially superior to their analyst, and therefore their catathetic signalling to him would have been increased rather than reduced.

Ritual Agonistic Behaviour (RAB) in Females. RAB is so much more conspicuous in males than in females that, in his classification of aggressive behaviour, Moyer (1976) labelled it "inter-male aggression". How can this be reconciled with the hypothesis that depression, which is more common in females, evolved from the yielding component of RAB?

1. *Female-female RAB.* Since Moyer's time it has become recognised that RAB between females is important in many species (Kevles, 1986) and affects the reproduction of low-ranking females in a number of ways, such as prevention of copulation, suppression of ovulation, inhibition of implantation, abortion and infanticide. The method of fighting may not be as dramatic as between males, but that is no reason why the yielding should be any less effective.

2. *Dependent Rank.* In primate society female rank is more variable

than that of males and therefore the mechanism for falling in rank needs to be more easily triggered. Females tend to enjoy rank that is dependent on that of the male they are consorting with, and if the consort relationship is dependent on the presence of oestrus, there may be a rise and fall of rank with each sexual cycle. In polygynous species the females have to cope with both within-harem and between-harem variation of rank.

3. *Formal rank.* In human society, occasions for contesting male rank are limited. In situations in which confict between males is likely to arise, formal ranks are allocated and differences are settled by outside adjudication, so that the interpersonal settlement of conflict by RAB is not required. In the armed forces, for example, the formal hierarchy of males gives rise to less problems than the less well-defined hierarchy of wives. Fall in formal rank is rare, and even when demotion does occur, as with retirement, the individual concerned seldom remains in the same social group as those who have been allocated his former rank.

4. *Male-female RAB.* Competition between males and females has become common in Western society (Weisfeld, 1986), and it occurs in its most intense form in the marital relationship. Dominance in the marital relationship increases genetic fitness because it allows control over extramarital conception, and over the allocation of resources to collateral relatives and to children not shared with the spouse.

It is not known what proportion of marriages are symmetrical, but we know from clinical experience that switches into asymmetry, and reversal of asymmetry, are not uncommon. We also know that down-hierarchy catathetic signalling is a common cause of depression in subordinate spouses. McLean (1976) has recorded this phenomenon, which consists in some cases of an almost continuous stream of criticism and abuse, and he made the interesting discovery that much of what is received as catathetic signalling is not conciously transmitted as such. It is probably true to say that more RAB occurs within marriage than anywhere else in human society.

Doubts have been raised about the reliability of dominance relations in marriage. Various measures of dominance are poorly correlated (Grey-Little and Burks, 1983). These experimental results are at variance with everyday observation, our professional experience, and with what we know of human life through fiction (see, for instance, the marital relationship depicted by Daphne de Maurier in *Jamaica Inn*).

I think the answer to this problem lies in the distortion introduced by an investigator. If a couple is asked which is dominant, their answers are not objective statements but may be part of their marital struggle, because to make a statement about whether or not one is dominant may affect one's dominance. It is necessary to ask people outside the marriage questions about where the power lies, and preferably to ask informants who rank lower in the hierarchy than the two individuals whose relative rank is in question, because the fine gradations of hierarchical rank are more visible from below. In clinical work, for instance, the replies of patients and their spouses to questions about dominance are very evasive, but asked about the power relations in their parents' marriages, they can give immediate and definite opinions.

5. *Reverted escape.* In many primate species the males move freely from group to group, and can thus escape from a disputed dominance relationship. Females tend to stay in the same group, and therefore must contest each conflict about rank with the risk of being forced into the role of yielder. In marriage, wives more than husbands feel trapped by the need of their children for a home and a parent of the opposite sex.

Depression After "Exit" Events. It is a matter of common clinical experience, and has been confirmed by careful research (Paykel, 1978) that depression frequently follows the loss of a loved person, whether by death, or rejection, or in some other manner. On the face of it the present hypothesis would suggest the opposite, because the exit of a group member offers an opportunity for rise in rank, as often happens when a parent dies; whereas the entry of a new group member threatens loss of rank. Therefore depression should be more common following "entry" events.

I suggest that the answer to this apparent problem lies in the fact that, in man and other social primates, rank depends largely on alliances. Therefore the loss of a loved one has profound implications for the rank of the bereaved individual. This is seen most conspicuously in widows, in whom, in some social groups, there is an obligatory submission to the wife of the eldest son. Or, again, a son who maintains his dominance over his brother because of his father's support may fall below his brother in rank after his father's death.

Depression has no adaptive value in the activities of the co-operative mode. When a group member dies, there is more work to be done by the survivors and decisions about reallocation of work and social roles have to be made, so that an increase in energy and decisiveness (superimposed on normal grief), rather than depression in the bereaved individuals, would be adaptive. But depression is adaptive according to the agenda of the underlying (but often unacknowledged) competitive mode, effecting a reduction in RHP of those individuals who have been supported by the lost person, and tailoring their dominance level to the new social situation.

Modification of RAB in Human Beings

Since the neural circuits for ritual agonistic behaviour (RAB) were laid down in our ancestors' reptilian brains, much further evolution has occurred. In particular, we have developed parental behaviour, pair-bonding and the capacity for affiliation between members of the same sex which allows for co-operative behaviour including alliance formation. Superimposed on these evolutionary developments there is the pervasive influence of culture. It is not surprising, therefore, that human RAB is somewhat different from the lizard's.

Cultural Organisation of Competition. Most areas of human competition are now prescribed by culture. RAB has been either modified or replaced by tournaments, duels, sporting contests, legal actions, gambling, examinations, selection panels and elections. RAB is competition for territory or status as guarantees of access to resources, but most human competition is for money, which can buy not only resources but also territory and status. Referees and law enforcement agencies ensure that losers lose, and there is no need for the yielding subroutine (YS) in these culturally ritualised contests. The primitive yielding of the YS is only required now in societies that are above the law (such as oligarchies), or below the law (such as street gangs), or beyond the law (such as the pioneers of the wild west), or deliberately ignored by the law (such as marriage and the family). In practice, therefore, and certainly in medical practice, RAB and its complications are encountered largely in marriage and the family, and in other areas in which society decrees that competition should not be taking place.

Language. Language enormously increases the scope for catathetic signalling, such as veiled criticism, sarcasm and the subtle induction of shame and guilt. Moreover, it allows people to put each other down without the recipient being aware of what is happening. I have already mentioned unilateral definitions of relationship and the use of asymmetrical neutral and anathetic signals; there is also constructive criticism, in which an attack on a person's ritual RHP is presented as an attempt to improve them. These are all technically double binds (Bateson, 1972) in which a signal appears neutral or anathetic but is catathetic at a different logical level. Individuals who experience these paradoxical put-downs during their formative years are likely to anticipate catathetic experiences even in co-operative settings, and will probably prefer situations in which the co-operative mode is definite such as a group task or inter-group conflict, rather than ambiguous activities such as "small talk".

Disparagement of an individual to a third party, an extension of catathetic signalling impossible in pre-linguistic societies, is beyond the scope of this discussion.

Alliance formation. In any current agonistic concern or rivalry, other people are likely to be either for one or against one. Therefore in our appraisal of others there will be a tag of "for," or "against" in addition to the rating of relative RHP. Those "for" increase our RHP and those "against" reduce it, particularly if they are higher ranking. This fits in with the fact that socially anxious people are improved by the presence of some higher-ranking people, but made worse by others.

In the case of subordinates, alliances set a limit to catathesis. Among reptiles there is no limit to the amount of catathetic influence which it pays one animal to exert on another, apart from the effort of doing so, for that other is never going to be of any use to them. But if a conspecific is, or is likely to become, an ally there *is* a limit to how put-down we want them to feel. We want them to be sufficiently depressed not to challenge us, but we do not want them to be too depressed to help us challenge someone else, or to help us ward off someone else's challenge. And so we adjust their mood carefully, by putting them down with catathetic signals and lifting them up with anathetic signals until their mood is such that they lack the initiative to fight on their own behalf but do not lack the confidence to fight on our behalf, given sufficient encouragement or coercion.

I mentioned above that the expression of hostility appears to be inhibited in depressive states, but it would be more true to say that it is hostile *initiative* that is inhibited, and that does not include bullying aggression towards lower-ranking individuals or aggression carried out on behalf of someone else (particularly on behalf of a higher-ranking person). In fact, there are occasions when a patient may be too depressed *not* to fight; for instance, a hen-pecked husband who is coerced by his wife to join her affray with a neighbour and to make a complaint to the neighbour's husband, and who, if he were less depressed, would firmly tell his wife that she must sort out her own quarrel.

Parent–Child Relations. Children may manipulate their parents' instincts subtly in order to get their own way (see above), but they may also confront them in a manner similar to a ritual agonistic encounter, but with an asymmetrical use of methods. For instance, a baby may have a series of encounters with its mother over its feeding schedule in which the baby cries and the mother uses different methods. If the baby wins this battle, it may become dominant to its mother, and it could be that this loss of control in the mother might precipitate a yielding subroutine which would be recognised as a post-natal depression.

However that may be, it is certain that the use of childish forms of both manipulation and confrontation occur in marital struggles to further complicate an already complex situation.

Pair Bonding. The human species appears to be the only mammal in which more than one pair forms bonds in a cohesive group. It is also unique in that parents have an influence over their children's bonding, and some parents may express their social aspirations in this way rather than in direct competition with other adults.

Inter-Group Conflict. Human social life is like a series of concentric circles, extending to ever more distant relationships. In one circle there is usually conflict, with co-operation in the circle immediately inside it. At some degree of social distance, probably around the stage at which people cease to know each other individually, RAB ceases to occur and gives place to warfare. The relation of the motivational bases of RAB and warfare is obscure; warfare does not appear to have evolved out of RAB, but the waging of war probably inhibits RAB in a group more than other forms of group task.

Anathesis. The capacity to boost other people's morale (raise their RHP) evolved after the mechanism for RAB had been laid down. Anathesis may have evolved out of parental behaviour, or possibly out of pair-bonding, or it could have evolved out of submission, one of the components of submission being the ritual acknowledgement of the other's higher RHP (see above). In the latter case, it may be that the reptilian brain treats anathetic signals as if they were negative catathetic signals, and the cessation of anathetic signals as catathetic signals. This would explain the mechanism for the precipitation of a yielding subroutine by loss or separation from a loved one (source of anathetic signals). But a catathetic signal contains a dominance vector: when it comes from a *subordinate* (or equal) it elicits a catathetic signal in return, and if it is overpowering it triggers a yielding subroutine of the change variety, the template for psychotic depression; a catathetic signal from a *superior* elicits a reduction of catathetic signals or an increase in anathetic signals, and if it is overpowering it triggers a yielding subroutine of the homeostasis variety, the template for neurotic depression. Now, if anathetic signals reach the reptilian brain with a dominance vector, and if this vector is maintained when their cessation is recognised as a catathetic signal, it follows that the loss of a lower-ranking source of anathesis should elicit hostility, possibly followed by a psychotic form of depression, whereas loss of a higher-ranking source of anathesis should cause an inhibition of hostility, possibly followed by a neurotic form of depression. It is recognised that most bereavement reactions are a confused mixture of hostility and depression (Parkes, 1986), but so far there has been no attempt to relate the balance of these affects or the type of the depression to the relative rank of lost person and bereaved individual.

Conclusion. In spite of culture, language and the evolution of various forms of affiliation, many human agonistic encounters are remarkably similar to those of lizards. Most married people have experienced pointless rows in which neither culture nor the higher centres of the brain seem to play much part. There are more formal similarities, such as the Yanomamo chest punching in which rivals take alternate turns, not unlike the alternate striking of the opponent's head with the tail by lizards, or the alternate broadside presentation of Siamese fighting fish. In addition to these primitive encounters, the process of RAB affects us at other times, when RHP may be raised or lowered in an apparently non-competitive setting. This is an area where more research is needed. For instance, are the RHP-comparing mechanisms inhibited when the group is operating in the co-operative mode, or are they still active below the level of awareness?

The Negotiation of Asymmetry in Human Society

It is clear from work summarised elsewhere in this volume that human beings can exist for long periods in a state of social symmetry—and that this is true both for peaceful people, such as the Kalahari bushmen, and aggressive people, such as the Yanomamo—provided there is plenty of space for conflicts to be managed by group splitting (freedom from circumscription). There are three main ways in which relationships may lose symmetry:

1. *Verbally Negotiated Asymmetry.* Two formerly equal people may enter a negotiation whereby one becomes the master and the other the servant, or a group may elect one person leader, etc.
2. *Agonistic asymmetry.* As a result of an agonistic encounter one individual adopts a subordinate role and acknowledges their lower RHP. If the two started with equal RHP, the RHP of the loser falls as part of the yielding subroutine.
3. *Adulatory asymmetry.* As a result of some social process which may take the form of a display of adulation-eliciting behaviour, one individual comes to respect, love, hero-worship or otherwise adulate the other. The adulator revises upwards his estimate of the other's RHP.

Agonistic and adulatory asymmetry both involve the creation of an RHP gap between two individuals, in both cases the RHP adjustment occurs in the one who finally has the lower RHP, and in both cases the adjustment of RHP is associated with an emotional state (depression in the case of lowering own RHP, adulatory emotion in the case of raising one's estimate of other's RHP). The main difference is in the quality of the emotion experienced by the adjuster, and it would not be an exaggeration to say that of the two sets of instructions for inducing asymmetry, the adulatory subroutine is more "user-friendly" than the yielding subroutine.

Adulatory asymmetry is maintained by the passage of anathetic messages from the adulator to the object of adulation, so that the object's RHP is ritually raised to match the inflated estimate of the adulator. However, adulatory asymmetry is fundamentally unstable. As Kemper (1978) has pointed out, the object of adulation may become dependent on the anathetic signals (processual status) of the adulator to maintain (structural power), and so is vulnerable to their withdrawal. This is a weapon in the hands of the adulator, which paradoxically increases their own RHP, so that the balance of RHP (structural power) may end up in favour of the adultator. The object can avoid this situation in a number of ways: by using catathetic or asymmetrical signals to put the adulator down; by recruiting more than one adulator, so that the object is not dependent on anathetic signals from any one individual; and by matching the adulator's new power by reciprocating their anathetic signals. However, we should note that, whereas the anathetic signals of the adulator correspond to the up-hierarchy anathetic signals of the agonistic mode, in that they are signals of unfavourable relative RHP, the reciprocal anathetic signals directed down-hierarchy to the adulator cannot be defined in this way.

The complexity of social life depends on the interplay of these asymmetries, in that a group which is ostensibly egalitarian may be affected by all three forms of asymmetry at different times, or even at the same time. There are definite correspondences between the roles and processes of the different asymmetries. For instance, the individual who is dominant in the competitive mode is likely to be the object of adulation in the co-operative mode, adopting the role of support-giver, employer or elected leader. On the other hand, this may not be the case; for instance, a group may elect a non-dominant leader in order to avoid the domination of the one who is dominant in the competitive mode.

SUMMARY

Within the framework of evolutionary biology, human depressive states are postulated to have evolved as the yielding component of ritual agonistic behaviour (RAB), which is the main vertebrate mechanism generating social asymmetry (subserving the intra-sexual component of Darwin's sexual selection). Two related but distinct yielding subroutines are postulated. One subserves social *homeostasis* in the form of maintenance of low rank in spite of motivation to rise in rank, and is a chronic and relatively mild condition that may provide a phylogenetic template for *neurotic depression*; it is similar to Seligman's model of learned helplessness (Seligman, 1975). The other yielding subroutine subserves social *change* in the form of fall in rank, and is a severe but self-limiting condition that may provide a phylogenetic template for *psychotic depression*; it is related to Klinger's model of depression as a means of disengaging from unattainable goals and incentives (Klinger, 1975).

The ritual agonistic encounter is analysed in terms of signalling of resource-holding potential (RHP), a term derived from behavioural ecology (Parker, 1974). Two stages of the encounter are distinguished. The first, which constitutes the *assessment* stage of the encounter, is a mutual offer of information about each other (about absolute RHP), and co-operative co-evolution between sender and receiver has ensured that the information offered is extensive and accurate. The second, which constitutes the *engagement* stage of the encounter, is a mutual exchange of the fact that each has decided that they have at least as much chance of winning as the other (exchange of signals of favourable relative RHP), and competitive co-evolution has ensured that the information is intensive rather than extensive, so that in most species it takes the form of what is generally called fighting.

It is suggested that the "signal" given off by the depressed patient reflects these two kinds of signalling. First, there is a cessation of activity related to the engagement stage in that the patient keeps "out of action" and does not seek new engagements; in existing relationships the change in signalling depends on the patient's relative dominance: towards dominant individuals the signal of favourable relative RHP (hostility) is reduced, whereas towards subordinate individuals it is increased. Secondly, there is a signal of low RHP in which the patient presents as someone who is incapable of fighting back because of a generalised incapacity, which in its extreme form includes even the capacity for signalling. The mildly depressed person is signalling on two channels with the paracommunication "I submit (unfavourable relative RHP), and even if I did not wish to submit, I am too incapacitated to do anything about it (low absolute RHP)", whereas the more depressed person is sending a signal about signalling with the metacommunication, "I am too depressed even to signal my submission." These signals relate to the agenda of the agonistic mode, but in humankind are usually transmitted in the hedonic mode, where they are associated with an inhibition of adulation-eliciting behaviour.

Apart from suggesting new animal models of depression, the main heuristic value of the theory relates to the expression of hostility in depression. It is predicted that expressed hostility to dominant individuals, such as an employer or a dominant spouse, is reduced in depression, whereas expressed hostility to subordinate individuals, such as employees, children and a subordinate spouse, is increased.

Some difficulties facing the theory have been discussed. Depression is commoner in women than in men, tends to occur after "exit" events rather than after "entry" events, and sometimes appears to be used to get one's own way. None of these phenomena is predicted by the subroutine model as it occurs in animals, and it is argued that they are due to the modification of the manifestation of the agonistic mode in humans by the evolution of affiliative behaviour, and by the influence of culture in the human lineage.

REFERENCE NOTES

1. Price, J. S. & Sloman, L. (1986, June). *The expression of hostility in complementary relationships*. Paper presented at the meeting of the International Society of Human Ethology, Federal Republic of Germany.

REFERENCES

Adams, R. M. (1980). Nonvocal social signals and clinical processes. In J. S. Lockard (Ed.), *The evolution of human social behavior* (Pp. 239–256). New York: Elsevier.

Bandura, A. (1983). Psychological mechanisms in aggression. In R. G. Green & E. Donnerstein (Eds.), *Aggression: Theoretical and empirical reviews, vol. 1: Theoretical and methodological issues*. New York: Academic Press Inc.

Bateson, G. (1972). *Steps to an ecology of mind*. New York: Ballantine Books.

Beck, A. T. (1973). *The diagnosis and management of depression*. Philadelphia: University of Pennsylvania Press.

Betzig, L. L. (1986). *Despotism and differential reproduction: A Darwinian view of history*. New York: Aldine.

Brown, G. W. & Harris, T. (1978). *Social origins of depression*. London: Tavistock.

Callan, H. M. W. (1976). Attention, advertence and social control. In M. R. A. Chance & R. R. Larsen (Eds.), *The social structure of attention*. New York: John Wiley & Sons, Ltd.

Campbell, B. (1973). *Sexual selection and the descent of man, 1871–1971*. Chicago: Aldine.

Carpenter, C. C. (1978). Ritualistic social behaviors in lizards. In N. Greenberg & P. MacLean (Eds.), *Behavior and neurology of lizards*. Rockville, Ma.: National Institute of Mental Health.

Chance, M. R. A. (1977). The infrastructure of mentality. In M. T. McGuire & L. A. Fairbanks (Eds.), *Ethological psychiatry* (Pp. 180–195). New York: Grune & Stratton.

Chance, M. R. A. (1986). The social formation of personality systems: The two mental modes and the identity of recursive mental processes. *American Journal of Social Psychiatry*, *6*, 199–203.

Chance, M. R. A. & Jolly, C. J. (1970). *Social groups of monkeys, apes and men*. New York: E. P. Dutton.

Cochrane, N. & Nielson, M. (1977). Depressive illness: the role of aggression further considered. *Psychological Medicine*, *7*, 283–288.

Coopersmith, S. (1967). *The antecendents of self-esteem*. San Francisco: W. H. Freeman.

Darwin, C. (1871). *The descent of man and selection in relation to sex*. London: Murray.

Deag, J. M. (1977). Aggression and submission in monkey societies. *Animal Behaviour*, *25*, 465–474.

Durant, J. R. (1981). The beast in Man: an historical perspective on the biology of human aggression. In P. F. Brain & D. Benton (Eds.), *The biology of aggression*. Rockville, Maryland: Sijthoff & Noordhoff.

Freud, S. (1924). Mourning and melancholia. *Collected papers*, vol. 4. London: Hogarth Press.

Gardner, R. J. (1982). Mechanisms in major depressive disorder: an evolutionary model. *Archives of General Psychiatry*, *39*, 1436–1441.

Gaylin, W. (1983). Epilogue: the meaning of despair. In W. Gaylin (Ed.), *Psychodynamic understanding of depression*. New York: Jason Aronson.

Gilbert, P. (1984). *Depression: From psychology to brain state*. London: Lawrence Erlbaum Associates Ltd.

Gilbert, P. (in press). Psychobiological interactions in depression. In J. Reason & S. Fisher

(Eds.), *Handbook of life stress, cognition and health*. Chichester, England: John Wiley & Sons Ltd.

Gray-Little, B. & Burks, N. (1983). Power and satisfaction in marriage: A review and critique. *Psychological Bulletin*, *93*, 513–538.

Greenberg, N. & Crews, D. (1983). Physiological ethology of aggression in amphibians and reptiles. In B. B. Svare (Ed.), *Hormones and aggressive behavior* (Pp. 469–506). New York: Plenum Press.

Haley, J. (1963). Marriage therapy. *Archives of General Psychiatry*, *8*, 213–234.

Harris, V. A. (1964). *The life of the Rainbow Lizard*. London: Hutchinson.

Heard, D. H. & Lake, B. (1986). The Attachment Dynamic in adult life. *British Journal of Psychiatry*, *149*, 430–438.

Hartung, J. (1987). Deceiving down: conjectures on the management of subordinate status. In J. Lockard & D. Pulhus (Eds.), *Self-deceit: An adaptive strategy*. Englewood Cliffs, N. J.: Prentice-Hall.

Henderson, A. S. (1974). Care-eliciting behavior in man. *Journal of Nervous and Mental Diseases*, *159*, 172–181.

Hill, J. (1984). Human altruism and sociocultural fitness. *Journal of Social and Biological Structures*, *7*, 17–35.

Hinde, R. A. (1979). *Towards understanding relationships*. London: Academic Press Inc.

Hodes, M. (1986). Dominance hierarchies in psychotherapy groups. *British Journal of Psychiatry*, *149*, 520 (correspondence).

Hooper, D., Vaughan, P. W., Hinchcliffe, M. K., & Roberts, F. J. (1978). The melancholy marriage: an inquiry into the interaction of depression, V: Power. *British Journal of Medical Psychology*, *51*, 387–398.

Kaufmann, J. H. (1983). On the definition and function of dominance and territoriality. *Biological Reviews*, *58*, 1–20.

Kemper, T. D. (1978). *A social interaction theory of emotions*. New York: John Wiley & Sons Ltd.

Kennedy, J. L. & Mackenzie, K. R. (1986). Dominance hierarchies in psychotherapy groups. *British Journal of Psychiatry*, *148*, 625–631.

Kevles, B. (1986). *Females of the species: Sex and survival in the animal kingdom*. Cambridge, Mass.: Harvard University Press.

Kleinke, C. L. (1986). Gaze and eye contact: a research review. *Psychological Bulletin*, *100*, 78–100.

Klerman, G. L. (1974). Depression and adaptation. In R. J. Friedman & M. M. Katz (Eds), *The psychology of depression* (Pp. 127–145). Washington, D.C.: V. H. Winston.

Klinger, E. (1975). Consequences of commitment to and disengagement from incentives. *Psychological Review*, *82*, 1–25.

Krebs, J. R. & Dawkins, R. (1984). Animal signals: mind reading and manipulation. In J. R. Krebs & N. B. Davies (Eds.), *Behavioural ecology: An evolutionary approach*, 2nd Edition (Pp. 380–402). Oxford: Blackwell.

Kummer, H. (1971). *Primate societies: Group techniques of ecological adaptation*. Chicago: Aldine-Atherton.

Lockard, J. S. (1980). Studies of human social signals. In J. S. Lockard (Ed.), *The evolution of human social behavior* (Pp. 1–30). New York: Elsevier.

Lorenz, K. (1981). *The foundations of ethology*. New York: Springer Verlag.

Lott, D. F. (1967). Threat and submission signals in mature male American bison. *Proceedings, 75th Annual Convention*, American Psychological Association, 121–122.

MacLean, P. D. (1985). Evolutionary psychiatry and the triune brain. *Psychological Medicine*, *15*, 219–221.

McLean, P. (1976). Depression as a specific response to stress. In I. G. Sarason & C. D. Spielberger (Eds.), *Stress and anxiety*. New York: John Wiley & Sons Ltd.

Maslow, A. H. (1937). Dominance feeling, behavior, and status. *Psychological Review*, *44*, 404–429.

Maynard Smith, J. (1982). *Evolution and the theory of games*. Cambridge, England: Cambridge University Press.

Moyer, K. E. (1976). *The psychology of aggression*. London: Harper & Row.

Parker, G. A. (1974). Assessment strategy and the evolution of fighting behaviour. *Journal of Theoretical Biology*, *47*, 223–243.

Parker, G. A. (1984). Evolutionarily stable strategies. In J. R. Krebs & N. B. Davies (Eds.), *Behavioural ecology: An evolutionary approach*, 2nd Edition (Pp. 30–61). Oxford: Blackwell.

Parkes, C. M. (1986). *Bereavement: Studies of grief in adult life*, 2nd Edition. Harmondsworth, England: Penguin.

Paykel, E. S. (1978). Contribution to life events to the causation of psychiatric illness. *Psychological Medicine*, *8*, 245–253.

Power, M. (1986). The foraging adaptation of chimpanzees, and the recent behaviors of the provisioned apes in Gombe and Mahale National Parks, Tanzania. *Human Evolution*, *3*, 251–264.

Price, J. S. (1968). The genetics of depressive behaviour. In A. Coppen & A. Walk (Eds.), *Recent developments in affective disorders*. London: Royal Medico-Psychological Association.

Price, J. S. (1972). Genetic and phylogenetic aspects of mood variation. *International Journal of Mental Health*, *1*, 124–144.

Price, J. S. & Sloman, L. (1987). Depression as yielding behavior: an animal model based on Schjelderup-Ebbe's pecking order. *Ethology and Sociobiology*, *8*, 85(S)–98(S).

Raush, H. L., Barry, W. A., Hertel, R. K., & Swain, M. A. (1974). *Communication, conflict and marriage*. San Francisco: Jossey-Bass.

Richards, S. M. (1974). The concept of dominance and methods of assessment. *Animal Behaviour*, *22*, 914–930.

Rippere, V. & Williams, R. (1985). *Wounded healers*. Chichester: John Wiley & Sons Ltd.

Rowell, T. E. (1974). The concept of social dominance. *Behavioral Biology*, *11*, 131–154.

Savin-Williams, R. C. (1987). *Adolescence: An ethological perspective*. New York: Springer-Verlag.

Schiff, M. & Lewontin, R. (1986). *Education and class: The irrelevance of IQ studies*. Oxford: Clarendon Press.

Schenkel, R. (1967). Submission: its features and function in the wolf and dog. *American Zoologist*, *7*, 319–329.

Schjelderup-Ebbe, T. (1935). Social behaviour of birds. In C. Murchison (Ed.), *Handbook of social psychology* (Pp. 947–972). Worcester, Mass.: Clarke University Press.

Seligman, M. E. P. (1975). *Helplessness*. San Francisco: Freeman.

Sinclair, A. R. E. (1977). *The African Buffalo*. Chicago: University of Chicago Press.

Sloman, L. & Price, J. S. (1987). Losing behavior (yielding subroutine) and human depression: proximate and selective mechanisms. *Ethology and Sociobiology*, *8*, 99(S)–109(S).

Snaith, R. P. & Taylor, C. M. (1985). Irritability: definition, assessment and associated factors. *British Journal of Psychiatry*, *114*, 1325–1335.

Totman, R. (1985). *Social and biological roles of language: The psychology of justification*. London: Academic Press Inc.

von Andics, M. (1947). *Suicide and the meaning of life*. London: John Wiley & Sons Ltd., p. 94.

Watzlawick, P., Beavin, J. H., & Jackson, D. D. (1967). *The pragmatics of human com-*

munication: A study of interactional patterns, pathologies and paradoxes. New York: W. W. Norton.

Weisfeld, C. C. (1986). Female behavior in mixed-sex competition: a review of the literature. *Developmental Review*, *6*, 278–299.

Weissman, M. M. & Paykel, E. S. (1974). *The depressed woman: A study in social relationships*. Chicago: University of Chicago Press.

Wood-Gush, D. G. M. (1955). The behaviour of the domestic chicken: a review of the literature. *British Journal of Animal Behaviour*, *3*, 81–110.

8 Psychiatric Syndromes as Infrastructure for Intra-specific Communication

Russell Gardner, Jr.
Department of Psychiatry and Behavioral Sciences,
University of Texas Medical Branch,
Galveston, U.S.A.

The most obvious is the hardest to fathom
—Edgar Allen Poe

This chapter's central idea is that distinctive human communicational states may be naturally occurring guides to the infrastructure of normal intra-specific communication in many species. Psychiatric symptoms and syndromes are maladaptive and prevalent, and at present they are inadequately explained in biological terms at either the whole-organism or cellular-molecular level. But from their phenomenology and from their response to psychotropic drugs, they may teach us about normal patterns of communication, similar to the way in which disorders of the blood cell instruct us about the normal workings of the blood. Primitive adaptation may be evident from "communicational pathology", which in turn may provide us with an indication of how to divide communicative experience into components or units, i.e., into an alphabet of communicational infrastructure.

SETTING THE BIOLOGICAL QUESTION

For communication to occur, there have to be signals, senders, receivers, and methods of decoding (Green & Marler, 1979). But we know little of the brain structures involved in signal choice or choice of response, or of why senders and receivers play their particular roles when they do, or of how brains have become programmed over evolutionary time to construe the playing of these roles. Overall, more is known about telencephalic cortical structures because these are larger in mammalia and so more accessible for examination. They clearly relate to such characteristics as language, fine motor co-ordination and perceptual analysis.

But what about "motivational" brain systems *below* the level of the cortex? There are conceptual problems in the "hydraulic" metaphor associated with "motivation" and "sexual and aggressive drives" because there is no evidence that the brain operates in a manner similar to those hydraulic mechanics that were newly exciting in the 19th century. This is now widely understood, but the same terms and implicit metaphors persist in discussion of basic concepts. Fleck (1979) has pointed out that prevailing and underlying "thought collectives" determine the nature of "scientific facts". In my presentation, no assumptions are made about unitary drives of sex and aggression, the mental and behavioural outcomes of which take on varied shapes depending on vicissitudes of ontogeny. Drives are reified concepts. To modernise the metaphor by alluding to a machine that is exciting in the 20th century, the brain more resembles a computer with interactive terminals than the water pipes of a steam engine.

Other common psychiatric terms are essentially reifications as well. For example, "affect" or "mood" disorders are implicitly assumed to be correlated with disordered neuronal systems, which are "corrected" by psychotropic drugs. Yet these terms refer in fact only to subjective and communicational experience (including empathetic identification with other individuals who also experience these feeling states), not to definitive neuronal structures or connections. Affects and moods have no independent referents: they may be "reliably" identified via questionnaires or interviews, but there is no index of "validity" connected to a biological system comparable to the way in which a blood disorder can be related to variations in blood composition, e.g., how the cause of pain and the reduced red blood cell count in sickle cell anaemia can be understood through the haemoglobin sickling phenomenon. This chapter tries to ignore reified concepts, and starts with this observation: Communicative acts pervade the animal kingdom and function in the human species to an extraordinary degree; this implies that there is considerable body machinery for communicational states in animals including humans.

Do psychiatric disorders, these strange behavioural states for which there is no explanation (Weiner, 1978), reflect the selective or overriding operation of primitive structures independently of, in spite of, and/or via brain structures more evolutionarily recent? Thinking through potential answers to this question led me to the importance of dissecting intraspecific communication as a general biological question.

We have gathered preliminary data that show resemblances between manics and high-profile normal persons, both of which are in turn significantly different from effectively treated manics and non-high profile persons (Gardner, Gustavson & Gustavson, Note 2). But even before those data are all collected, we now need to frame the issue differently, partly so that the data can become as meaningful as possible, and partly because so doing now has support from existing knowledge that merits further intensive examination. "Intra-specific communication" has obviously determined many biological phenomena and many somatic structures—from the human larynx, to bird feather patterns, to pheromone receptors. Intra-specific communication is not a frequently used term at present, but other descriptions of its meaning (e.g., social relations, animal communications) are not sufficiently precise nor inclusive of humans. Therefore, I have used it here as a first component in the reframing task.

I assume in this chapter that to posit psychiatry's role in this intellectual endeavour is an important bridging effort. But regardless of psychiatry's ultimate usefulness in this issue, the question now summarised remains a very interesting one: What components of communication can be understood as the infrastructure to specific communicative acts; how can these components be isolated from the on-going flow of communicative behaviours; and what parts and systems of the brain are involved in their generation and mediation?

INTRA-SPECIFIC COMMUNICATION DELINEATED

Intra-specific communication is necessary to social interaction but the two are not synonymous. Although they label similar phenomena, the first term can be more readily used to focus on individual communications, whereas social interaction refers to the overall process, to the results of communications (McGuire & Essock-Vitale, 1981). The first uses the individual whole-organism as the basic unit of analysis, and the other instead more naturally refers to varied, multiple-individual groupings as basic units. Because an ultimate purpose of my work is to stimulate investigation of now unknown, or barely known, cellular–molecular mechanisms in communicational processes, the individual as the basic unit of comparison is an important distinction that requires emphasis.

This distinction is a major reason why such work is not labelled "sociobiology", which analyses biological determinants of social behaviour more broadly (Wilson, 1975). For example, inclusive fitness, sociobiology's important paradigm, has fostered the mathematical modelling of individual behaviours in populations (Hamilton, 1964; Queller, 1985). This paradigm implicitly assumes the individual organism to be a "black box" or unitary vehicle for its genes (Dawkins, 1976). In contrast, my approach here is to use the metaphor of the individual as a functioning machine, the parts of which can be analysed and understood. This metaphor has been used most profitably in the basic sciences of the other specialities of medicine, and can be applied in psychiatry as well.

However, the terms "neuroethology" or "ethology" will not serve either, although these areas are obviously much concerned with animal communication (Eibl-Eibesfeldt, 1970). These fields have not restricted their focus to cross-disciplinary study of intra-specific communication—predator–prey relations are just as relevant to them, for example. Also, I avoid use here of such key concepts in ethology as fixed action patterns, releasers, search images, and innate releasing mechanisms, in the hope of establishing relevance to the human species as well.

Definition of Intra-specific Communication. This term refers to any kind of information sent and registered among or between conspecifics (members of a same species) in order ultimately to achieve individual survival and survival of genes. Such information includes systems of messages and meanings that each member of the animal species under examination spontaneously gives and receives to accomplish a variety of proximate adaptive functions. These may include brief encounters with information exchange of varying specificity, or long-lasting states involving the propensities of how the individual relates to others, e.g., group membership, leadership, or isolation from the group through choice or exclusion. Reproductive and nurturant ends are often served by intra-specific communication.

Intra-specific Communication in Humans and Related Species

There are major differences amongst taxa concerning the amounts of ontogenetic learning necessary to achieve communicative function in the adult animal (Bonner, 1980). Signal systems and their meanings seem to be largely inborn for invertebrates and, to a major extent, for many vertebrates as well. However, work with birds has shown that early ontogenic experience takes part in the normal development of avian song (Baker, Spitler-Nabors, & Bradley 1981; Bottjer, Miesner & Arnold, 1984; Marler & Tamura, 1964. Of course, humans across the world learn many different languages, indicating great plasticity, and spend much time during youth mastering this highly flexible and specific communicative mode. Indeed, multiple languages can be learned, although the capacity to do this easily and without accent lessens with age (Cazden & Brown, 1975).

Human "emotional" communication, including posture, is less plastic, variable, and flexible, and its code seems more or less universal throughout the species. A smile, frown, eye-flash, or crying have very similar meanings to the sender and to the receiver even when they are strangers to each other, and regardless of their culture (Darwin, 1873; Eibl-Eibesfeldt, 1972;

Ekman, Friesen & Ellsworth, 1972). Apparently, they have meaning from the earliest days of life as shown by the infant's responsiveness to particular facial expressions (Meltzoff & Moore, 1977) even within 36 hours of birth (Field, Woodson, Greenberg, & Cohen, 1982). What is learned later is appropriateness of timing, audience, and nuance, not the basic emotional coding itself.

Components of such communication seem to have carried over from ancestral species into various radiations that have developed since. Indeed, many emotional messages are sufficiently "hardwired" that they allow *inter*-specific communication between related species. Humans and non-human primates seem to understand mutually a number of congruent feelings. Such inter-specific communication may in part determine the existence of pets, as dogs and cats, for example, seem to be "related" sufficiently to human individuals or families to learn responsiveness and interact with them, extensively and powerfully. Moreover, monkey babies can relate to their canine adopted mothers (Mason & Kenney, 1974), so this phenomenon is not limited to humans. Still, the capacity to send and receive meaningful messages, and to relate to significant others, is mostly limited to conspecifics.

Returning then to *intra*-specific communication, the role of telencephalic structures in the recognition of a large number of faces, visually and with great specificity, is clear in humans. Cerebrovascular accidents involving parts of the cerebral hemispheres not only cause language dysfunction (several kinds of aphasia) but also can abolish the ability to recognise famous and familiar faces (prosopagnosia; Bruyer et al., 1983), thereby locating an important link for these several facets of the human communicational repertoire within the cortex, the largest, most recently evolved part of the brain. Of course, the cortex also mediates sensory and perceptual analysis, and fine-movement co-ordination. This level of brain structure may initiate, constrain, or otherwise modulate powerful emotional expression but probably is not critical for such communications. Aphasic stroke patients with cortical damage routinely still exhibit emotional expression. A symmetrical smile was shown in a photograph of the face of a patient whose voluntary facial movements were unilaterally paralysed from cortical brain damage (Monrad-Krohn, 1924, cited in Reynolds, 1981, p. 92).

Intra-specific Communication in the Animal Kingdom

Let us now take a different tack. I have briefly reviewed some communicative components that distinguish humans and some other mammalians, and I have located an important anatomical site for these functions in the recently evolved (expanded) telencephalic structures. But it would be a mistake to view intra-specific communication as arising anew from structures that emerged so late in evolution. Indeed, inter-organism messages have probably been important from the time organisms became individuated. Certainly, with the evolution of sexual reproduction, such messages became obligatory. Unless males and females recognise each other and their gametal materials are exchanged, their genes die when they die (Dawkins, 1976). We assume that evolution, via diversity of phenotype, natural selection, and genetic transmission (Mayr, 1982), has influenced the derivation of communication in many taxa, and that, indeed, species are *defined* by their communicative specificity, especially as required for mating and reproduction.

Across the wide spectrum of the animal kingdom, message systems vary as to the dominant sensory modalities for transmission. Pheromones were probably the first such mode. For single-cell organisms, some pheromones were apparently so useful in the ancient evolutionary past that they were retained as hormones when multicellular organisms evolved, so that the one-time pheromones are used now to "communicate" between organism parts (Feldman et al., 1984). Examples are corticosteroids and oestrogens; receptors for these "messenger molecules" are found in present-day, single-celled organisms, the yeasts, and these "hormones" are also produced by these same cells. Significant components of what we may think of as the biochemistry of the body apparently were once intra-specific communicative devices!

Noting the antiquity of intra-specific communication lessens our surprise on finding that there are many levels of communicational production in such complex creatures as humans. Language and facial recognition exist partly as a function of cortical structures but some facets of our communicative interactions surely involve neuro-anatomical structures on subcortical levels as well. If the brainstem is involved, such mechanisms reflect very primitive vertebrate origins (McLean, 1973; Sarnat & Netsky, 1981).

Functions of Intra-specific Communication

Of course, intra-specific communication serves adaptive patterns other than information exchange about sex. For example, formation of groups is an observable phenomenon in many species. Foraging, protection from predators, reproductive aims, and nurture of the young are among the probably benefits of this affiliative process (Bonner, 1980).

In some species, adult animals live in isolation except at mating times. Examples include prototherians, the most primitive mammals to have survived to the present time (Eisenberg, 1981). However, despite such isolationist tendencies, animals in these species seem very involved in the detection of whether a conspecific has been present, including when, e.g., how long ago was a smell-trace left? In part, this may assist in the most efficient spacing or distancing of individuals one from another, so that those with shared genes can exploit the ecosystem the most effectively. These (and more gregarious) animals make considerable use of specialised glands that secrete species-specific substances to be left as distinctive traces, meaningful to conspecifics (Colgan, 1983). Linking together for them is through time more than through congregation. Spacing functions are clearly important throughout the animal kingdom. For example, defended territories or aggression fields can be detected at many levels of that kingdom and, in part, can be seen as a "spacing" mechanism, distributing and perpetuating conspecific genes as much as possible (Waser & Wiley, 1979).

Formation of groups in some species also implies that membership is restricted. In-group membership versus out-group persecution are documented for many species (see Wenegrat, 1984, for review of such considerations, particularly in humans and similar species), and these are related to aggression fields. For species that congregate, the formation of social rank hierarchies within the group is also adaptive. Particular emphasis has been put on the individual in the group who is dominant or who, in many mammalian species, behaves as an alpha member in ways other than aggressive dominance activity alone (Bernstein & Gordon, 1974). However, those acceding to this group member are also exhibiting adaptive behaviours. Certainly, group influences, e.g., following the group, are powerful determinants of individual human behaviours. Crowd behaviour, where the leader may or may not be important, is an example. At times, groups may "elicit" leadership behaviour from one of the members; this is true at least, of humans. Finally, nurturing (parenting) behaviours seem to perpetuate the genes of a species, as does the eliciting of such behaviours from parents or other adults by the young.

Thus, intra-specific communication stems in its evolutionary history from single-cell organisms, some of which still exist in a relatively unaltered form. At the other extreme, humans exemplify the use of both highly specific and plastic forms (language and recognition of faces), and of forms that require either no or little learning (emotions). But, two basic messages are conveyed by all intra-specific communications: conspecifics tend to link themselves or to space themselves in optimal ways in physical space or time.

HOMOLOGY AS A RELATIVE CONCEPT

Can Humans Truly be Compared to Other Animals?

Examples given have been selected freely from a variety of species ranging from humans to single-celled yeasts. But can this be done legitimately? Easy extrapolation from animal to human is justifiably viewed with suspicion, and many have considered the relating of human to non-human animal behaviour as especially fraught with conceptual hazards. For example, criticism prevails of any assumption that traits similar in humans and animals have come to be so from homologous evolution when convergent or analogous pathways may be an equally likely explanation, and indeed the more parsimonious hypothesis. That is, may communicational patterns not be *homologous* between species but evolved independently?.

However, these terms refer to slippery concepts and, when used, they require that the user should clearly conceptualise and qualify the characteristic in question, and designate the starting point (Campbell, 1976). Differentiating between functional and anatomical aspects may be useful. The question then turns out to be not whether phylogenetic homology exists but to what degree. Indeed, I propose that specifying the level of homology for a particular characteristic should be an important question for research in intra-specific communication.

For example, do the songs of birds compare to the songs of man as either homologous or analogous melodic communications? Since Darwin, homology has implied that a trait that two taxa have in common was also once a trait of their common ancestor; analogy, in contrast, means that the trait was elicited by environmental circumstances from separate origins (Ospovat, 1981). According to Mayr (1982), eyes evolved independently more than 40 times in the animal kingdom. In the case of melody and song, independent evolution almost certainly occurred (this obviousness caused the choice of example). If bird and human song *were* homologous, then the pre-avian, pre-mammalian (probably reptilian) ancestor would have sung also. But if one broadens the function to that of making audible signals meaningful to conspecifics, and then considers that the structures involved need only to have made sound and deciphered the sounds of others, it follows that the ancestral creature probably did share the more general trait.

Interestingly, current research on auditory communication concerns exactly these issues. There is a motor theory not only of human speech but also of avian song perception as well. The motor neural structures responsible either for the articulation of speech phonemes or of avian song syllables receive information about the acoustic details of what is heard from differently evolved telencephalic sources (Williams & Nottebohm, 1985). These telencephalic structures for both avians and mammals seem to have been expansion areas, mediating ever more elaborate specific codes, which developed after the ancestral radiations branched apart. Phonemes and birdsong syllables are both "articulatory gestures", each probably stemming from unmelodic and very ancient articulatory gestures.

Another example of "depth of homology" comes from the musculoskeletal system. Bird and bat wings are *not* homologous if the structures are considered as highly elaborated forelimb bone and muscles (with the function of flight exploiting the ecosystem of the air). Separate digits become extended independently in developing the increased surface area necessary for flight, indicating separate "invention" through natural selection. But if the structures are considered generic forelimbs instead of wings (with a redefined function, that of "forward motion"), then "homology" emerges as an accurate descriptor for the trait possessed by the two species. The definition depends on the starting point, and on the descriptions of the structure and its counterpart function.

Delineation of "depth of homology", "degree of phylogenetic canalisation", or how "highly conserved" is the trait or characteristic, allows one to designate more precisely what is meant by any statements about homology or analogy. The broader the definition of the function and the greater the evolutionary age of the structures underlying it, the deeper the homology, the greater the phylogenetic canalisation and the more highly conserved the structural components of the characteristic in question.

Whatever they may be called, we readily accept that some components of bodily systems are "more basic" than others. That the vertebrae of vertebrates exist is more basic than their exact number in a particular species. That most terrestrial vertebrates have two limb girdles and four limbs is more basic than the exact form of these anatomical structures in example taxa. The exact form often depends for its details on the ecosystem that the species inhabits, and to which it has conformed.

This is also true of intra-specific message systems. Some components are very basic; others, such as the precise message codes used by animals of a particular taxon, have been elaborated from more basic components. That auditory and olfactory spheres of communication are used is more basic than the particular phonemes or pheromones used by a particular species. That an animal is a part of a group is more important than any of the myriad ways through which it communicates its sense of position, e.g., by submissive gestures, escaping from a dominant, or recursive maintenance of a certain distance from the dominant. That language is used for seduction is less important than the fact that sex is on the minds of the two involved; their "communicative state" or "sexual psalic" (S)—as described later—is an infrastructure for the words spoken.

As molecular genetics blossoms still further and the D.N.A. coding of various traits, such as body structures and learning propensities, becomes gradually understood, then classifications such as the ones mentioned here will become more important. Codings for the more deeply homologous characteristics will probably be seen to act more like primary or master computer programs, and the less basic characteristics like subprograms.

I emphasise here a particular viewpoint for comparing animals, one aimed at quantifying "depth of homology" on the dimension of intraspecific communication. Operational definition of the communicative communalities demonstrated by individuals of different species will allow us to determine the cellular–molecular mechanisms of these functions. To detect such communalities, appropriate cross-species comparisons are important. The determining of homological branching points should constitute an important scientific activity.

PSYCHIATRIC SYNDROMES DESCRIBED

The psychiatric syndromes and symptoms of the major mental illnesses are stereotyped and prevalent. Bipolar or manic-depressive disorder and schizophrenia each exist in approximately 1% of the population. Paranoid symptoms are part of a number of psychiatric conditions stemming from a variety of causes. Mania, depression, and paranoia are best understood as "final common pathways", i.e., stereotyped syndromes caused by varied brain conditions or arising for no easily discernible reason. The likelihood of bipolar disorder or schizophrenia illness is greater when first-degree relatives have the same disorder, indicating a diathesis transmitted genetically.

Some psychiatric illnesses, the organic brain syndromes, are caused by changes in general physiology or brain degeneration. These are not directly considered here although in their study in the 19th century (Hughlings) Jackson (1884) and his neurologist followers have provided a model of the nervous system as evolved through natural selection and hierarchically arranged. Neurological symptoms result from release phenomena (a latent function usually held in check by active inhibition), from direct stimulation, e.g., epileptic motor seizure, or from deficit states, i.e., the results of absent, impaired, or inhibited neurones. That the so-called functional mental disorders may in fact be caused or exacerbated by subtle organic features is something taken for granted by present-day psychiatry. Indeed, the efforts to postulate psalics, which are detailed and defined later, may allow different neurological hypotheses of mental illnesses to be tested.

Psychiatric Disorders as Communicational Propensity States

Most psychiatric syndromes without a clear organic aetiology are states lasting days, weeks, months or even years, which usually involve social, communicational postures and propensities. Precise moment-to-moment predictions of behaviour are not possible but more general ones or the prediction of propensities are. The manic is often boisterous with real or with imagined others. The schizophrenic's "negative symptoms" ensure that he[1] actively avoids other persons, showing his involvement through indifference. The depressed person typically describes pessimism about her place or her future in the world of people. Because the paranoid is sure others will hurt or persecute him, he overreads potentially harmful messages, as in "ideas of reference" in which he thinks some behaviour or cue refers to him and is more significant than would non-paranoid persons perceiving the same stimulus.

These major mental illnesses are further notable because the usual ways of verbally or otherwise persuading someone to change their behaviour fail to change the ill person's communicative-propensity state. Traditional talking therapies are ineffective for this purpose, and psychiatric patients persist in their behaviour patterns regardless of what others say or do. In contrast, psychotropic drugs, known also to affect brain neurotransmitter systems, do have the capacity to change the state, as well as to create other body effects, such as movement disorders and blood pressure changes.

This is not to say that such patients do not register the communications of others. The manic claims to control others and to be in charge; she is filled with energy and ideas for others. Opposition is quickly detected and countered (Janowsky, el-Yousef & Davis, 1974). The severely depressed person communicates a state of extreme submission or self-debasement. That the other person is reassuring may be registered but the reassurance is not believed and the person continues to debase himself and to claim worthlessness. The paranoid patient assumes the status of an out-group member, well advised to be wary, cautious, and to expect persecution. She may relate to a new interviewer temporarily with seeming trust, but small cues may make her soon distrust the new person as well. With those who exhibit the "negative" symptoms of schizophrenia, the pattern is that of a distancing or spacing from other persons. Acknowledgement of the other person occurs to varying degrees, but autistic fantasies often seem preferable to interactions with real persons.

Descriptive Psychiatry as Whole-organism Biology

Recent advances in consensually agreed descriptions can be found in the Research Diagnostic Criteria (Spitzer, Endicott, & Robins, 1978), and in the third edition of the Diagnostic and Statistical Manual (D.S.M–III) of the American Psychiatric Association (1980). These now widely accepted categories allow clinicians from different locations and kinds of training to understand in congruent fashion the problems under consideration. Hard

[1]Pronoun gender is alternated in this chapter.

work over recent decades has produced consensus on the basic descriptions of many disorders, although fine-tuning on the basis of empirical data will occur in future editions of the D.S.M.

Because psychiatrists have put this enormous research and organisational effort into getting congruence and agreement about these descriptions of behaviour, they might perhaps be considered as "clinical" biologists of the natural-history variety. The ethological tradition in biology involves making empirical observations of the naturally occurring behaviours of animals in their natural habitat. Mental illness occurs naturally in all cultures in similar, stereotyped forms—this has been particularly well-documented for schizophrenia (Murphy, 1985)—and it seems to have existed in past millenia (Mora, 1980), so that observations of patients (as they are now called) and classifications of their behaviour do constitute a variation of natural-history biology.

We have taken the line that this enormous effort may have some investigational pay-off in demonstrating communicational infrastructure. Thus, we have used the official listing of the core characteristics of mania as the methodological starting point for our comparisons of mania and high-profile normals (Gardner et al., Note 2). We will use the same ploy in other investigations.

To be fully an ethologist, however, implies that the worker will make cross-species comparisons, especially of the functional consequences of observed behaviours (McGuire, Essock-Vitale, & Polsky, 1981). Price (1972) pioneered such ideas, and McGuire and colleagues have tried to put psychiatric disorders into the context of evolutionary biology. In general, however, such comparisons have been few and unsystematic. The recent sociobiological review by Wenegrat (1984), guided by the "inclusive fitness" paradigm, is another interesting exception.

Psychopharmacology and Cellular–Molecular Biology

Development of drugs effective in the treatment of major mental disorders and discovery of their metabolic and brain effects has dominated psychiatric research effort over the past three decades. These agents include antipsychotic drugs, which counter the active symptoms of psychosis from various causes, e.g., hallucinations, delusions, and agitation; antidepressants, which counter serious depressive states and panic attacks; and lithium carbonate, which treats mania and, when taken regularly, is prophylactic against future episodes of either mania or depression in bipolar illness.

The antipsychotic and antidepressant classes of drugs have powerful effects on the monoamine central neurotransmitters—dopamine, serotonin, and norepinephrine. The neurones in which these metabolites originate are in brainstem structures that have ancient vertebrate origins, e.g., in fish

(Parent, Dube, Braford, & Northcutt 1987). Midbrain structures include dopamine neurons, originating in the substantia nigra, and serotonin neurones in the raphe nuclei near the central canal. Norepinephrine neurones chiefly originate in the locus coeruleus of the dorsal pons, a hindbrain structure (Feldman & Quenzer, 1984). These are highly conserved brain systems; their primitive origins, and the effect of psychotropic drugs on both them and on psychiatric disorders, argue indirectly but saliently for an evolutionarily ancient, rather than recent, basis for the behaviours that they affect.

The Three Components of Psychiatric Disorders

In an earlier paper, I postulated that the pathogenesis of psychiatric syndromes may eventually be shown to have three components: (1), an overly sensitive onset trigger; (2), deeply homologous (or canalised) communicational states (which are triggered), and; (3), reaction components to the plights caused by the fixity of the state (which may include the stimulation of another psalic, as described in the next sections).

Examples of the first include greater ease of onset in vulnerable patients. Persons with a family history of illness who themselves more easily incur the same disorder do so for unknown reasons apparently related to factors transmitted genetically. In psychopharmacology, "disregulation" of neurotransmitters is a popular incurrent speculation about pathogenesis (Siever & Davis, 1985). The model presented here suggests that such disregulation involves a threshold to illness.

As mentioned earlier, the central idea argued here emphasises the second (listed) component in the conjecture that similarities exist between manics and normal high-profile persons, between depressed and low-status, humbled, submissive supplicants, between paranoid persons and individuals truly persecuted, and between other pathological and normal human communicational states. Differences between pathology and normality are obvious and are a common focus of investigation. But these differences may be functions of components (1) and (3) which in turn cause problems with timing, nuance, and audience (or with secondary effects—see third component), *not* with the communicational state itself. The capacity to engage any of a number of psalics may be indispensable to the maintenance of life; as mammalia became gregarious, relating to conspecifics may have become necessary for bodily homeostasis. Even if physically alone, we are never separated from persons remembered in our past or contemplated in our future.

Returning to the third of the three components listed here, if one feels conviction about one's role in relation to others, then perceptions tailored for that communicational set may augment one's "faulty" interpretations, or impair still further "reality testing". The manic may develop persecutory delusions or pathological denial when his "leadership" is rebuffed.

Thus, psychiatrists have been "biologists" of two varieties (Ledley, 1983) in their quest better to understand and treat illnesses, especially the most severe, recalcitrant, and crippling disorders. They have been natural-history (whole-organism) biologists in describing naturally occurring mental disorders as much "without theory" as possible. But further, guided by observation of the therapeutic action and side effects of psychotropic drugs, psychiatrists have become cellular–molecular biologists, investigating the interactions of various hormones and hormone systems, neurotransmitters and their cell origins, and even cell membrane physiology, including receptor sites. To foster a coherent, biological, basic science of psychiatry, I propose that these two domains of biology be integrated by focusing on the communicational-state aspect of psychiatric disorder. This becomes practically significant in developing rational, animal models of various psychopathologies.

DELINEATION OF PSALICS

This new term seemed necessary because there was no other to describe the central idea being developed here, that the second component (see list in previous section) of psychiatric states may be an index to deeply homologous communicational states, mediated by highly conserved neuronal systems. Mayr (1982) advises that a new term should be devised when no existing term adequately provides the specific meaning, or where other candidate terms carry them too many misleading connotations.

Problems with Other Candidate Terms

I earlier referred to an already extensively used term, "fixed action pattern" (Gardner, 1982), but this term has many other meanings in biology, and especially in ethology, and fails to convey some of the meanings required by the framework described here (M. Hofer, personal communication, 1983). The terms "schema", as used by psychologists, and "sensory template", as used by biologists, are closely similar to each other but connote the developmental trajectories of detailed learning (Green & Marler, 1979) rather than the infrastructure of communication, as required here.

Two recent books dealing with similar concepts refer to "psychobiological response patterns" (Gilbert, 1984) and genetically transmitted "response rules" (Wenegrat, 1984). The states we are dealing with here display not only responsivity but initiative. "Psychobiological" also perpetuates the dysfunctional Cartesian split between "mind" and "body"

(Bateson, 1979; Engel, 1977) and does not indicate the critical role of intra-specific communication. The response rule, with its sociobiological "black box" implications, fails to suggest the necessary integration between the whole-organism and cellular–molecular spheres of biology. However, both terms refer to related concepts and the idea of "rules" may eventually be useful in description of the operational effects of psalics.

Derivation and Etymology of Psalic

The requirements for a new term included emphasis on the importance of intra-specific communication, which antedates in its evolutionary development signal systems that are specifically human, such as language. That is, functions that intra-specific communications accomplish in all species should be conveyed by the term; when reduced to final common denominators, these are "linking" and "spacing" functions. Further, as discussed earlier, the psychiatric states used in this derivation are best described as propensity states, not predictive in detail of the person's next behaviour or perception, but conveying greater likelihood of particular categories. The term should be phonemically similar to psychiatry and psychology, the clinical sciences that sparked its origin. It was hoped there would be helpful connotations stemming from roots in Greek.

The Greek term "psalis", as found in *Dorland's Medical Dictionary* (1957), refers to the fornix, a subcortical limbic structure. Fornix is Latin for "arch", which describes the shape of this nerve tract; psalis was the Greek counterpart term used by Galen. This term thereby concerns anatomy (of an evolutionarily ancient brain structure) as well as metaphorically referring to an architectural foundation structure (the arch) through which communication is possible. Therefore, psalis was itself adopted as the foundation for two acronyms that condensed many of the foregoing concepts. So, psalic comes from **P**rogrammed **S**pacings **A**nd **L**inkages **I**n **C**onspecifics, and from **P**ropensity **S**tates **A**ntedating **L**anguage **I**n **C**ommunication. Other Greek meanings of psalis (Jones, 1940) augment the potential usefulness of the term: These include "plucking a stringed instrument" (giving rise to psaltery and psalm), which connotes non-verbal communication, underlying and more basic than verbal communication alone; "Scissors" or "clipping, as with a shears", which connotes a separating and thereby spacing function; "a part of a bridle, a curb-chain, formed of links . . .", which refers back to linkage functions.

To help define the concept further, let us consider what psalics are *not*. What motivational states are comparable to the postulated psalics but not in the same subset? Examples include states basic to survival but not requiring (necessarily) communication with conspecifics, such as appetitive, consuming or foraging states, and defensive states, as from predators from other species, whether the defence is passive or active. However, psalics may occur in conjunction with such states, although how such conjunctures occur, and what brain mechanisms these states share, are important research questions.

For example, panic states are active defence states, maladaptive in our present-day world, which obviously are not limited to stimuli arising only from one's own species; Nesse (1984) argues that panic is an active defence state that had adaptive significance over evolutionary time. To speculate on a possible sequence (in order to illustrate the point), an *inter*-specific defence may, in its earliest versions, have antedated *intra*-specific communicative states. Then as new species evolved, defences against, and adaptations to, one's conspecifics also became important. Jacob (1982) compares evolution to a tinkerer rather than an engineer, as it remodels devices from the past rather than starting anew, even if new functions flow from old structures.

Formal Definition of Psalic. A psalic is a primitive communicational state, mediated by deeply homologous neural structures, that, when stimulated and activated, causes the organism to demonstrate an unusual readiness to assume distinctive roles relating to functional activities involving one or more of its conspecifics. Psalics can overlap (i.e., two or more psalics can be simultaneously active in the same individual at the same time), and can override to varying extents other factors influencing the organism's behaviour and perceptions.

Algorithm for Deriving Eight Psalics

The purpose in deriving psalics was not to cram all psychiatric illnesses into new categories but rather to exemplify how the concept might work, in order then to try out its utility. Carl and Joan Gustavson and I decided on eight Psalics through trying to account for some of the major psychiatric states and, for the major functional communicational activities observable in normal humans and animals. (see Table 8.1). In this "first-draft" effort, psalics were formulated with an eye to their parallelism with each other and with the appetitive and defensive states that do not necessarily rely on intra-specific communication. It is conceivable that there are many more psalics, but we decided to designate a number sufficient to demonstrate well the algorithm for any future research efforts. We wanted to derive enough psalics for the concept to be extensively explored but, at the same time, we hoped to keep the number manageable for practical usage.

The following tripartite formula generated the eight psalics so far: (1), a psychiatric disorder or symptom can be described succinctly as a distinctive communicative state; (2) a normal human counterpart also exists, such that the syndrome/symptom describes it also, except that timing, nuance,

TABLE 8.1
Eight Psalics

Psalic	*Abnormal Human*	*Normal Human*	*Non-human Animal*
Nurturant (N)	Kidnapping without financial aims	Parenting Caretaking	Nest-tending Parenting
Nurturance Recipience (NR)	Dependency Anaclitic depression depression	Normal offspring behaviour with parents (and other caretakers)	Normal care-seeking behaviours
Sexual (S)	Perversions Rape	Being in love Sexual intercourse	Oestrous behaviour Male sexual behaviour
Alpha (A)	Mania ? Early stages of alcohol intoxication	High-profile normal Charismatic leader	Dominant member of a social grouping
Alpha-Reciprocal (AR)	Loss of values as in a mob or with an inappropriate leader Conversion disorder	Followers Audiences Hypnotic trance	Non-omega subordinates in a grouping
In-group Omega (IGO)	Melancholic depression	Very low ranker who exhibits submissive behaviours to others in the group	Lowest ranker in a social hierarchy
Out-group Omega (OGO)	Paranoia delusions Persecutory delusions	Persecuted member of an out-group	Persecuted member of an out-group (non-member of an in-group)
Spacing (Avoidant) (SA)	Schizoid behaviour Autism	Hermit Isolated living	Avoidant behaviour towards conspecifics

audience and intensity variables cause it to be normal; (3), counterparts are detectable in other non-human animal species, albeit with differences stemming from taxon-specific communicational repertoires.

Specific Psalics

Alpha, Sexual and Alpha-reciprocal Psalics. The alpha (A) psalic is the model psalic first characterised, seen in its most powerful and maladaptive form in the bipolar patient during manic psychosis. Thirteen characteristics of mania listed in the D.S.M.–III Training Guide (Webb et al., 1981) also describe high-profile normals but not euthymic bipolar patients or normal-profile controls (Gardner et al., Note 2). Despite the high degree of malfunction that manic patients display, they seem typically to have little insight, characteristically to display high spirits and, in a proportion of cases, they refuse lithium medication in order to keep their "highs".

Interestingly, a comparison of the social behaviour of persons who have ingested alcohol with that of non-drinkers, as reviewed from 34 studies involving about 2000 subjects (Steele, 1986), has shown that alcohol causes them to make more extreme responses as follows: They exhibit more aggression and more conflict behaviour; increased drinking, eating, gambling, and risk-taking; engage in more self disclosure; and display greater sexual interest. This is congruent, of course, with common knowledge about early intoxication with alcohol and is replicated if the subject *thinks* she has drunk alcohol but has not; in contrast, lesser inhibition is an alcohol-specific effect not duplicated by the control condition.

Of course, my interest in these behaviours in the context of this chapter is because they are closely similar to those displayed by the manic (including the manic's lesser inhibition) and seem to characterise the normal non-intoxicated alpha as well. This reinforces the idea of psalics as final common pathways, stimulated in a variety of ways, which in turn depend on the state of the nervous system.

What was so powerfully adaptive about the A psalic that it should be perpetuated in the genes of so many distantly related animals? Why is it "deeply homologous"? Holding forth in aggression fields, competing successfully for sexual reproduction, and controlling subordinates in social rank hierarchies obviously seem to be good adaptive devices. All seem characterised in humans by "feeling good"; we can surmise that the so-called "reward" sites in the central nervous system—which also exist at lower brain levels (Feldman & Quenzer, 1984)—may indicate systems involved in generating and sustaining A psalics.

Sexuality also makes one "feel good", and also seems mediated by an ancient programme for determining intra-specific communicative behaviour. Interestingly, the psychiatric representatives of the sexual S psalic are disorders that are "ego syntonic"—which means that the person with the disorder does not complain about it. Persons with sexual perversions, exhibitionists, or paedophilics may complain bitterly about the consequences of their behaviour but generally not about the state itself; in this they are similar to the manic.

Furthermore, heightened sexual activity is often a component of the manic's behaviour, and of the high-profile normal's behaviour as well. This emphasises the notion that the psalics are not independent of one another but can overlap. This lack of orthogonality has the disadvantage of making psalics harder to study, but their simultaneous presence also means that the underlying neuronal systems must be independent to some degree.

In humans and other species, group behaviours are observed in the form of crowds, flocks, schools, herds, and other designations. Many members of these are "subordinates" who demonstrate considerable attention to an alpha animal (Chance, 1975). In our thinking about the adaptive significance of A psalic, the known elicitation of such behaviour by group membership caused us to conjecture about an alpha-reciprocal (AR) psalic.

As we observe humans, we note that sometimes group behaviours can be most maladaptive for those in the group. These are exemplified by destructive mobs, or the arsenic ingestion of the Jonestown residents. Not all such behaviours follow a leader's command but may be communicated from individual to individual less specifically. At the individual level, those who are very susceptible to suggestion, such as the "follower participant" in a folie à deux, or someone who is very hypnotisable, may also be experiencing the same or a very similar psalic. Animals congregate in groups, and some gatherings are unrelated to obvious leaders, such as schools of fish or flocks of birds. However, we have lumped these varied "responsiveness" behaviours into the AR psalic as a first approximation, recognising that later work with actual neuronal systems and subsystems may subdivide this global designation.

Other In-group and Nurturance Psalics. Bipolar patients at times exhibit severe depressions. Do depressed persons express a "communicative state" in which the message is that of exaggerated, self-humbling, submissive display? Price (1972) noted that low-ranking chickens, as in Schjelderup-Ebbe's initial observations of the "peck order" of barnyard fowl, exhibited phenomenological depression. If such parallel behaviour in these very separate taxa does turn out eventually to stem from activation of homologous anatomical structures, it will provide evidence of an early vertebrate branching point, and of a relatively greater depth of homology.

D.S.M–III, the manual mentioned earlier, also provides the category of a bipolar mixed state in which the patient is *both* depressed and manic, i.e., meets the operational criteria for both conditions. Again, such lack of orthogonality makes these states harder to study, but they also imply separate neuronal mechanisms in their generation and continuance. Such mixed states could be examples of the third reactive component of psychiatric disorders in the list shown earlier. Perhaps it relates on the one hand to the clinical aphorism that mania is a "defence" against depression or, on the other hand, to the idea that the rebuffed manic has a "depressive reaction" to the rebuff ("I am your leader but why aren't you responding?"). Both relations are speculative explanations for simultaneously present psalics.

Price and Sloman (1984) raised the idea that a "competition model of depression", into which category the in-group omega (IGO) psalic falls, is separate from the "loss" type that stems from the model in *Mourning and Melancholia* (Freud, 1915), as well as from the observations of Spitz (1946), Harlow and Harlow (1962) and others, as summarised by Bowlby (1969) in regard to the anaclitic depression of abandoned infants, human and monkey. Price and Sloman point out that the two models of depression may overlap to some extent and ought not be seen as mutually exclusive.

Through deployment of the algorithm for psalic formation, we suggest that in non-human animals, as well as humans, there is adaptive pay-off for neonates in making demands on maternal and other adult resources for nurturance. If these demands are not met, or not met to satisfaction, then a problem ensues for the offspring, and a normal intra-specific communicational state of continued elicitation by them may occur. Loss-induced depression and states of abnormal dependency are the abnormal counterparts. Dependent personalities are life-long in many individuals so these states are not a function of developmental conditions only. We have called this the nurturance–recipience (NR) psalic. The fact that both the IGO and NR psalics are dysphoric and unpleasant ("ego dystonic"), and have other characteristics in common, suggests that they too are mixed but less dramatically so than the A and IGO psalics in mixed bipolar states. Their further study, and the eventual dissection of the degree to which they are different—or overlap—may be done in congruence with an exploration of the separate—or/and overlapping–neuronal mechanisms.

The nurturance (N) psalic barely meets the requirements of the first listed criterion. That is, its pathology is less obviously psychiatric than the others, perhaps because nurturance is an extremely welcome trait in mammalia, including humans, (indeed through a mother's offering her mammary glands to her infant, this psalic *defines* mammalia!) Mothering and parenting are seldom societally inappropriate. However, despite this algorithm problem, nurturance seemed to demand definition because it obviously exists and must operate in conjunction with its NR reciprocal.

Out-group Omega Psalic. In-group omega status requires involvement within a group; much submissiveness will bring continued survival of oneself (and one's genes). But out-group status also occurs and may be reflected in out-group omega (OGO) displays. For example, there seems to be a communality between paranoia and discriminated-against-outgroup-status for both normal humans and non-human animals. Extreme "normal" human examples are the Jews in Hitler's Germany, or blacks newly moved into a segregated neighbourhood; less extreme examples are near at hand in any community, as conflict is generated by being members of different religious communities, competing companies, ethnic groups, etc. These designations distinguish in- and out-groups. For individuals with characteristics "alienating" them from territory holders, i.e., defining them as out-group members or enemies, it is highly adaptive to pay great attention to one's potential persecutors, to be wary and cautious in general, and to *expect* persecution from these conspecifics.

Paranoid psychosis is characterised by the patient who defines himself as an out-group member ("I've got enemies"), by being wary and cautious, and by expecting persecution—which he gives concrete definition through persecutory delusions. Delusions, i.e., false but fixed ideas, involve expectation of harm from a person or a group, e.g., "the F.B.I. has poisoned my food". Wenegrat (1984, p. 188) argues similarly: he notes that, ". . . In every case, the delusional paranoid experiences feelings appropriate only if he or she were potentially exposed to outgroup enemies. . . . Between individuals, the camaraderie and trust that normally distinguish in-group relations are notably tenuous, and in-group bonds of reciprocal altruism are never taken for granted."

Non-human animal counterparts also occur, as when a stranger gerbil is introduced into a cage of gerbils already resident there. The stranger may not survive as it will be attacked by its conspecific hosts. Such strangers are therefore naturally wary and cautious on entry into the area and from their behaviour seem to expect persecution (Thiessen & Yahr, 1977).

When the paranoid discusses or acts on his persecutory delusions, he exhibits, in a highlighted form, the OGO psalic. We speculate that the idiosyncratic delusional content is based on cortical interpretation of a subcortical influence. Being delusional is more basic than the details of the person's conviction at any moment (even "fixed" delusions change subtly in content over time).

In an initial and informal review of psalics, P. Gilbert (personal communication, 1985) queried whether the out-group omega psalic truly represented "omega" status because paranoid patients are often rather energetic in their own defence, i.e., more "alpha" than omega. My response is that they are omega with respect to another group; in normal circumstances, this other group is a hostile and actual entity; for the patient, the other group is often imaginary, either in fact or in terms of its hostility towards them. However, Gilbert's point is well taken, although with the recognition of simultaneous overlapping psalics, there may be no final contradiction. The paranoid can simultaneously demonstrate A and OGO psalics, as in a normal human variant; Spartacus did so in leading his maligned group in rebellion against their persecutors.

Some of the most striking examples of mixed psalics include the OGO psalic; persecutory delusions characterise psychosis of several varieties, including both manic and depressed, and organic states. We may conjecture that the OGO psalic may have a lower threshold if the other psalics are fully in force.

Spacing (Avoidant) (SA) Psalic. The broadness of the psalic classification is both an asset (for the reasons given) and a problem. In the Popperian approach to experimental science, as articulated by Platt (1964), hypotheses need to be excluded to provide the most persuasive scientific argument. Truth does not advance by being demonstrated; rather, when clearly articulated alternative hypotheses can be excluded, truth gets ringed in. Do psalics represent such a broad category that they are impossible to exclude?

In general, the implicit neurological hypothesis of psalics is that of release of inhibition: a psychiatric illness results when a psalic programme is inappropriately released from the forces that otherwise keep it at bay. Thus, in the face of stress, mania is known to occur more readily (Kennedy et al., 1983). It may be that the A psalic gets stimulated in a manner once required by group circumstances under primitive conditions, as in the face of turmoil and leaderlessness. Perhaps "stress" elicits various psalics for similar reasons, and with lack of present-day appropriateness.

This release may also involve the tendencies to avoid others (SA psalic), as seen in schizophrenia (and at lower gain in schizotypal personality disorder—compare to Wenegrat, 1984). Stress also exacerbates these disorders (Lukoff, Snyder, Ventura, & Nuechterlein, 1984; Nuechterlein & Dawson, 1984) although the "positive" symptoms of psychosis are usually the defining element of the abnormal state (in our parlance, OGO psalic is often simultaneously stimulated). But is the "negative" symptomatology, including the prominent feature of avoidance behaviour, similarly disinhibited? An alternative neurological hypothesis is that of deficit: work is underway to determine if the negative symptoms of schizophrenia are in fact a dementing illness (Kraepelin first called it "dementia praecox"), characterised by reduced ability to calculate and orientate to surroundings, and by other indices of neuropsychological deficit (Carpenter, Note 1). If so, then the idea could be rejected that such avoidant behaviour is a "positive" psalic rather than an absence of expected function similar to paralysis of a limb. At present, it is clear that 20 to 35% of schizophrenic patients have brain impairment (Seidman, 1983), but there is as yet no clear-cut answer to the deficit hypothesis.

Thus, the psalic hypotheses may be proven or not by the exclusion of alternative hypotheses; they are theoretical structures, accessible to new information. They may also serve an additional function, that of generating derivative hypotheses; so, at times they may be less hypotheses and more framework constructs. For example, when mania and paranoia are simultaneous, there may be different sets of brain stem activity than if one or the other were present alone.

A CLINICAL POPULATION CHARACTERISED BY PSALICS

In an effort to determine whether psalics could be used for rating clinical populations, I evaluated a consecutive series of patients presented at "debriefing rounds" during the autumn of 1985, at the University of Texas Medical Branch in Galveston. Residents in psychiatry who had been on emergency duty presented for supervision patients seen during their just-ended tour (4 of the 68 patients were presented at other teaching conferences). I then rated each patient who had sufficient clinical data according to the presence or absence of each of the eight psalics. Each history was evaluated for evidence of either normal or abnormal psalic components (as listed in Table 8.1). The results are regarded as pilot study data.

Table 8.2 shows psalics of the 68 patients, 33 females and 35 males, average age 33.4 years (S.D. = 13.3). On average 1.9 psalics were scored per person, and fewer than 50% had only a single psalic. All psalics were scored from 5 to 38 times (average = 16.2; S.D. = 11.8), with two (N and S) never seen except in combination. NR was scored in more than 50% of the patients, which may reflect the heavy proportion of drug abusers encountered on the Emergency Service, most of whom registered sufficient neediness to be scored positively on this "nurturance recipience" psalic. Pairings were tabulated (Table 8.3) for those patients registering more than one psalic. The greatest juxtaposition of pairings included the NR and IGO psalics. These pilot data may develop the hypothesis that depression does indeed often represent a "mixed" set of psalics.

This work exemplifies the data that can be gathered in relation to psalic concept. As these were rated by one person on an implicit or qualitative scale, the construct has yet to be verified clearly. Indeed, a small negative correlation was calculated between week of study and the number of

TABLE 8.2
Tabulation of Psalics on 68 Patients

Psalic	*Single*	*Double*	*Triple*	*4 or 5*	*Sum*	*Percentage*
A	4	3	3	2	12	9.2
AR	1	5	3	4	13	10.0
IGO	4	8	5	4	21	16.2
N	0	0	3	2	5	7.4
NR	11	13	9	5	38	29.2
OGO	9	7	9	3	28	21.5
S	0	3	2	2	7	5.4
SA	2	1	2	1	6	4.6
Total						
Psalics	31.0	40.0	36.0	23.0	130	99.9
%Psalics	23.8	30.8	27.7	17.7	100	
Total						
Subjects	31.0	20.0	12.0	5.0	68	
%Subjects	45.6	29.4	17.6	7.4	100	

TABLE 8.3
Psalic Pairings on 37 Patients

Psalic	*AR*	*IGO*	*N*	*NR*	*OGO*	*S*	*SA*
A	2	1	2	3	7	1	1
AR		6	2	10	3	2	1
IGO			2	14	6	1	2
N				3	3	1	0
NR					10	7	2
OGO						3	3
S							0

psalics scored per patient (Pearson $r = -0.22, P < 0.1$), indicating a hint of bias toward scoring multiple psalics earlier in the series.

With the help of medical students, K. Christian and S. Ferguson, rating scales have been drafted that we hope will measure the existence and intensity of each of the eight psalics, based on detailed clinical or other present-state information. In large part these scales are derived from symptom lists in D.S.M–III and from other clinical work.

PSALICS AND ORDINARY COMMUNICATIONAL EXCHANGES

An early commentator on the psalic concept (himself a noted and articulate psychiatrist-leader and administrator) pointed out that, in contrast to the manic, he was alpha enough at work but not when he returned home in the evening, when he was glad to relax and let his wife give orders (R. Michels, 1983; personal communication). Part of the problem in the psalic proposition is that it interferes with our set ideas about usual behaviour, which is not "driven" in the manner outlined, but instead is flexible and responsive. A relaxed ambience seems incongruent with extreme psalics.

As a way of briefly rationalising this, and to sketch in an account of ordinary non-extreme interactions, let me invoke the idea of "graded" psalics—states in which the underlying "psalic master program" is strongly modified by subprograms, so that it seems less evident in behaviour and less completely commanding of the organism. The psalic less completely overrides other influences on behaviour. If, in addition, psalics overlap or are mixed, the "gain" that is evident for any one of them may seem much lower. Small deferences ("excuse me", instead of the massive submissive manoeuvres deployed by the depressive—"I am terrible"; "I deserve to die") may be a quick euthymic non-extreme variation to facilitate in-group function. A gracious acceptance of such deference may be a euthymic equivalent of the manic's assumption of high status with presumption, condescension, and control. Flirtation may be a similar version of sexual psalic; low-key warranted suspicion of another, the parallel and normal version of OGO; and tolerant interest in one's students a lesser version of N than the mother invests in her infant. Forming an audience at a concern may be a milder version of intense crowd behaviour (AR). Acceptance of a cup of coffee from one's host is a less intense version of NR than is the begging of a dependent patient. Wishing to be alone to read a book is an easily reversible equivalent of SA.

With lower gains on one's psalic mechanisms, humour and pleasure in another's company can occur. Play represents "non-serious" versions of adult inter-individual activities, and continues as a neotenised activity in human adults. Play and the "hedonic mode" (Chance, 1984, and this volume) share many characteristics in common and may represent such low-gain, evanescent surges of psalic activity, albeit highly controlled by subprograms to avoid injury. These contrast strikingly with the tense attention to the alpha animal that is characteristic of "agonic mode" (Chapter 1), and they may be the key to how we benefit from our larger cerebral hemispheres.

SUMMARY

This chapter aimed to present:

1. A rationale for psychiatric clinical science as a branch of the biology of behaviour on both whole-organism and cellular–molecular levels.
2. A focus on intraspecific communication as a method of making cross-species comparisons of functional behaviour.
3. An emphasis on social rank hierarchies and in- and outgroup behaviours as powerful determinants of human as well as non-human animal interactions.
4. A method of dissecting basic from species-specific levels of communication by exploring homology as a relative concept and by using some psychiatric syndromes as indicators of deeply homologous "master programs" for organising behaviour.
5. A concept to concretise and exemplify these points. Called "psalic", this concept was provided rationale and etymology. Eight examples of such primitive communicational states were described and preliminary attempts of using them to typify patients were presented.
6. A brief speculation in which "ordinary" behaviour is characterised as modified outcomes of "master" psalic programs by species-specific "subprograms" in which the overriding nature of the psalics gets reduced, as in humour and play.

ACKNOWLEDGEMENTS

I owe many thanks to the thoughtful considerations of Carl and Joan Gustavson as they heard with great patience many of these ideas in their nascent stages; also a seminar at Thomas Jefferson University during my sabbatical there was very exciting, with Gail Zivin, Herbert Adler and Rachel Schindler particularly helpful; Karl Christian and Scott Ferguson were participants in a recent and very useful seminar sessions on psalics. Finally, I appreciate the recent conversations and/or letter exchanges with J. Price, L. Sloman, P. Gilbert, J. Feierman, A. Wassef, R. Good, E. Barratt, M. Amadeo and M. R. A. Chance. The ideas expressed here are my own, however, and do not necessarily reflect the final conclusions or endorsements of these others.

REFERENCE NOTES

1. Carpenter, W. (1985, November). *Current research on schizophrenia*. Presentation at Department of Psychiatry Research Conference, University of Texas Medical Branch, Galveston, Texas.
2. Gardner, R., Gustavson, J. C., & Gustavson, C. R. (1985). *Alpha behavior in manics as a model communicational state*. Paper presented at: The ethology of psychiatric populations, a joint meeting of the Animal Behavior Society and the International Society of Human Ethology, Raleigh, N. Carolina.

REFERENCES

American Psychiatric Association. (1980). *Diagnostic and statistical manual* (3rd edition). Washington, D.C.: APA Press.

Baker, M. C., Spitler-Nabors, K. J., & Bradley, D. C. (1981). Early experience determines song dialect responsiveness of female sparrows. *Science*, *214*, 819–821.

Bateson, G. P. (1979). *Mind and nature: A necessary unit*. New York: E. P. Dutton.

Bernstein, I. S. & Gordon, T. P. (1974). The function of aggression in primate societies. *American Scientist*, *62*, 304–309.

Bonner, J. T. (1980). *The evolution of culture in animals*. Princeton, N.J.: Princeton University Press.

Bottjer, S. W., Miesner, E. A., & Arnold, A. P. (1984). Forebrain lesions disrupt development but not maintenance of song in passerine birds. *Science*, *224*, 901–903.

Bowlby, J. (1969). *Attachment and loss, vol. 1: Attachment*. New York: Basic Books Inc.

Bruyer, R., Laterr, C., Seron, X., Feyereisen, P., Strypstein, E., Pierrard, E., & Rectem, D. (1983). A case of prosopagnosia with some preserved covert remembrance of familiar faces. *Brain and Cognition 2*, 257–284.

Campbell, C. B. G. (1976). Morphological homology and the nervous system. In R. B. Masterson, W. Hodos, & H. Jerison (Eds.), *Evolution, brain, and behavior: Persistant problems*. Hillsdale, N.J.: Lawrence Erlbaum Associates Inc.

Cazden, C. B. & Brown, R. B. (1975). The early development of the mother tongue. In E. H. & E. Lenneberg (Eds.), *Foundations of language development: A multidisciplinary approach (vol. 1)* (Pp. 299–309). New York: Academic Press Inc.

Chance, M. R. A. (1975). Social cohesion and the structure of attention. In R. Fox (Ed.), *Biosocial anthropology*. New York: Malaby Press.

Chance, M. R. A. (1984). Biological systems synthesis of mentality and the nature of the two modes of mental operation: hedonic and agonic. *Man–Environment Systems 14*, 143–157.

Colgan, P. (1983). *Comparative social recognition*. New York: John Wiley & Sons Ltd.

Darwin, C. (1873). *The expression of the emotions in man and animals*. London: John Murray.

Dawkins, R. (1976). *The selfish gene*. Oxford University Press.

Dorland's illustrated medical dictionary, 23rd edition. (1957). p. 1120.

Eibl-Eibesfeldt, I. (1970). In E. Klinghammer (Translator), *Ethology: The biology of behaviour*. New York: Holt, Rinehart & Winston.

Eibl-Eibesfeldt, I. (1972). Similarities and differences between cultures in expressive movements. In R. A. Hinde (Ed.), *Non-verbal communication* (Pp. 297–312). Cambridge, England: Cambridge University Press.

Eisenberg, J. F. (1981). *The mammalian radiations: An analysis of trends in evolution, adaptation and behaviour*. Chicago: University of Chicago Press.

Ekman, P., Friesen, W. V., & Ellsworth, P. (1972). *Emotion in the human face: Guidelines for research and an integration of findings*. Elmsford, N.Y.: Pergamon Press.

Engel, G. L. (1977). The need for a new medical model. A challenge to biomedicine. *Science, 196*, 129–136.

Feldman, R. S. & Quenzer, L. F. (1984). *Fundamentals of neuropsychopharmacology*. Sunderland, Mass.: Sinauer Associates, Inc.

Feldman, D., Tokes, L., Stathis, P. A., Miller, S. C., Kurz, W., & Harvey, D. (1984). Identification of 17 beta-estradiol as the estrogenic substance in *Sacchamyces cerevisiae. Proceedings of the National Academy of Science, U.S.A., 81*, 4722–4726.

Field, T. M., Woodson, R., Greenberg, R., & Cohen, D. (1982). Discrimination and imitation of facial expression by neonates. *Science, 218*, 179–181.

Fleck, L. (1979). In T. J. Trenn & R. K. Merton (Eds.), *Genesis and development of a scientific fact*. Chicago: University of Chicago Press.

Freud, S. (1917 [1915]). Mourning and Melancholia. In J. Strachey (Translator), *The standard edition of the complete psychological works of Sigmund Freud* 57 (Pp. 243–258). London: The Hogarth Press.

Gardner, R. (1982). Mechanisms in manic-depressive disorder: an evolutionary model. *Archives of General Psychiatry, 39*, 1436–1441.

Gilbert, P. (1984). *Depression: From psychology to brain state*. London: Lawrence Erlbaum Associates Ltd.

Green, S. & Marler, P. (1979). The analysis of animal communication. In P. Marler & J. G. Vandenberg (Eds.), *Handbook of behavioral neurobiology, vol. 3: Social behaviour and communication* (Pp. 73–158). New York: Plenum Press.

Hamilton, W. D. (1964). The genetical evolution of social behavior, I and II. *Journal of Theoretical Biology* 7, 1–52.

Harlow, H. F. & Harlow, M. K. (1962). The effect of rearing conditions on behaviour. *Bulletin of the Meninger Clinic, 26*, 213–224.

Jackson, J. H. (1884). Evolution and dissolution of the nervous system: Croonian lectures delivered at the Royal College of Physicians. In J. Taylor (Ed.), *Selected writings of John Hughlings Jackson* (vol. 2, 1958). (Pp. 45–75). London: Staple Press.

Jacob, F. (1982). *The possible and the actual (Jessie and John Danz Lectures)*. New York: Pantheon Books.

Janowsky, D. S., el-Yousef, M. K., & Davis, J. M. (1974). Interpersonal maneuvers of manic patients. *American Journal of Psychiatry, 131*, 250–255.

Jones, H. S. (1940). *A Greek–English lexicon compiled by Henry George Liddell* (9th edition). London: Oxford University Press (Pp. 2017–2018).

Kennedy, S., Thompson, R., Stancer, H. C., Roy, A., & Persad, E. (1983). Life events precipitating mania. *British Journal of Psychiatry 142*, 398–403.

Ledley, F. D. (1983). Review of E. Mayr, *The growth of biological thought: Diversity evolution and inheritance*, (Harvard University Press, 1982) *New England Journal of Medicine, 308*, 1174–1175.

Lukoff, D., Snyder, K. Ventura, J., & Nuechterlein, K. H. (1984). Life events, familial stress, and coping in the developmental course of schizophrenia. *Schizophrenia Bulletin, 10*, 258–292.

Marler, P. & Tamura, M. (1964). Culturally transmitted patterns of vocal behavior in sparrows. *Science, 146*: 1483–1464.

Mason, W. A. & Kenney, M. D. (1974). Redirection of filial attachments in Rhesus monkeys: dogs as mother surrogates. *Science, 183*, 1209–1211.

Mayr, E. (1982). *The growth of biological thought: Diversity, evolution and inheritance*. Cambridge, Mass.: Belknap Press of Harvard University.

McGuire, M. T. & Essock-Vitale, S. M. (1981). Psychiatric disorders in the context of evolutionary biology: a functional classification of behavior. *Journal of Nervous and Mental Disease, 169*, 672–686.

McGuire, M. T., Essock-Vitale, S. M., & Polsky, R. H. (1981). Psychiatric disorders in the context of evolutionary biology: an ethological model of behavioral changes associated with psychiatric disorders. *Journal of Nervous and Mental Disease, 169*, 687–704.

McLean, P. (1973). A triune concept of the brain and behavior. In T. J. Boag & D. Campbell (Eds.), *A Triune concept of the brain and behavior: The Clarence M. Hincks Memorial Lectures, 1969* (Pp. 4–66). Toronto: University of Toronto Press.

Meltzoff, A. N. & Moore, M. K. (1977). Imitation of facial and manual gestures by human neonates. *Science, 198*, 75–78.

Mora, G. (1980). Historical and theoretical trends in psychiatry. In H. I. Kaplan, A. M. Freedman, & B. J. Sadock (Eds.), *Comprehensive textbook of psychiatry III* (3rd edition). Baltimore: Williams & Wilkins.

Murphy, J. M. (1985). Cross-cultural psychiatry. In R. Michels, J. O. Cavenar, H. K. H. Brodie, A. M. Cooper, S. B. Guze, L. L. Judd, G. L. Klerman, & A. J. Solnit (Eds.), *Psychiatry* (vol. 3, Pp. 1–15). Philadelphia: J. B. Lippincott Co.

Nesse, R. (1984). An evolutionary perspective on psychiatry. *Comprehensive Psychiatry, 25* 575–580.

Nuechterlein, K. H. & Dawson, M. E. (1984). A heuristic vulnerability/stress model of schizophrenic episodes. *Schizophrenia Bulletin, 10*, 300–312.

Ospovat, D. (1981). Analogy/homology. In W. F. Bynum, E. J. Browne, & R. Porter (Eds.), *Dictionary of the history of science*. Princeton: Princeton University Press.

Parent, A., Dube, L., Braford, M. R., & Northcutt, R. G. (1978). The organization of monoamine containing neurons in the brain of the sunfish (*Lepomis gibbosus*) as revealed by fluoresence microscopy. *Journal of Comparative Neurology 182*, 495–516.

Platt, J. R. (1964). Strong inference. *Science, 146*, 347–353.

Price, J. S. (1972). Genetic and phylogenetic aspects of mood variation. *International Journal of Mental Health, 1*, 124–144.

Price, J. S. & Sloman, L. (1984). The evolutionary model of psychiatric disorder. Letter to the editor. *Archives of General Psychiatry, 41*, 222.

Queller, D. C. (1985). Kinship, reciprocity and synergism in the evolution of social behavior. *Nature, 318*, 366–367.

Reynolds, P. C. (1981). On the evolution of human behaviour: the argument from animals to man. Berkeley: University of California Press.

Sarnat, H. B. & Netsky, M. G. (1981). Evolution of the nervous system (2nd edition). New York: Oxford University Press.

Seidman, L. J. (1983). Schizophrenia and brain dysfunction: an integration of recent neurodiagnostic findings. *Psychological Bulletin, 94*, 195–238.

Siever, L. J. & Davis, K. L. (1985). Overview: toward a dysregulation hypothesis of depression. *American Journal of Psychiatry, 142*, 1017–1031.

Spitz, R. (1946). Anaclitic depression. An enquiry into the genesis of psychiatric conditions in early childhood. *The Psychoanalytic Study of the Child, 2*, 313–342.

Spitzer, R. L., Endicott, J., & Robins, E. (1978). Research diagnostic criteria: rationale and reliability. *Archives of General Psychiatry, 35*, 773–782.

Steele, C. M. (1986). What happens when you drink too much? *Psychology Today, 20*, 48–52.

Thiessen, D. & Yahr, P. (1977). *The gerbil in behavioral investigations: Mechanisms of territoriality and olfactory communication*. Austin, Tex.: University of Texas Press.

Waser, P. M. & Wiley, R. H. (1979). Mechanisms and evolution of spacing in animals. In P. Marler & J. G. Vandenbergh (Eds.), *Handbook of behavioral neurobiology, vol. 3: Social behavior and communication* (Pp. 158–223). New York: Plenum Press.

Webb, L. J., DiClemente, C. C., Johnstone, E. E., Sanders, J. L., & Perley, R. A. (1981). *The DSM-III training guide*. New York: Brunner/Mazel.

Weiner H. (1978). The illusion of simplicity: the medical model revisited. *American Journal of Psychiatry, 135*, 27–33.

Wenegrat, B. (1984). *Sociobiology and mental disorder: a new view*. Menlo Park, Calif.: Addison Wesley Publishing Co.

Williams, H. & Nottebohm, F. (1985). Auditory responses in avian vocal motor neurons: a motor theory for song perception in birds. *Science, 229*, 279–282.

Wilson, E. O. (1975). *Sociobiology: The new synthesis*. Cambridge Mass.: Harvard University Press.

9 The Therapeutic Use of an Ethogram in a Drug Addiction Unit

Peter Scott Lewis
Broadmoor Hospital, Berkshire, U.K

Michael Chance (1976; and see Introduction) has written about the characteristics of rank in the social system of a group of monkeys, and it was my first encounter with him and his video tapes that prompted me to compare the phenomena that he was describing with human groups in captivity in a prison setting.

Here, I decided that in order to practice proper group psychotherapy, there were certain preconditions that had to be laid down and, as these pre-conditions did not exist, I would have to manipulate the social group to make them more favourable. What I did not know until later was that I was unconsciously aware of the structure of the unit (Schottstaedt, 1963), and that this structure, which I had tampered with and reformed, could be understood in ethological terms, hence "Ethogram" in the title of this chapter.

I was introduced to the drug addiction unit in a prison as a visiting psychotherapist. I found the prevailing ethos very puzzling but this may be explained in part by the fact that there had been no psychotherapist available to the unit for some months. This had allowed certain social structures to form that were interfering with therapy.

On reflection I now begin to understand the implications of my arrival on the unit and the hostile reaction I received when I first met its inmates. My visit allowed me to assess them, and to a certain extent, for them to assess my acceptability. Whilst it seemed I was, in some way, acceptable to them, I also experienced a good deal of hostility about my intrusion, which threatened to upset the unit's social order (Buirski, 1973; 1975; 1980).

It was apparent that there was one dominant member supported by two major sub-leaders. This dominant was aggressive both in physical and verbal terms, with a long history of crime and a long history of drug abuse. All other members of the group were submissive to her. This submissiveness (i.e., deferring to her in matters of opinion) was not only in terms of their general behaviour, but in sexual terms also. The two main sub-leaders were facultative homosexuals in the prison setting but had longstanding heterosexual relationships outside the prison and had in fact made a living from prostitution. Below these two in hierarchy were relatively passive people who would never speak out of turn and who could not, it seemed, speak for themselves.

Maintenance of the dominance and power of the hierarchy had very much to do with listening to the dominant members and not communicating with them except in ways that maintained the hierarchy. Below these experienced supporters in the hierarchy were a number of young offenders who were in prison for the first time and were at their most impressionable. I have named these "newcomers" or potential supporters (Fig. 9.1)

During initial observations it was discovered that there were interesting alternative hierarchies that related to the different members of the group and may have been additional factors in determining the hierarchical arrangements of the individual members. These various divisions of drug takers had some effect in determining rank, but one also had to take into account personality factors.

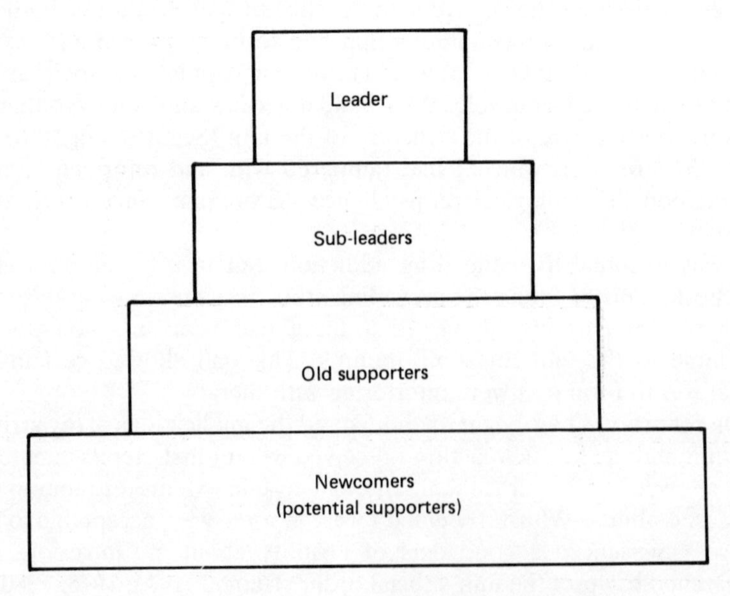

FIG. 9.1. Social structure of inmates in a drug unit.

It seemed that the arbitrary division into registered or non-registered addicts afforded the non-registered to appear more dominant and higher in rank because of their proven capacity to survive without taking the "soft option" of going to a clinic and having a small supply of drugs on a regular basis. It also implied that people who were non-registered had the opportunity of taking more drugs by virtue of using their wits. Those people who took the harder types of drugs tended to be regarded as higher in the social order than those who took the so-called soft drugs. The taking of natural derivatives rather than the synthetic opiates also accorded higher status. People who took soft drugs were deemed to be lower in the social order.

The mode of administration of the drugs had an influence on hierarchical status, in that "main lining" (intravenous injection) seemed to be superior to "skin popping", i.e. a practice of injecting drugs into the skin. "Sniffing" or "snorting", the nasel inhalation of drugs, was deemed to be higher in hierarchical order to "dropping", the oral ingestion of drugs, which was lowest in hierarchy.

Another phenomenon that needed understanding was the position of each individual in the hierarchy of the drug scene outside the prison (Fig. 9.2). Street addicts were at the bottom of the hierarchy, whilst the

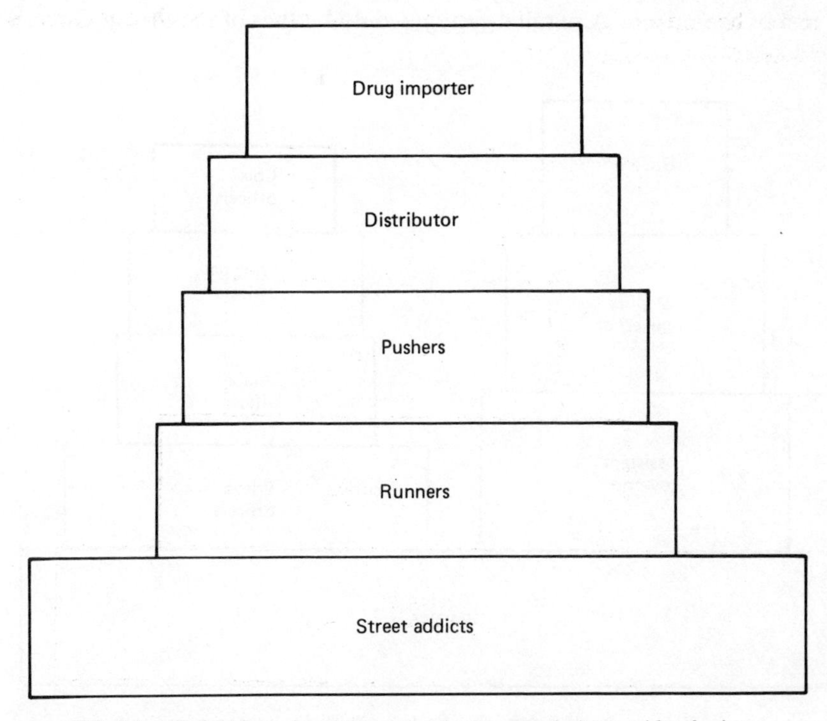

FIG. 9.2. Social hierarchy of drug users in the community outside of prison.

so-called runners, i.e., people who work for the people who are "pushing" the drugs, were in the next layer. People might ascend the ladder from being street addicts to runners and pushers but it was more likely that the pushers might descend the ladder to end up as street addicts. It is unlikely, however, that the importer and the distributor would move from their particular perch unless toppled by the law. It seems that the people who had been pushers and/or runners outside the prison had, whilst in prison, an unusual relationship with street addicts. They were envied because of the previous availability to them of drugs, but they were also deprecated because they had not made drugs available to the street addicts in the past. If they had made drugs available to persons outside and not provided them with useless substitutes, this conferred a degree of popularity and social dominance.

The idea that drug addicts work reasonably together in the unit was based on the fact that they do appear to have a degree of social cohesion and social order of their own outside of prison.

There are a number of groups of people that the inmates come into contact with both in the unit and in the prison generally. The persons who have primary contact with the inmates are, of course, the prison officers, who have their own hierarchy within the unit and a further hierarchy in the rest of the prison. A parallel system would be that of the Prison Governor

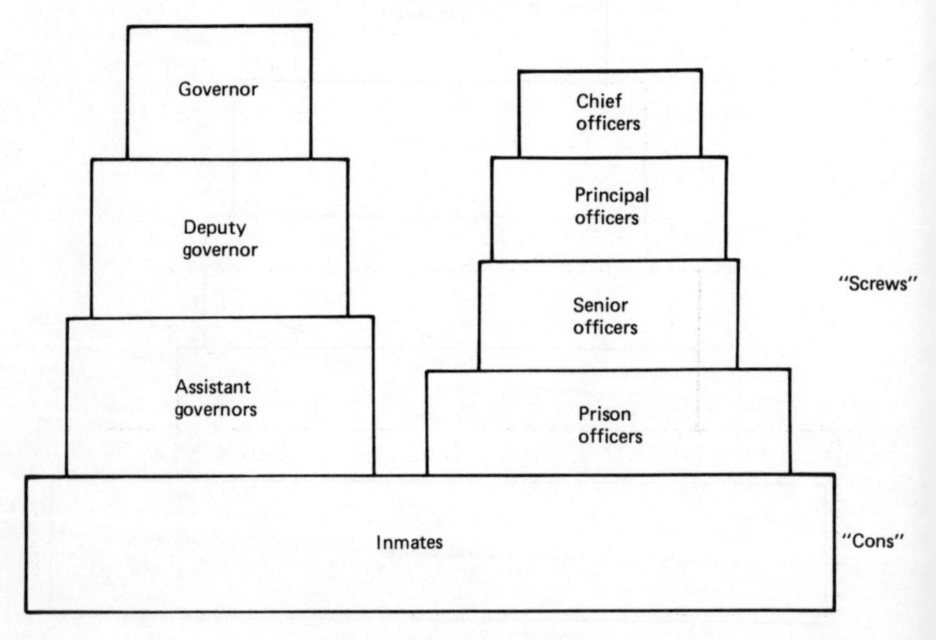

FIG. 9.3. Prison hierarchies.

grades. (Fig. 9.3). The nurses who visited the unit would also have some relationship with a nursing hierarchy, as would the unit's medical officer with his own medical hierarchy. The social and welfare department too would have its own individual representative who would relate to inmates as well as to that department. Similarly, the education department and the psychology department would have a particular person in contact with the unit.

There is thus a very complicated series of relationships within the prison, which has, as well as intra-professional hierarchies, a tendency to implicit social hierarchy. It is virtually impossible to draw all the interrelationships

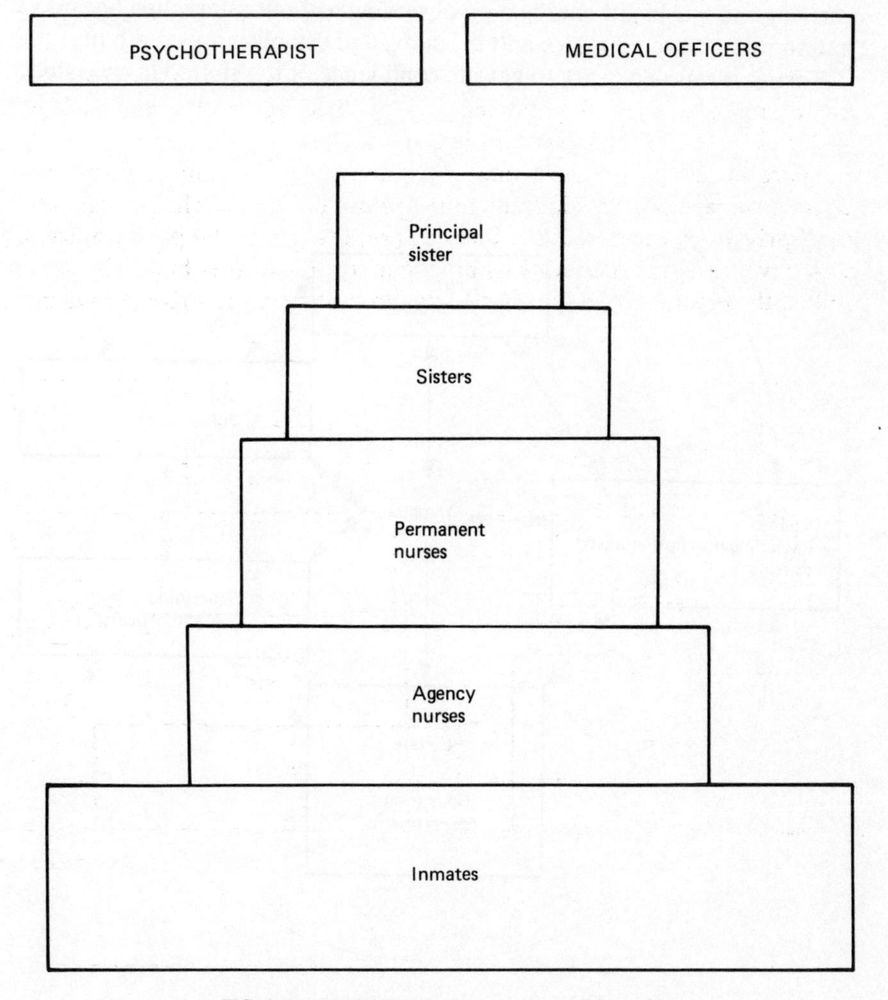

FIG. 9.4. Medical services—consumer hierarchy.

that may exist in an institution; a diagrammatic representation of the way in which people in different categories impinged on the inmates is represented in Fig. 9.5. Although an observer may become preoccupied with the impingement of personnel upon the inmates, it is important to remember that at each point of contact of each relationship there is an opportunity for disturbance of a relationship, i.e. the possibility of tension. When one examines in detail the hierarchical relationship structure of institutions it is difficult to understand how institutions can possibly run smoothly.

After observing these complicated interrelationships, which posed many problems, I was doubtful whether I could effect any change. This new culture was a way of functioning that I did not entirely understand, yet it was not one in which I could go on observing without interacting because I had to relate to the inmates and the staff, and establish my credibility.

Firstly, it was necessary to get the confidence of the staff. This was done

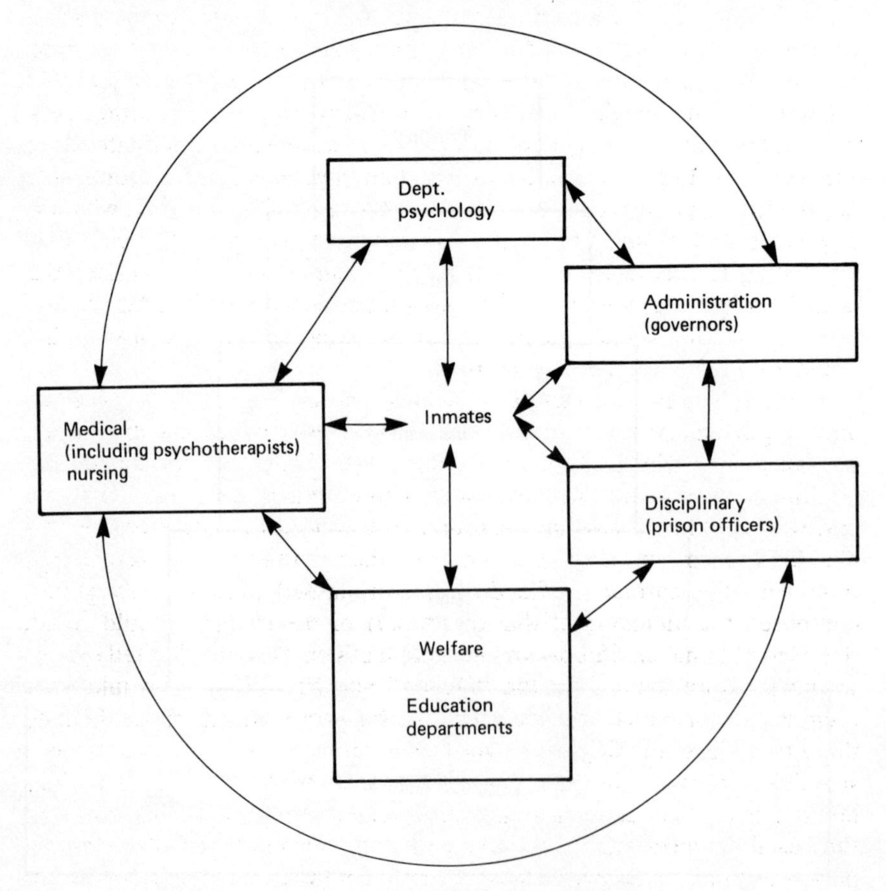

FIG. 9.5. Group interaction in prison.

by: (1), making myself available to them for general discussions about difficulties that arose; (2), responding to some of their needs and difficulties both at a personal and professional level; (3), offering some explanation of some of the problems; and (4), giving them support. I would sympathise with them and encourage them to do different things until they were able to accept the suggestions I made. When they tried to put them into operation and found them successful, they became encouraged.

Later the decision was made to undermine the credibility of the maintainers of the inmate culture that kept life as comfortable as possible for the inmates and that was not dealing with their difficulties with drugs or with their repertoire of social responses. It became clear to me that there were a number of people who had been badly selected for the unit and who did not contribute to its aim, which was to help people in their attempt to give up drugs when they left prison.

At the apex of the inmate culture was the person, already described, who presented herself as a leader to whom everyone behaved deferentially and to whom a number of the members submitted themselves sexually. Interestingly she regularly had her hair brushed and combed by others. All this occurred in a unit reputed to be egalitarian, in which all inmates were expected to get up for breakfast by a certain time and all should contribute to preparing it; but where she was, to say the least, the "Queen", who was waited on and brought breakfast, albeit unobtrusively, in her cell. It is interesting in this context to reflect that Chance (see Introduction) has talked about the phenomenon in a monkey troop of agonistic testing and re-affirming the social system in the morning. It seems that the phenomenon I have described is similar.

Below this dominant person, as I have outlined, were the sub-leaders and supporters. At the bottom of the triangle were the chronic drug users; passive people who go to maintain the ethos. There was, of course, the additional lower layer, the newcomers, who were in prison for the first time and who could so easily have succumbed to that prevailing ethos. It was decided to form two subgroups, knowing that I would have a pretty tough time with the hardened offenders. It was hoped that the new group, composed the members of the lowest part of the triangle, would attach themselves to me and follow an alternative ethos. This, to a limited extent, would separate them from the "Queen" and her retinue. I would treat them as new patients, educating them in the group analytic mode, helping them to understand the way in which they transferred their feelings about significant persons in their past to persons in the present and their behaviour to them as such (e.g. transference), and also the way in which they used defences. As I would have them in a group for a relatively long period of time, I would attach them to me and would be able, to a certain extent, to erode the hierarchy that previously prevailed.

I was encouraged by the fact that the other group, comprising the leader, the sub-leaders and the old supporters would, because their sentences were expiring, be leaving the prison. With this encouragement, and having found this divide-and-rule method to be advantageous, I decided to begin to erode the hierarchy that was maintaining the culture. As it was inappropriate to remove the group leader, it was more important to replace her by my presence as a dominant person; i.e., in Chance's terms (see Introduction), to provide them with an alternative social focus of attention and so bring about a split in the cohesion of the group, as well as in the ethos.

I began by questioning the tenets on which the ethos was based, eroding the credibility of the dominant leader and encouraging the more passive members to begin to speak up and assert themselves by offering them support and protection. In this was I also acted as a positive social referent; see Chance, Emory and Payne (1977). This group, however, still contained people who had not fulfilled my expectations of therapeutic work, and I therefore decided to question their non-adherence to the aims of the unit and get them to accept that they were not matching up to these obligations. I was then in a strong position to manipulate the administrative hierarchy of the prison so that I could get those who were not adhering to the aims of the unit in my terms transferred to another institution, thus depriving the dominant members of their passive, supportive, relatively non-vocal members.

By means of personal selection of new patients for the unit, I replaced the submissive members with newcomers who were more appropriately endowed with an acceptance of the new ethos. We questioned and examined the taking of drugs and explored the possibility of managing their lives without them. By this selection process I therefore supplanted the support provided in the hierarchy with questioning, more comfrontational, less passive persons. These would now support my ethos rather than that of the original leader, thus further eroding her position and that of her minions.

Another fortuitous experience that helped me in achieving my objectives was that one of the members of the newcomers group had formed an alliance with a member of the more hardened group. This girl, who had taken drugs over many years, had remained repetitively hostile to me over some weeks. This hostile reaction was out of all proportion to her character and to the way in which she related to prison officers. I recognised this difference to be due to the phenomenon of transference. I suspected that she was treating me in a way that derived from the way in which she dealt with another person in her past. The way in which she dealt with me was seen by her friend (who was one of the newcomers in the other group, and with whom she had a close alliance), as different from the way in which her friend perceived me. Clearly, over a period of time, and not necessarily only in the groups but in the on-going interaction of the unit, she became increasingly aware that she was seeing me in a distorted way. Through her alliance with her friend, she was able to learn from her ally (who was, in a way, my ally) that she was transferring on to me hostile feelings that she had had towards her step-father.

I was very touched when finally "the penny dropped" and she was able to acknowledge this and see me in a different light. One morning she presented me silently with a flower, which was really saying, in the parlence of that time, "Love and Peace". A few days later she was able to explain the difficulty that she had had. She thereafter became a valuable therapeutic ally in the difficult group and by virtue of that change, moved ranks and worked very hard in giving up her long-standing drug addiction. She subsequently married and had another baby, and then took back the baby she had had some years before and built a new life for herself. She and others helped to maintain a therapeutic ethos and, in that way, also helped to maintain my position as a dominant leader dictating that ethos, which assisted in enhancing the capacity of the inmates to deal with their addiction.

REFERENCES

Buirski, D. (1973). A field study of emotions, dominance and social behaviour in a group of baboons (*Papio anubis*). *Primates*, *14*, 67–78.

Buirski, D. (1975). Some contributions of ethology in group therapy dominance and hierarchies. *International Journal of Group Psychotherapy*, *25*, 227–235.

Buirski, D. (1980). Towards a theory of adaptation of analytic group psychotherapy. *International Journal Group Psychotherapy*, *30*, 447–459.

Chance, M. R. A. (1976). Social attention: Society and mentality. In M. R. A. Chance & R. R. Larsen (Eds.), *The social structure of attention*. New York: John Wiley & Sons Ltd.

Chance, M. R. A., Emory, G. R. & Payne, R. G. (1977). Status referents in long-tailed macaques (*Macaca fascicularis*): Precursors and effects of a female rebellion. *Primates*, *18*, 611–632.

Schottstaedt, W. W. (1963). Social interaction on metabolic ward—the relation of problems of status to chemical balance. *Journal of Psychosomatic Research*, 7, 83–95.

10 Social Interactions of Young Children with Peers and their Modifications in Relation to Environmental Factors

H. Montagner
Laboratoire de Psychophysiologie, Faculté des Sciences, Université de Franche-Comté, Besançon, France
Unité 70, Institut National de la Santé et de la Recherche Médicale, Montepellier, France

A. Restoin, D. Rodriguez, V. Ullman, M. Viala, D. Laurent
Laboratoire de Psychophysiologie, Faculté des Sciences, Université de Franche-Comté, Besançon, France

D. Godard
Unité 70, Institut National de la Santé et de la Recherche Médicale, Montpellier, France

Blurton Jones (1967; 1971; 1972) and McGrew (1969; 1972) were probably the first to study systematically the behaviour of young children and their peers in British nursery schools. Their work is a basic reference for biologists and psychologists who study the behaviour and interaction of young children (see the example of "rough-and-tumble play" described by Blurton Jones, 1972). From 1970 our laboratory in Besançon started doing research work in child day-care centres (children under three years of age) and kindergartens (children from two to six years of age) to try and list those behaviour sequences of young children in their peer groups that remained stable over a period of weeks and months (Montagner, 1974; 1978; 1979; 1983; Montagner & Henry, 1975; Montagner, Restoin & Henry, 1982; Montagner et al., 1978; 1979; 1981). The groups only varied in size at certain times of the year with the arrival or departure of a child or a small number of children for social or medical reasons. After the necessary descriptive stage at the beginning, a resolutely functional and ontogenetic approach was adopted:

1. Possible correlations were sought, in children aged from 18 to 36 months, between the behaviour of a child as an emitter and the behaviour of one or more receivers during activities that were created by the children themselves or by their nurses. In this way it was possible to identify

progressively the most probable consequences or functions of a type of behaviour according to age, sex, context, etc.

2. The evolution of the exchanges and activities was studied for each group in relation to the structure, duration and frequency of the interactions throughout the year.

3. The interactions between children of less than 18 months of age were observed according to whether they were lying down, sitting, squatting, kneeling, or standing.

It was thus possible to follow children from the age of 12 weeks, i.e., as from the time (in France) when the mother must return to work after her maternity leave. In this way we are able to see at what age, how, and under what influences, the communication behaviour emerged and changed, while at the same time observing the temporary and lasting groups that were formed.

POPULATIONS AND METHODS

Since 1970, studies have been made in 2 local government, child day-care centres receiving children aged from 12 weeks to 36 months and, according to the protocol and years, in 2 to 15 local government kindergarten classes receiving children from 2 to 6 years of age. These studies have included more than 1200 children (it is difficult to give the exact number because all the information concerning all the children could not always be used: in fact the number of children studied is far higher than 1200). Some research was also done with children from 6 to 9 years of age in primary schools.

All types of socio-professional groups were represented in the centres and in the kindergartens. At a day-care centre the entrance conditions are that the child must be less than three years of age and that both parents must be at work (the parents have to pay a fee that is proportional to their annual salaries). At the kindergarten all children over two years of age can attend free of charge.

From the outset, the research workers observed and filmed the children in their usual peer group. The filming was done with Super 8, 16mm or video cameras according to the protocol, availability of cameras, and the financial means of the research team. The children were usually observed on their arrival at the centre or kindergarten in the morning between 7.30 and 9 a.m., then systematically followed from 9 to 10 a.m., and then often from 10 to 11 a.m. In this way three of four workers could observe and film from one to three times per week each of the children who evolve, during the whole day, in the same group of 10 to 15 children of the same age. These observations were made from September to June and were only interrupted by week-ends and holidays (in some cases the children were also observed during these periods). The children were thus observed and filmed in four different situations:

1. Free activities in the playroom or yard when they are free to use their own initiative.
2. Situations imposed or suggested by the staff of the centre or kindergarten.
3. Experimental situations suggested by the research workers and created by the staff. The creation of new situations (one low table upside down on another; the introduction of large cardboard boxes inside which the children could go and huddle together; the introduction of normally attractive and sought after objects, etc.) enabled the workers to study and compare interactions and competitions in relation to the other, more common situations (Montagner, 1978).
4. Situations at table during meal time (Kontar, 1981; Laurent, 1983).

The behaviour recorded on film or video cassette was analysed for each child in relation to previous and subsequent situations and, at the same time, in relation to the behaviour of other children. This established concomitant and causal relationships between the behaviour of a child and that of a peer (and also, in many cases, that of the nurse in the centre and the teacher in the kindergarten). By comparing the child's behaviour in different situations experienced from one day to another, or at two- or three-day intervals, the relative frequency of the different types of behaviour and interactions with peers in the same group could be calculated. Even the probability of the child going from one behaviour to another could be estimated according to the sex and the behaviour characteristics of the child with whom they interacted, and the context. When the behaviour of a child is compared from one month to another, or every two or three months, the evolution of the behaviour can be followed in relation to age, time spent in the same peer group, events that occur in the home environment (birth, absence of mother or father, change in work rhythms of the mother or father, etc.), and events in the centre of kindergarten (arrival of a child, departure of a child, absence of a member of staff, etc.). Thus it was possible to define progressively the modifications of structure, the frequency and duration of behaviours, their combination and linking in relation to internal factors (growth, illness, emergence or modification of biological rhythms including the sleep-awake rhythm, etc., Montagner, 1983), and in relation to events experienced by the peer group and modifications of the home, centre, and kindergarten environments.

The investigators were able to define several categories of elementary behaviour by using as a basis the responses of the receiver and the context (see Table 10.1). For example, linking and appeasing behaviours are most likely to be followed by motor, vocal, and/or verbal responses bringing about an interaction with smiles, offering, light body-touching (fingers or palm of the hand on the cheek, chin, or head of the other), or non-rejecting behaviour (hand-in-hand), or reciprocal imitation. The result will be a strong probability of activity together, often in co-operation (construc-

TABLE 10.1
Categorisation of Elementary Behaviours

Appeasement and Linking Items	Agonistic Items: Threat	Agonistic Items: Aggression	Fearful and Withdrawal Items	Isolations
Smiling	Wide opening of the mouth	Striking a blow with arm, hand, or fist	Enlargement of the uncovered part of cornea	Sucking one's finger
Bending the head on the shoulder	Clenching the bared teeth	Kicking	Putting the arms in front of the face (protective gesture)	Sucking toy, cloth, etc.
Bending the bust on one side	High-pitched vocalisation	Striking a blow with an object	Withdrawal of the head or chest	Standing aside
Nodding one's head	Loud vocalisation "Ah"	Scratching	Moving back	Lying down
Waddling	Frowning	Pinching	Running away	Curling up
Swinging	Clenching fiists	Biting	Blinking	Sobbing or crying independently of any interaction with other children
Offering	Stretching out the arm with one hand opening downwards	Pulling the other child's hair	Crying after having received a threat or an aggression	
Stretching out the arm with one hand opening upwards	Throwing out the arm in front of oneself	Pulling the other child's clothes by force		
Light touching	Forefinger pointing to the other child	Knocking over		
Taking the other child by the hand	Sudden opening of both arms	Shaking the other child		
Stroking	Sudden bending of the head forward	Grasping an object in the other child's hand		
Kissing	Sudden bending of the chest forward			
Squatting down	All sudden movements forward			
Jumping	Breaking a blow			
Hopping				
Turning oneself round				
Clapping the hands				
Vocal exchanges with no threatening vocalisation				

NOTE Combined data from 18-month to 6-year-old children in free-play situations at the day-care centre (18- to 36-month-old children), and the kindergarten (2- to 6-year old children). A behavioural sequence is a combination of 2 or more items. The frequency of items or sequences is calculated for each child every one, two and three months or every year (see Table 10.3). This Table does not give: items rarely observed; attraction items (the child goes towards or imitates a peer); non-determined items, i.e., those that appeared to have different meanings or functions in a given context (putting arms ... another child's back; gently hitting another child's head, etc.); simulations (for example barking

tion with Lego building blocks, movement of objects that are relatively heavy for a child to move alone, etc.).

After these categories of behaviour had been distinguished, other groups of children were observed by other research workers in order to verify that the behaviours in each category do actually correlate, i.e., a given behaviour is even more likely when the other behaviours of the same category also appear either for a given child or for a peer group. For example, it was observed that the behaviour of grasping an object (taking an object without prior body or verbal solicitation) preceded, accompanied, or often followed an overt aggression (biting, pinching, scratching, hitting the body of the other, etc.) both in the individual child and in a group of three to five children. Grasping behaviour and overt aggressions had significant, positive correlation coefficients in all structures studied. For example, in 14 children of less than two years of age who were studied for three consecutive years at the day-care centre the correlation coefficient between these behaviours was 0.84, $P < 0.01$ (Ullmann, 1981). Again, in each of the kindergarten classes receiving children from three to

TABLE 10.2
Correlation Coefficients Between Behaviours

Behaviours / *Kindergarten Classes*	*Overt Aggressions vs. Grasping*	*Overt Aggressions vs. Offering*	*Overt Aggressions vs. Isolation*	*Grasping vs. Offering*	*Grasping vs. Isolation*	*Offering vs. Isolation*
A La Bouloie	0.90	−0.18	−0.24	−0.22	−0.26	−0.41
B Bourgogne–Planoise	0.80	0.22	−0.01	−0.06	−0.27	−0.04
C Rosemont I	0.53	0.25	−0.19	0.56	−0.29	−0.28
D Rosemont II	0.49	−0.08	−0.28	−0.15	−0.42	0.11
E Ile de France–Planoise	0.75	−0.08	−0.01	−0.03	−0.29	−0.11
F Champagne–Planoise	0.80	0.17	−0.31	0.25	−0.34	−0.21
G Les Sapins–St. Ferjeux	0.81	−0.16	0.18	−0.43	−0.13	−0.42
H Saone	0.84	0.09	−0.31	0.06	−0.30	0.03

NOTE Combined data from kindergarten classes receiving children from 3 to 4.5 years. The coefficients were calculated from data collected every Monday and Friday from September 1976 to June 1977.

four-and-a-half years of age, studied from September 1976 to June 1977, the correlation coefficients between these behaviours went from 0.49 (Rosemont II class) to 0.90 (La Bouloie class) $P < 0.01$ (Viala, 1980; Table 10.2). A relationship therefore appeared to exist between these behaviours. However, no significant correlation was found between behaviours belonging to different categories; for example, between grasping and offering, except in one case (Rosemont I class; see Results), or between overt aggressions and offering.

It was then possible to calculate for each child the relative frequencies of the different categories of behaviours shown in Table 10.1 from one month to another, every two or three months, or from one year to the next, in accordance with the particular cases and the child populations (Table 10.3). Then the frequency of the linking and appeasement behaviours could be divided by that of the agonistic behaviours (threats and aggressions) or of just the aggressive ones; the frequency of linking and appeasement behaviours by that of fearful behaviour; or the frequency of aggressions by that of fearful behaviour, etc. In this way, individual coefficients were obtained, and behavioural profiles for a given period (one, two or three months, a year) could be built up. The fluctuations of each coefficient could then be studied in relation to time. For example, in October 1975 some children had a significantly higher frequency of aggressions and combined threats and aggressions than of linking and appeasing behaviours. In January or February the difference was even higher; in May or June it increased again. However, at the same time but in other children, the coefficient obtained by dividing the frequency of linking and appeasement behaviours by that of aggressions was significantly higher in January than in October and even higher in April. These children became more and more appeasing.

The behaviour profiles can also be seen when factoral analysis of correspondences is used (Fig. 10.1). This gives a visualisation of the relative proximity of the axes that define different categories of behaviour according to whether they are emitted or received. The behaviours emitted and not received enable children to be differentiated one from another (Ullmann, 1981). Thus we could see at what age(s), how, and under what influences, each child is able to reinforce, modify, or enrich the range of their interacting behaviours in relation to age, sex, and events being experienced or experienced in the past, etc.

At a later stage the list and the categories of behaviour in Fig. 10.1 can be used as a reference grid for new research. For example, children and groups of children can be compared one with another in relation to cultural and socio-professional factors, ecological conditions (mountains in relation to plains; mountains in relation to sea-level; forests in relation to savannah; etc.), changes in environment (adoption, holiday camps), sensory defects or

TABLE 10.3
Relative Proportions of Behavioural Items

Name	*Date of Birth*	*Appeasement/ and Linking (percentage)*	*Agonistic (percentage)*	*Fearful/ and Withdrawal (percentage)*	*Participation in Competitions*	*Behaviour Profile (see Text)*
Jerome	09.11.75	28.8	7.6	63.4	–	Dominated–fearful
Karine C	30.09.75	48.7	18.4	32.7	–	Dominated with leader mechanisms
Frederic	18.07.75	51.9	34.6	13.4	+	Dominant–fluctuating
Nicolas	20.06.75	21.0	7.8	71.0	–	Dominated–fearful
Katia	27.04.75	44.6	17.8	37.5	–	Dominated with leader mechanisms
Sebastien	21.04.75	25.9	40.7	33.3	+	Dominant–aggressive
Elizabeth	07.04.75	36.0	26.4	37.3	+	Dominant–fluctuating
Damienne	10.03.75	70.4	11.4	18.0	+	Leader
Karine H	01.03.75	68.9	9.2	21.7	+	Leader
Stephanie	28.01.75	68.9	10.0	20.9	+	Leader
Kaouthar	03.07.75	21.0	34.0	45.0	–	Dominated–aggressive
Marianne	16.02.75	66.6	20.2	13.0	+	Leader
Gerald	20.12.74	31.2	54.1	14.5	+	Dominant–aggressive

NOTE Children were in the same peer group, aged from 2 to 3 years, at the day-care centre in 1977 and were observed in free activities throughout the year (from September to June). + high level of participation; – low level of participation; agonistic items = threat and aggression.

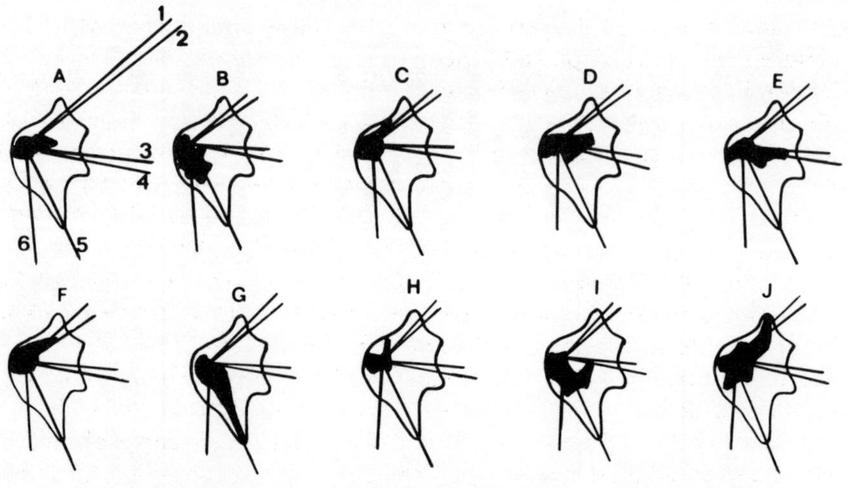

FIG. 10.1 Graphic representation of the multifactorial analysis of the different behaviour characteristics in each child (A to J) as related to the same characteristics in the total peer population at the day-care centre (n = 10). The large area is where all observations concerning the 10 children are recorded; the dark-shaded area is where all observations concerning each child are recorded. The axes 1 to 6 define the behaviour characteristics of Table 10.1, except that threatening behaviour and isolation are not represented, fear and withdrawal are analysed separately, and the "non-determined" interactions and "attraction" categories have been added. Non-determined interactions are those without any particular meaning or function; attraction is the number of times that the child goes towards or imitates another child. Ages are from 8 to 26 months (7 girls and 3 boys). Each individually shaded area goes towards the child's "privileged" characteristics. Thus, the most marked behavioural tendency or tendencies of each can be visualised. For example, children C and J were the most aggressive ("dominant–aggressive" profile), children D and E the most linking and appeasing, and G the most fearful ("dominated–fearful"). Note the proximity of the non-determined interactions and aggression axes, and the proximity of the linking/appeasement (appeas.) and attraction axes.

anomalies (partially blind or blind children, partially deaf or deaf children, etc.). Thus communication studies can have a true functional value.

RESULTS

The Emergence of Communication Systems Between a Child and its Peers

Observation of children in the day-care centre from the age of five or six months showed that, varying from child to child, the behaviours in Table

10.1 appear between the seventh and fourteenth month at least when the children can move about and meet each other without being seriously hindered by their motor activities. This list of behaviours has been observed in all the child populations studied in the two day-care centres, and in the fifteen classes in kindergartens where our research team have been working since 1970, and in two kindergartens in Brazzaville (Popular Republic of Congo) studied by H. Didillon and H. Gremillet. The migrant children (Algerian, Moroccan, Tunisian, Yugoslavian, Portuguese, Spanish and others from Africa) observed in the day-care centres and kindergartens in Besançon since 1970 have also shown the behaviours listed in Table 10.1 in the same contexts as children with a French culture and background, whatever the composition of the peer groups. Finally, most of the items in the list have been described by Blurton Jones (1972), McGrew (1972), and Lewis (1978) in English and American children. All the recent ethological studies would suggest that the behaviours in Table 10.1 actually do appear during the second half of the first year, at least when there are no pathological symptoms in the development of the child. Thus, they could be "universals" for the human species. This, however, does not mean that they are necessarily of genetic origin. There are three hypotheses:

1. A genetic origin.
2. A differentiation under the effect of the same type of social influences (aside from exceptional cases, all children from all cultures are influenced by the same configuration, "two eyes, one nose and one mouth", the same somaesthetic and proprioceptive influences when the child is in the mother's arms, the same intonations and vocalisations by people who are familiar to them, etc., even if there are obvious differences in morphology, colour of face, texture of skin, social habits, etc.).
3. A double influence: genetic and social.

We believe that at the present time there is no method of deciding between one or the other of these hypotheses and thereby knowing with certainty the influence of the genetic code in the behavioural manifestations of the human being. Even comparative study of the behaviour of children born blind, deaf, etc., with that of children born without any sensory defects cannot provide us with a conclusion in this field. Replacement mechanisms could exist between the different sensory channels and these could even be prepared by the conditions of fetal life. However, the most recent research only gives an indication and does not prove there are combinations and replacement of stimuli that could be received by the fetus.

Several conclusions and hypotheses can be drawn from the studies that have been done on the emergence of those behaviours listed in Table 10.1.

The first is that vocalisations play a particularly important role in communications between children of less than one year of age. Thus, D. Rodriquez (manuscript, in preparation) has shown that the acceptance of an offering by a child from six to twelve months is facilitated when a vocalisation accompanies the offering gesture (the frequency of the fundamental, the duration, the form of the vocalisation, and the number of harmonics are not characteristic). In actual fact the offering is accepted in 76% of the cases where vocalisation accompanies the gesture; there are only 34% of acceptances when the offering is silent. This is not the case between the age of two and three years, during which period the offering gesture brings about the same percentage of acceptances, whether the offering is accompanied or not by vocalisations. It is as if, during the second year, sound reinforcement of the offering becomes less and less necessary for the offering gesture to be recognised as a signal, with a minimum of ambiguity. This appears to be the case for many other signals; research is being done to try and confirm this point.

Everything would suggest that among the vocalisations emitted by children from seven to fourteen months there is only one that is characteristic—that which is combined with sudden opening of the mouth, raising the eyebrows, lifting and then lowering of one of the arms, and often with the sudden throwing forward of the chest (when the child is standing), and often which appears in conflictual and competitive situations between children from eighteen to thirty-six months (Montagner, 1978; Montagner et al., 1982). This vocalisation, which often brings about the dropping of an object, the turning of the head or chest, and/or running away, can be considered as having mainly a threatening function in the ethological meaning of this term. When this vocalisation is emitted there is a reduction in the probability of overt (biting, scratching, etc.) aggression resulting from both the emitter and the receiver. When this vocalisation is analysed on the sonograph there is a typical bell shape (Fig. 10.2), which cannot be confused with other sound forms. A child from seven to fourteen months is thus able to decode one of the signals that announces the possibility or the beginning of a conflict or a competition between the emitter and itself or another child, for example, when it turns its back towards the emitter. It is as if one of the priorities of the young child is to recognise and differentiate from the beginning of its life in a peer group, one of the signals that governs many conflicts and that enables it to participate without overt aggression in competitions, so letting it have access to a limited number of situations and objects. The parallel differentiation, during the first year, of smiles, offering gestures, and solicitation gestures (in lying down, sitting, squatting, or standing positions, according to age and situation, the child bends its head sideways, some times also its chest, and tries to catch the eye of the other person) enables the young child of less than one year of age to

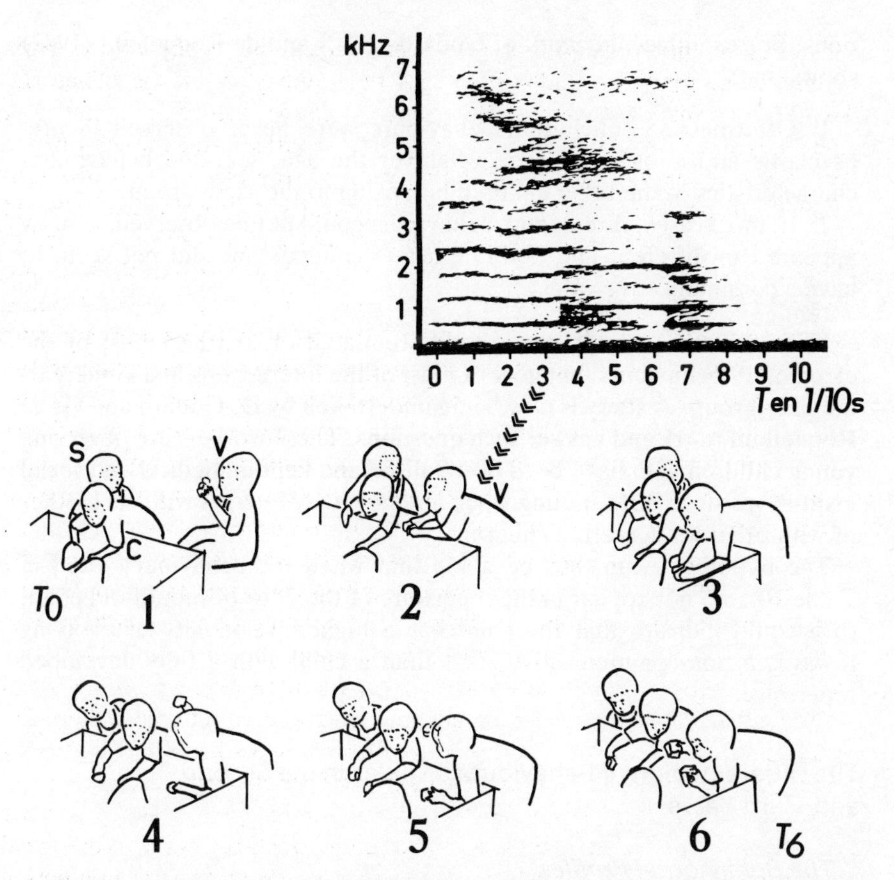

FIG. 10.2 Sonagram of a typical vocalisation, y : frequency in kHz ; x : time in one-tenth seconds. This vocalisation was emitted by the 11-month-old girl Virginie (V) when she reached the cardboard box where two other children were playing. The vocalisation is combined with sudden opening of the mouth, raising of the eyebrows (pictures 2 and 3), lifting (picture 4) and then lowering (picture 5) the right arm. There were 6 seconds between the first and the sixth picture. The 6 pictures are drawn from a film that was analysed frame by frame.
T_0 to T_6: time scale in seconds.

set up, reinforce, and complicate, communication systems. Consequently, reciprocal imitation, activities in common, and co-operative activities begin to outnumber the aggression, crying, and rejection, even in competitive situations.

A second conclusion is that although the behaviours listed in Table 10.1 were also observed in groups of peers at kindergarten and primary school, most of them would never be seen in children living in psychiatric institutions. For example, the work of Godard (1982) and de Roquefeuil (1982) shows that:

1. Offering and solicitation behaviours were never observed in pre-psychotic and autistic children, whatever the age, sex, and behavioural characteristics of the other children belonging to the same group.
2. In most cases, threatening behaviours could not be observed, or they appeared in different and non-interactive contexts and did not seem to have a communicative value.

It is as if these children never differentiated or had lost most of the elementary behaviours that govern most of the interactions in a kindergarten peer group. A study is now being undertaken by D. Godard and G. de Roquefeuil to try and answer such questions. These workers are observing young children who have been hospitalised and kept in medical and social institutions since birth, noting when and how they interact with each other or with other kindergarten children.

The hypothesis can thus be made that when the behaviours listed in Table 10.1 do not appear in the repertoire of the 7- to 14-month-old child, this would indicate that the child has a higher probability of evolving towards a non-communicative state than a child with a fully developed repertoire.

The Differentiation of Behavioural Profiles in the Second and Third Years

The Behavioural Profiles

Several behavioural profiles that differentiate during these years originate from the behaviours (Table 10.1) that have been set up during the second half of the first year and the beginning of the second year. There are seven of the more marked profiles, enumerated and explained as follows.

1. Leader Children. These participate much in competitions, such as for the legs of an upside-down table, cardboard boxes, and new attractive objects (see Populations and Methods). They impose themselves in these competitions and have a frequency of linking and appeasing behaviours much higher than that of agonistic behaviours. These children communicate by varied, non-aggressive and non-fearful, behavioural sequences. They do not confuse threat and aggression as they rarely mix these two behaviours in the same sequence, and rarely go from a threat to a marked and sustained aggression. In most cases they react positively to an offering (they accept it in more than 80% of cases) as well as to a soliciting behaviour (they give the object they are holding in more than 75% of cases). When they become really aggressive, this aggression is more gentle, less sustained, and not repeated. The frequency of threats is most often higher than the frequency of overt aggressions. These children are imitated and followed the most often, and during longer periods, than the others, with the exception of the dominated children with leader mechanisms (see category 6). These children are called leaders because of the large number of times in which they lead the others in free activities. It is as if more effective attention is paid by peers to their behaviour than to that of the other children, with the exception of the dominated children with leader mechanisms. There may be a relationship between the predominance of linking and appeasing behaviours in the behavioural repertoire of the leaders and the development of attention structures (Chance, 1967; Chance & Jolly, 1970) in these children and also in the peers of the same group. Consequently, the groups that form around the leaders (there can be one or several leaders in the same group) could be compared to the hedonic social groups that were described by M. R. A. Chance in non-human primates (Chance & Jolly, 1970; Chance, 1974).

2. Dominant–Aggressive Children. These participate much in competitions, impose themselves, and have a much higher frequency of aggressive and agonistic behaviours (according to whether the aggressions and threats are quantified separately or together) than linking and appeasing behaviours. The analysis made week after week and month after month showed that these children are differentiated from the others by their more frequent aggressive behaviour. They are imitated and followed less often and over shorter periods than the leaders. They can change to marked and repeated aggressions "without any apparent reason" during free activities, activities organised by an adult and, less often, during meals. They often disorganise activities and the interactions between other children. The combinations of their behavioural sequences are less varied and, to a significant degree, include more threats and aggressions than those of the leaders. They give more attention to conventional games, and appear to be less creative in activities and new games than the leaders. The groups that form around these children are more quickly disorganised than those formed around leaders. The disorganisation of a group is often accompanied by crying in children who leave it. It is noteworthy that these groups are usually composed of children with the same "dominant–aggressive" profile, of fluctuating children who participate a lot in competitions (category 3), and of fearful children who often receive the aggression of the others. It is as if the attention structures were very different in these groups as compared with the leader group, whether the children in these groups are dominant–aggressive or not. Consequently, the dominant–aggressive groups could be compared to the agonistic social groups described by M. R. A. Chance in non-human primates.

3. Fluctuating Children. These participate much in competitions, impose themselves, and vary greatly from one day to the next or one week to the next in the relative frequency of their different categories of behaviours. Some days or weeks they have a profile that is near to or non-dissociable from that of leaders. Other days or weeks their profile is that of "dominant–aggressive" children. Most often they fluctuate in their behaviour during the whole year.

4. Dominated, Fearful Children. These participate little or not at all in competitions, only rarely impose themselves, and have a higher frequency of crying and fearful behaviour than the other behavioural profiles, with the exception of some of the most isolated. They have a linking and appeasing behaviour that is equal to or higher than that of the leaders; a significantly higher frequency of crying and fearful behaviour, and a significantly lower frequency of threats and aggressions. Compared to the dominant–aggressive children they have a significantly higher frequency of linking and appeasing behaviours and, of crying and fearfulness, and a significantly lower frequency of threats and aggressions. Usually these children are the followers of children with the three previous profiles. When there is a conflict, these fearful children often receive the threats and aggressions from the others. They seek out the help and support of the staff. They take longer than those with the previous profiles to adapt to new structures (the difficulty of adaptation is measured by the duration and frequency of crying, and of isolation) when they have just changed structures.

5. Dominated, Aggressive Children. These participate little or not at all in competitions, only rarely impose themselves, and have alternating periods of crying, fear, aggression, and isolation. Their aggression can appear and develop without any apparent reason and without being aimed at the appropriation of an object. It is difficult to quantify the behaviour of these children from one day to the next because the duration and the frequency of their isolation can be so high that the other behaviours do not appear or, if they do, they are short and undeveloped. These children are amongst the least attractive, the least imitated, and the least frequent imitators. Their aggression is amongst the strongest. Thus, the example of these children clearly shows that thc beginning and reinforcement of aggression are not necessarily related to the situations of conflict and competition.

6. Dominated, Leader Children. These participate little or not at all in competitions, only rarely impose themselves and have, in every case, a significantly higher frequency of linking and appeasing behaviours than agonistic, crying, and fearful behaviour. Their combinations of behavioural sequences are the same as those of leaders. The frequency of their agonistic behaviour is always very low. They are imitated and followed by a small number of children, usually from two to four. Like the leaders they appear to create situations and new games more often than the dominant–aggressive children. Thus, it appears that the ability to lead peers and to create new events is not necessarily related to dominance factors. These children are typed "dominated with leader mechanisms".

7. Isolated Children. These isolate themselves often and for long periods. Their linking and appeasing behaviour is infrequent compared with the other dominated children. Their agonistic behaviour is infrequent or non-existent. Some of these children come out of their isolation more and more often during the third or fourth year. Others maintain their isolation and its duration increases. Thus this profile appears to be very heterogeneous. Other, longer studies are required to define more accurately the characteristics of these children.

In most of the cases these profiles appear to be reinforced and thus more obvious in the younger classes in the kindergarten. However, they can change considerably from one month to another and sometimes from one week to another. In all the cases where significant modifications in the behavioural profile were observed between the age of two and four we also noted, at the same time, significant changes in the behaviour of the parents, especially in the mother in relation to the child [the parents' behaviour in relation to the child could be studied when they came to the day-care centre and, in the late afternoon, when they welcomed and dressed their children before taking them home (Montagner, 1978)]. The behavioural profiles are thus not rigid, set once and for all during early childhood. Although these profiles often appear to be reinforced from three to six years and beyond, in most cases this is probably because of reinforcement of the child's behaviour by the parents' attitude in relation to the child and, in particular, by reinforcement when the child returns to the family environment at the end of the day and during the week-end. For example, it was clear that children, from 18 to 36 months, tended more and more to behave more aggressively to their peers both in the day-care centre and in the kindergarten when their mothers increased her aggressions in relation to them. Such changes often occur when a mother becomes tired, usually through an imposed change in work rhythm (workers on production lines, in offices, private secretaries, etc.) or through voluntary action, e.g., medical doctors. The increase in the frequency of aggression by the child is often translated into an increased tendency for aggressions to be received by them in the day-care centre, in the kindergarten, and in the primary school. Thus, a mutually reinforcing mechanism is set up between the child and its family, whereby aggressions tend to outnumber the other modes of behaviour. It must be added that the child tends to differentiate and reinforce a "dominated–aggressive" profile when receiving a high level of aggressions from both parents. In these children, an outburst of marked and sustained aggressions can alternate with a period of isolation during which the child does not interact with peers and adults, whatever the context.

Modulation of Aggressive Behaviour Profiles

The behaviour of the family appears to be the most obvious and frequent cause of the differentiation of the most aggressive behaviours, but the behaviour of the staff, the situations experienced, and the physical and social environment also play a role in the reinforcement or modification of these profiles.

The Role of the Staff

We observed a significant decrease in the frequency of aggression in the most aggressive children when a female member of staff individually received each child in the morning at the day-care centre or kindergarten and gave value to the child by her behaviour and words. Then, by suggesting to the child a manual and usually attractive activity (assembling blocks together, pictures to complete, etc.), the member of staff actually channelled the motor activity of the child and, at the same time, reduced the probability that the child would be aggressive to one of its peers. Psychomotor exercises, which channelled the motor activities of the child's limbs, continued the channelling of the child's corporal expression. Finally, the member of staff did not behave aggressively to a child after it had been aggressive to another. This brought about a decrease in the probability of the appearance of aggressive behaviour. For example, this was the case in Rosemont I Kindergarten class (Table 10.2): because of the teacher's behaviour, the frequency of overt aggressions became lower as the frequency of offering behaviours by peers became higher; at the same time, the frequency of grasping behaviour tended also to decrease, but more slowly than did overt aggressions, and grasping behaviour was more often followed by an offering from the grasping aggressive child than in the other classes. Consequently, there was a positive correlation between grasping and offering behaviours, and a relatively low positive correlation between grasping and aggression, as compared to the other classes under study (with the exception of Rosemont II class, where the teacher was also appeasing in relation to the children, especially the most aggressive ones, but did not facilitate the exchange of objects and co-operation during the manual and usually attractive activities).

To summarise, all the studies that have been carried out since 1970 show that a diversified solicitation of motor activities in the most aggressive children, combined with praising and non-aggressive relations on the part of the staff, result in a decrease in the frequency of aggressive behaviour by those children in relation to their peers. During all the time that aggression is channelled, the usually aggressive child can develop other modes of relationships with its peers, during which reciprocal imitation, appeasement, and co-operation can develop. This way the child can "situate" itself more often and more clearly in relation to the others, at least at certain times. Then the child can pay more attention to the locomotion, interactive processes, school activities, etc., of its peers and, at the same time, receive attention from them. Thus it can be seen that situations created by the teacher that channel aggression can play an important role in allowing the development of attention structures in relation to peers, in the sense implied by Chance (1967) and so help the most aggressive children, at least partially, to control the frequency, force, and extent of their aggression, and then to develop a greater diversity of strategies in the peer groups in which they move.

The Role of Situations

In our research group, Laurent (1983) has shown that although the frequency of linking and appeasing behaviours is not significantly changed by the table arrangement created at the day-centre at lunch time, there is a significant decrease in the frequency of aggressive behaviours in this situation (Table 10.4: the frequency of threatening behaviour remains unchanged; the decrease of agonistic behaviours is from a decrease in aggressions). However, there is no significant change in the frequency of fearful and withdrawn behaviours, and of isolations.

It must be added that the frequency of aggressive behaviours is especially decreased at the meal in the most aggressive children. As these are among the most active in the free activities, and as they participate in the longest interactions at the meal, they can, in this particular situation, appear to be leaders like the most appeasing, soliciting, and usually attractive leaders in free activities. This study confirms that certain situations bring about a significant reduction in the frequency of aggressions and that the children differentiate, especially during the second and third year, from aggressive behaviours that are induced and reinforced by the home environment. At the same time, all the children can express linking and appeasing behaviours each time the situation warrants them. Thus, it would appear to be important, in both the day-care centre and the kindergarten, to plan for alternating free activities where the children can move about at will, and where they can develop exchanges and communication in relation to their behavioural profile and activities.

La Prairie kindergarten, which has just been built in Baume les Dames, 40km from Besançon, has been designed to take into account the indi-

TABLE 10.4
Quantity and Coefficients of Behaviours

Behaviour	1978–1979						1979–1980					
	Linking and Appeasing		*Agonistic*		*Linking and Appeasing/ Agonistic*		*Linking and Appeasing*		*Agonistic*		*Linking and Appeasing/ Agonistic*	
Activity	*F.A*	*M*	*F.A*	*M*	*F.A*	*M*	*F.A*	*M*	*F.A*	*M*	*F.A*	*M*
Day-Care Centres												
Planoise	34	50	25	7	1.3	7.1	93	95	36	17	2.6	5.6
Montrapon	32	45	26	20	1.2	2.2	70	55	41	26	1.4	2.1
Planoise + Montrapon (Mean)	33	47	25	13	1.3	3.6	81	75	39	22	2.0	3.4

NOTE Quantity of linking and appeasing behaviours and of agonistic behaviours and coefficients obtained by dividing linking and appeasing behaviours by agonistic behaviours (linking § appeasing/agonistic) in 18 children (9♂, 9♀) observed for 150h in 1978–1979 and in 16 children (8♀, 8♂) observed for 120h in 1979–1980. The 2- to 3-year old children were observed in free activities (F.A) for 30 min. before lunch, which started before 10.30 and 10.45 a.m. and then during meal times (M) which lasted 30 min. Planoise day-care centre receives children from very different socio-professional classes; Montrapon day-care centre receives a majority of children coming from lower socio-economic classes.

vidual behavioural profiles and to allow the most aggressive and the most fearful children to interact with their peers in situations such that aggression is most often channelled. Many different motor activities can be performed in this kindergarten by all children at all times (climbing onto a wall and up to a platform using the arms and legs; reaching the same place by steps; diving into a bed full of cushions; throwing polystyrene objects, etc.), and each child is able to select isolation or communication with one or several peers, to sleep or doze off in a darkened or lightened area, and to have a conventional (painting, pottery, etc.) or personal kindergarten activity (Y. Durin, manuscript in preparation). It is obvious that in these conditions there is a very low frequency of aggressive behaviours between children, even among the most aggressive (a precise quantitative study is being carried out; the first results show that there is a significantly lower frequency of marked and sustained aggressions than in the other kindergarten classes being studied by our research team). It is thus possible for the children to achieve a certain degree of control over the behaviour that causes breaking off from the others if the day-care centre and the kindergarten create conditions that enable each child to invest its motor activity and turbulence in activities so varied that the frequency of conflict and competition is decreased.

DISCUSSION AND CONCLUSIONS

The continuous study of agonistic behaviours (threats and aggressions) in children who go to the day-care centre and kindergarten shows that these behaviours are part of the usual behaviours of "non-pathological" children. The emergence of these behaviours, at least the interactions with peers, takes place between the seventh and fourteenth month. The most characteristic vocalisation appearing in the young child's repertoire is the threat when there is conflict and competition. It is as if one of the priorities of the less than one-year-old child in a group of peers is to emit and decode one of the signals that enables it to participate, without overt aggression, in competitions enabling access to a limited number of situations and objects. However, although overt aggressions can be observed in all children from one to three years of age who move in a peer group, their frequency and the circumstances of their appearance vary from one child to another. Thus, some children have, in free activities, higher frequency of aggressions than of linking and appeasing behaviours. When analysing the temporal patterning of their behavioural sequences, we found that these children could suddenly go from an offering, or any other linking and appeasing behaviour, to hitting, biting, scratching, or any other aggressive behaviour, even after they had succeeded in getting the object or the situation that gave rise to a competition, and even when there was not an obvious conflict (Gerald; Fig. 10.3). If these aggressive children are imitated and followed by peers of the same social group, it is over shorter periods of time than the most appeasing and soliciting children (the leaders and the dominated children with leader mechanisms). The groups that form around the most aggressive children are more quickly disorganised than those around the most appeasing and soliciting. By comparison, the leaders (Damienne; Fig. 10.3) and the dominated children with leader mechanisms have a low frequency of aggressions (even if they are able to react aggressively at any moment when they are the target of an aggression), and a high level of linking and appeasing behaviours. The leaders, and dominated children with leader mechanisms, do not confuse threat and aggression as they rarely mix these two behaviours in the same behavioural sequence, and rarely go from a threat to a marked and sustained aggression. More generally, they appear to show the elementary behaviours of Table 10.1 in such ways that they are the most often imitated and followed, and this from one day to the next. They are also the ones who create the most new situations and games.

There is thus no correlation between the level of aggression and the level of dominance in free activities and experimental situations, nor between the level of aggression and the level of leadership in a peer group. However, there is a positive correlation between the level of linking and appeasing behaviours, and the level of leadership, whatever the level in participation and success in competitions for usually attractive objects and new situations (high in leaders and very low in dominated children with leader mechanisms).

When the frequency of aggressions appeared to increase significantly from one month to the next, when they occurred more and more often without obvious cause (without being obviously linked with a competition or conflict), and when they were more and more marked and sustained, the home influence could be clearly seen. The continuity of work rhythms, a poor state of parental health, in particular, of the mother, and whether the parents themselves were receiving or had received marked and repeated aggressions in the past, especially in childhood, played a particularly important role in the high frequency of the aggressions that they gave to their child, and then in the high frequency, and marked and sustained, aggressions that the child gave to peers in the same group. This observation was also made in kindergartens and primary schools.

Institutional factors (behaviour of staff, composition of the peer group, lack of physical space, etc.) can also bring about an increase in the frequency, and in the marked and sustained (?violent) nature of the aggressions that result in rejections by peers, staff, and others. This reinforces the tendency for the aggressive child to give a larger place to repeated and marked aggression whether these aggressive actions be spasmodic or continuous. However, when the staff adopt a praising and

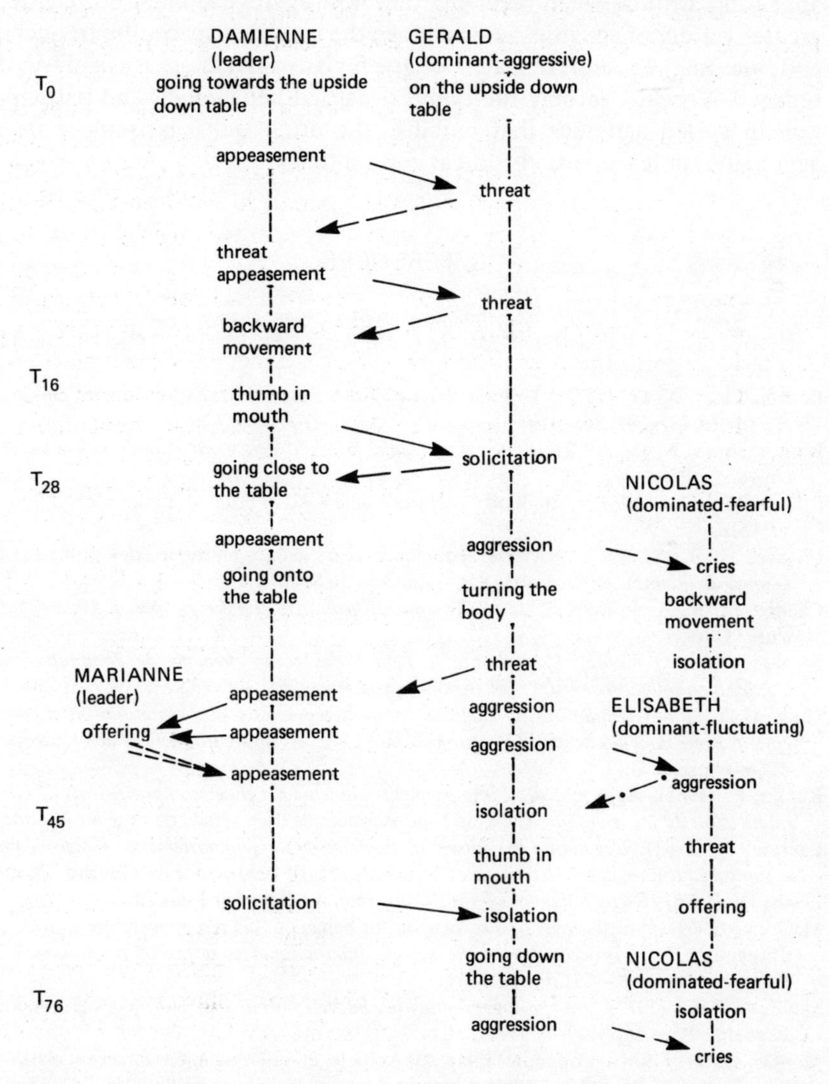

FIG. 10.3 An example of the linkings of behavioural sequences of 30- to 36-month-old children in an experimental situation that often gives rise to competition (a low table upside down on another). Damienne (see Table 10.3) was the most appeasing child and one of the leaders of the peer group; the frequency of her aggressions was always very low. Gerald (see Table 10.3) was the most dominant–aggressive child of the same peer group; the frequency of his linking and appeasement behaviours was always low. Nicolas (see Table 10.3) was the most dominated, fearful child of the same peer group; the frequency of his agonistic behaviours (threats and aggressions) was always very low.
T_0 to T_6: time scale in seconds.

appeasing attitude when receiving the most aggressive children, create a greater variety of activities, and arrange the available space, the frequency and the marked and repeated character of aggression is significantly reduced. Threats, turning the body, ritualised behaviours, and participation in varied activities that mobilise the arms and legs, replace overt aggression, at least partially and at certain times.

REFERENCES

Blurton Jones, N. G. (1967). An ethological study of some aspects of social behaviour of children in nursery schools. In D. Morris (Ed.), *Primate ethology* (Pp. 347–368). London: Weidenfeld & Nicolson.

Blurton Jones, N. G. (1971). Criteria for use in describing facial expressions of children. *Human Biology*, *43*, 365–413.

Blurton Jones, N. G. (1972). *Ethological studies of child behaviour*. London: Cambridge University Press.

Chance, M. R. A. (1967). Attention structures as the basis of primates rank orders. *Man*, *2*, 503–518.

Chance, M. R. A. (1974). Sociétés hédoniques et sociétés agonistiques des primates. In *L'unité de l'homme* (Pp. 83–91). Paris: Editions du Seuil.

Chance, M. R. A. & Jolly, C. J. (1970). *Social groups of monkeys, apes and man*. New York: Dutton.

de Roquefeuil, G. (1982). *Contribution à l'étude des comportements de communication d'enfants en situation de jeu*. D. E. A. de neurosciences de l'Université d'Aix-Marseille II.

Godard, D. (1982). *Etude des interactions comportementales et des communications entre enfants présentant des déficits d'insertion sociale*. D. E. A. de Neurosciences de l'Université d'Aix-Marseille II.

Kontar, F. (1981). *Approche des comportements alimentaires et des comportements de communication du jeune enfant*. Thèse de Neurosciences de l'Universite de Franche–Comté.

Laurent, D. (1983). *Régulation des entrées alimentaires et des comportements chez les jeunes enfants fréquentant la crèche*. Thèse de Neurosciences de l'Université de Franche–Comté.

Lewis, D. (1978). *Secret language of your child*. London: Souvenir Press Ltd.

McGrew. (1969). An ethological study of agonistic behaviour in preschool children. In C. R. Carpenter (Ed.), *Proceedings of the second international congress of primatology, I: Behaviour* (Pp. 149–159). Basel: Karger.

McGrew, W. C. (1972). *An ethological study of children's behaviour*. New York & London: Academic Press Inc.

Montagner, H. (1974). Communication non-verbale et discrimination olfactive chez les jeunes enfants–Approche éthologique. In *L'Unité de l'Homme* (Pp. 235–253). Paris: Editions du Seuil.

Montagner, H. (1978). *L'enfant et la communication*. Paris: Stock.

Montagner, H. (1979). Ethologie humaine. In *Encyclopaedia Universalis* (Suppl. Pp. 569–578). Paris.

Montagner, H. (1983). *Les rythmes de l'enfant et de l'adolescent*. Paris: Stock.

Montagner, H. & Henry, J. C. (1975). Vers une biologie du comportement de l'enfant. *Revue de Questions Scientifiques (Bruxelles)*, *146*, 481–529.

Montagner, H., Restoin, A. & Henry, J. C. (1982). Biological defense rhythms, stress and communication in children. In W. W. Hartup (Ed.). *Review of child development research*, *6*, (Pp. 291–319). Chicago: University of Chicago Press.

Montagner, H., Henry, J. C., Lombardot, M., Restoin, A., Bolzoni, D., Durand, M., Humbert, Y., & Moyse, A. (1978). Behavioural profiles and corticosteroid excretion rhythms in young children. Part 1: Non-verbal communication and setting up of behavioural profiles in children from 1 to 6 years, in V. Reynolds & N. G. Blurton Jones (Eds.), *Human behaviour and adaptation* (Pp. 207–228). London: Taylor & Francis.

Montagner, G., Henry, J. C., Lombardot, M., Restoin, A., Benedini, M., Godard, D., Boillot, F., Pretet, M. T., Bolzoni, D., Burnod, J. & Nicolas, R. M. (1979). The ontogeny of communication behaviour and adrenal physiology in the young child. *Child Abuse and Neglect*, *3*, 19–30.

Montagner, H., Restoin, A., Schaal, B., Rodriguez, D., Ullmann, V., Ladouce, I., Guedira, A., Viala, M., Godard, D., Hertling, E., Didillon, H., & Gremillet H., (1981). Apport éthologique à l'étude ontogénétique des systèmes de communication de l'enfant. *Médecine et Hygiène (Genève)*, *39*, 9–18.

Ullmann, V. (1981). *Contribution à l'étude ontogénétique du comportement et des rythmes biologiques chez le jeune enfant à la crèche*. Thèse de Neurosciences de l'Université de Franche–Comté.

Viala, M. (1980). *Etude comparative des effets de changements de rythme de vie contrôlés à l'école maternelle chez les enfants de 3 à 5 ans*. Thèse de Neurosciences de l'Université de Franche–Comté.

11 The Structure of Politically Relevant Behaviours in Pre-school Peer Groups

Carol Barner-Barry
Department of Political Science, University of Maryland, Cantonsville, U.S.A.

A "group" can be defined (Berkowitz, 1975, p. 433) as, ". . . a collection of individuals who react to each other in some way, however indirectly". In face-to-face groups, such as those under consideration in this chapter, the individual members perceive themselves as interrelated. It is this perception that gives the group its "social" character. Once a group has been defined as a "social group", new dimensions come into play. According to Smith (1945, p. 227), a "social group" is ". . . a unit consisting of a plural number of separate organisms . . . who have collective perception of their unity and who have the ability and tendency to act and/or are acting in a unitary manner toward the environment". This ability to act in a co-ordinated or unified way is a direct function of group cohesion.

The term "group cohesion" refers to the extent to which a group of individuals coheres or "hangs together". Shaw (1981) has identified at least three meanings that researchers have attached to the term "cohesiveness". First, it can refer to the inter-personal attraction that exists between the members of the group. Second, it can refer to the level of morale or motivation shown by group members. Finally, it can refer to the extent to which the activities of the group are co-ordinated. Cohesion is based on an infrastructure of customary or habitual interactions between and among the members of the group. These interactions may be related to any of the various meanings of group cohesion. The first and the third meaning will be emphasised in this study because it is easier to identify, with some degree of confidence, the behavioural manifestations of inter-personal attraction and co-ordination of effort than it is to indentify the more subjective behavioural manifestations of morale.

Each member of a group that has been in existence for any amount of time will usually become identified with certain roles. That is, the other members of the group come to expect that they will usually act in certain ways in given situations. This confers on the individual a social position, and often is a manifestation of the division of social labour that exists within the group.

Such roles, as well as the behaviours associated with them, channel the energies of the members in ways that contribute to the group's purposes and have a function in the group's life (Berkowitz, 1975; Linton, 1936). There are also normative overtones: group members develop notions of how one ought to act when one is occupying a certain position or role. One of the key roles in any group is that of leader. There may, however, be more than one leader per group. In the artificial laboratory climate of much psychological research, there has been a tendency for groups to generate relational (socio-emotional) leaders and task leaders. The functions implied by these designations may be performed by two different people, or may be combined in the activity of a single person (Bass, 1981).

In naturally occurring groups, the filling of certain roles can be more or less fluid, depending on the personalities of the members and on the situation. Where there is no formal leader, or where the formal leader is weak multiple informal leaders with different scopes of authority and levels of activity can emerge. That is, in any particular area of group concern there might arise persons whose behaviour organises and guides the other members. Such persons, be they formal or informal leaders, can be regarded as constituting a focal point around which the group members can cohere, and may, in some instances, become a symbol of group loyalty. (Chance & Larsen, 1976; see also Chapters 4 and 10).

Within any naturally occurring group that remains relatively stable over a significant period of time, certain functions must be performed that can be regarded as "governance functions". A leader (or leaders) in these functional areas might then be regarded as one who is, in fact, governing the group. They are performing, for the group, types of service similar (though not identical) to those a government provides for a polity. This analysis will focus on such behaviours, using terminology and some general conceptual frameworks borrowed from the discipline of Political Science.

Although an exhaustive list of such functions, and the rationale for considering them governance functions, is outside the scope of this chapter, some notion of their general nature can be conveyed by listing a few that seem (to the author, at least) to be most basic. Most governments, ranging from the primitive to the most complex, perform the duties associated with basic social planning for the polity. They establish some apparatus for authoritative decision making, regulate those activities of members that have society-wide or at least very broad impact, and arrange for services that are designed to make the polity reasonably attractive to its citizens.

Group leaders, then, would be concerned with such matters as planning for group activities, making decisions that are binding or are perceived to be binding, resolving conflicts, taking care of those who (for whatever reason) need aid or protection, and arranging for emotionally gratifying experiences that allow members to identify in an affectively positive way with the group. The performance of such functions requires a relatively wide range of abilities and motivations; it is conceivable that the same person might not be able to do them all, or not at a consistently high level of competence. Thus, there is no reason why all such functions have to be performed by the same person, or why any one of them has to be performed exclusively by one person all the time.

This factor is balanced somewhat by the parallel possibility that leadership, which is initially based on differential contributions to the performance of some limited type of task within or on behalf of the group, may, in time, be generalised to other domains through some process of imagery development (Nimmo, 1974), or the accumulation of mutual satisfactions (Hollander & Julian, 1976; Larsen, 1976; Verba, 1961).

My research on relatively stable peer groups of young children suggests that, at least at early stages of social development, governance functions in peer groups can be subdivided into three inductively derived categories: (1), directive behaviour; (2), constituency behaviour; and (3), receiving behaviour. Directive behaviour is most closely related to the act of governing as it is commonly conceptualised. Constituency behaviour and receiving behaviour contribute to the growth and consolidation of group cohesion around a given leader or set of leaders.

GOVERNANCE FUNCTIONS

"Directive behaviour" involves interactions begun by the leader in which they attempt to direct or control the behaviour of one or more followers. The intent of the leader is manifest, and the behaviour to be influenced is occurring at that given point in time, or is about to occur in the immediate future. The types of behaviour falling into this category that I have observed in the field include:

1. *Regulation of On-going Activities*. Where an individual or the members of a group look to the leader to guide their behaviour and the leader performs this function.

2. *Inclusion or Exclusion*. Where the follower wants to join a group of peers and is overtly included or excluded by the leader; the follower accepts the leader's decision, as do the other members of the group.

3. *Conflict Resolution.* Where the conflict is settled to the satisfaction of all parties by a leader who is either a party to the conflict or an uninvolved third party; the criterion used is the amicable acceptance of the leader's solution.

4. *Distribution of Resources.* Where there are not enough desired objects to satisfy every person involved in the interaction and the leader's distribution is accepted by both the satisfied and deprived parties.

5. *Rule Articulation.* Where the leader states the rule and the follower alters his or her behaviour to conform to the requirements of the rule.

6. *Rule Interpretation.* Where there is a difference of interpretation between the leader and at least one other person, and the subsequent behaviour of all parties is in accordance with the interpretation of the leader.

7. *Rule Enforcement.* Where the leader makes some effort to see that a rule is obeyed (beyond simple articulation or interpretation of the rule), and elicits compliance with no effective challenge of the leader's right to enforce.

In all of these, the direction of influence is from the leader to the follower, and the interactions are initiated by the leader. The essential element is that the leader "controls more than he is controlled" (Eckstein & Gurr, 1975).

"Constituency behaviour" also involves behaviour initiated by the leader. Here, however, the main goal of the leader seems to be to create "political debts", or to elicit expressions of support or loyalty. Whereas directive behaviour has the effect of "testing" the viability of the dyad's leader/follower relationship within a given scope of activity, constituency behaviour has the effect of consolidating the relationship for that dyad or, alternatively, of extending its reach into a qualitatively different area. The attempt of the leader to control is less obvious, and the overall image projected by the leader is one of concern or nurturance. The types of behaviour I have observed that fall into this category include:

1. *Aid.* Where the offer of help comes from the leader and is accepted by the follower.

2. *Giving Information.* Where the leader decides that the follower needs a certain piece of information and proffers it to an accepting follower; some indication that the follower has actually assimilated the information and attached importance to it is required.

3. *Protection.* Where the follower perceives a need for protection and accepts an offer of protection from the leader.

4. *Retaliation on Behalf of Another.* Where the leader is informed of or apprehends a wrong done to the follower and "punishes" the wrong-doer.

5. *Giving Attention.* Where the leader indicates interest in the follower or in something associated with the follower, and the follower expresses gratification or happiness with the leader's show of interest.

6. *Showing Affection.* Where the leader initiates an affectionate interaction with the follower (e.g. holding hands) and elicits a positive response from the follower; these tend to involve situations where the follower verbally or non-verbally expresses gratitude for the affection shown.

Here, again, the direction of influence is from leader to follower, and the interactions are initiated by the leader. There is present, however, a "helping" or solicitous quality that is not present in directive behaviour.

Finally, "receiving behaviour" is initiated by the follower and directed toward the leader. In some cases, the followers behave in ways that induce the leaders to respond. To this extent, it could be said that the follower is controlling the leader. The interaction, however, has the effect of either validating the leader's position or of relieving the follower of the necessity of making a given decision. The types of behaviour falling into the categories that were actually observed include:

1. *Proximity Desired.* Either a verbal request (e.g., "Can I come with you?"), or following behaviour are included.

2. *Attention Solicited.* The solicitation (e.g., "Jane, look at me!") is crucial here, and this behaviour is scored whether or not the attention is forthcoming.

3. *Imitation.* The imitative nature of the behaviour has to be quite explicit and follow the behaviour imitated with only a brief time lapse.

4. *Appeasement.* The behaviour has to be placatory and is scored whether or not it was successful; situations involving physical violence or threat were eliminated from this category.

5. *Permission Solicited.* The solicitation (e.g., "Can I play too?") is crucial here, and this behaviour is scored whether or not the permission is granted.

6. *Deference Shown.* There is only one instance of this type of behaviour in the data to be presented here, and it involved one child spontaneously yielding a place in line to another.

Directive behaviour is important to group cohesion in that it leads to a reasonably orderly orchestration of group behaviour. Constituency behaviour and receiving behaviour tend to be conducive to the formation of socio-emotional bonds among group members, and the development of a set of tacit expectations among them regarding the relative roles they would be expected to play in any appropriate social interaction.

They establish a set of interactional "norms", which, although they are not followed invariably, do contribute to behavioural predictability, and to the consolidation of group structure and cohesion.

DATA COLLECTION

The data presented here were gathered from two groups. Group I comprised 38 children attending a 4-week summer playground programme. Their ages ranged from three-and-a-half to six-and-a-half years. Most of the group's time was spent out of doors in a large fenced-in playground of approximately half an acre in area. Group II comprised 20 children who were attending a nursery school. Their ages ranged from two-and-a-half to four-and-a-half years. This group spent most of its time indoors in a large room with a high ceiling and one full-window wall. The composition of both groups remained stable for the duration of the study. Although the groups met for different lengths of time, it was possible to adjust the amounts of observing time analysed so that they were approximately equal for both groups.

The method used to gather the data was non-participant observation with event sampling (Selltiz, Wrightsman & Cook, 1976; Wright, 1960). The events were all asymmetrical interactions that took place in the presence of the observer. Asymmetrical interactions were defined as dyadic interactions in which X exerted more influence over the development or the outcome of the interaction than did Y. As a practical matter, the direction of balance had to be inferred from the verbal and non-verbal behaviours of the subjects. In almost all cases, subject behaviour was given the most straightforward of the likely interpretations. In some cases, however, the observer's familiarity with the group allowed the identification of more subtle behaviours, although there were undoubtedly some that went unappreciated.

A large amount of data was collected in an effort to dilute the effect of inaccuracies on the aggregate results. In cases that seemed questionable, the interaction was recorded in detail, and the decision as to asymmetricality was made after the analyst had had time to reflect and to consider relevant contextual factors. All instances in which coercion was the sole basis of the influence exerted were eliminated for purposes of this analysis.

In both groups the three children who most often engaged in directive, constituency, or receiving behaviour were two males (M) and one female (F):

Group I	*Group II*
S (f) : 5.0 years	M(m) : 4.0 years
C(m) : 5.5 years	J(m) : 3.0 years
F(m) : 6.0 years	A (f) : 4.5 years

The 38 children in Group I were divided into 21 males and 17 females; Group II had 9 males and 11 females. Group I had one retarded child, and one normal child who was significantly older than the rest of the group; both were males. The normal older boy, who was eight years old, spent most of his time with the supervisors and played a very limited role in peer-group social relations. The retarded boy, who was seven, participated in all group activities and, according to the teacher in charge, had a mental age of approximately four. There were no older or handicapped children in Group II.

RESULTS AND DISCUSSION

The Groups

Table 11.1 gives the frequencies of directive, constituency, and receiving behaviour for each group.

The first thing that should be noted is that these behaviours were much more common in Group I than in Group II; one probable reason for this is that Group I had much less close adult supervision because the physical area to be kept under surveillance was much larger, and there were almost twice as many children, so diluting the amount of attention a supervisor could direct toward any individual child. Group I had a ratio of approximately 13 children per adult supervisor, and one of the supervisors was a maintenance person with no training in child care or education. Group II had a ratio of approximately 10 children per adult supervisor, and both adults had had professional training. Also, Group II frequently had several more adults in the room, because the classroom was used for the training of child-care students. In contrast, Group I was frequently left with less than its full complement of supervisors, and on several occasions only one adult was supervising approximately thirty children. Finally, Group II had a much more structured programme and much less free play time than Group I.

Because of these environmental factors, there were many more instances in which it was necessary for the members of Group I to "govern" themselves. This, in turn, created a situation in which the members of that Group became more independent and had less tendency that those of Group II to seek adult guidance or help in solving social problems.

The proportions of constituency behaviour were roughly the same for both groups. There was, however, a much higher proportion of directive behaviour in Group II than in Group I. Conversely, there was much more receiving behaviour in Group I than in Group II. Overall, this suggests that in the less structured social and physical environment of Group I there was more need for weaker or less socially competent group members to make a

TABLE 11.1
Frequencies of Directive, Constituency and Receiving Behaviour

	Group I		*Group II*	
	Number	*Percentage*	*Number*	*Percentage*
Total	538	100	403	100
Directive Behaviour	236	44	256	64
Constituency Behaviour	60	11	65	16
Receiving Behaviour	242	45	82	20
Directive Behaviour				
1. Regulation of activities	153	28.0	101	25.0
2. Inclusion or exclusion	43	8.0	33	8.0
3. Conflict resolution	17	3.0	26	6.0
4. Distribution of resources	13	2.0	18	5.0
5. Rule articulation	4	0.7	21	5.0
6. Rule interpretation	3	0.5	37	9.0
7. Rule enforcement	3	0.5	20	5.0
Constituency Behaviour				
1. Aid	31	6.0	27	7.0
2. Giving information	12	2.0	14	4.0
3. Protection	7	1.0	10	2.0
4. Retaliation for another	5	0.9	0	0.0
5. Giving attention	3	0.5	1	0.2
6. Showing affection	2	0.4	13	3.0
Receiving Behaviour				
1. Proximity desired	185	34.0	40	10.0
2. Attention solicited	26	5.0	19	5.0
3. Imitation	23	4.0	19	5.0
4. Appeasement	6	1.0	0	0.0
5. Permission solicited	1	0.2	0	0.0
6. Deference shown	1	0.2	4	1.0

concerted effort to form and maintain social ties with other group members (Hallinan, 1978/79).

There are many possible reasons for this. One of the most plausible is that the desire for affiliation was an outcome of the greater perceived need to create personal bonds, which yielded security and structure, because the environment of Group I was relatively unstructured and lightly supervised. Thus, individuals might attempt to form numerous ties with other members of the group. If successful, this would result in the formation of a protective subgroup that, in turn, would increase the order and predictability in an individual's social relationships. There might also have been an effort by many group members to become the protégés of stronger and more socially competent group members.

The frequencies of occurrence of the behaviours in each of the subdivisions of directive behaviour were in similar proportions or activity levels in the first two types, "regulation of activities" and "inclusion or exclusion". After that, though, the picture becomes more complex. There was more conflict resolution in Group II than in Group I. This difference is almost entirely attributable to one individual in Group II who had an exceptionally high level of skill in verbal conflict resolution, and who was not reluctant to utilise that skill when the opportunity presented itself. The difference in the amounts of behaviour involving the distribution of resources is probably too small to indicate anything.

The levels of rule articulation, interpretation, and enforcement were higher in Group II; the reasons for this difference are not clear. One possibility is that because of the greater structuring of Group II there was more of this type of activity by adults, and the higher level of child activity was imitative. Another possible interpretation is that with less physical space per child, the members of Group II might have felt a greater need to limit extremes of behaviour by calling on quasi-adult authority in the form of rules. Other possibilities might be raised; suffice it to say that this is an area that deserves closer scrutiny.

What should be noted in regard to constituency behaviour is the substantial agreement over categories. Why the amounts of this behaviour should be so much more similar than for the others is not clear. Further research is necessary to determine whether this happened by chance or was the result of some underlying dynamic.

The dramatic difference between Groups I and II in the amount of receiving behaviour is almost entirely a function of the difference in the levels of desire for proximity. There is also a small contribution in the category of appeasement. Both of these suggest that life was less predictable and more hazardous in Group I. Children of this age do not normally have a high level of verbal skills. Therefore, they are relatively physical in their approach to social interaction. If a problem arises, such as one child grabbing toys from another, and if there is no adult present, the children's response will probably be to use physical force to solve the problem. In light of this, it is a reasonable conjecture that the greater the extent to which children are left to their own devices in regulating their social life, the greater the need for the relatively weak or less socially competent children to become allied with other children, who will serve as a buffer against the vicissitudes of peer group life.

Finally, Table 11.1 indicates that there is substantial overall similarity in the relative amounts of each subtype of behaviour found in both groups. For example, although "proximity desired" has a much higher frequency in Group I than Group II, it is still the most frequent type of receiving behaviour manifested by the members of Group II. In fact, if one calculates a product moment, correlation co-efficient over the subcategory data, grouped together without reference to major categories, one gets a correlation of 0.75. This is a high correlation for social-science data, and it indicates that the importance of each subtype of behaviour (as measured by frequency of appearance) is quite similar for both groups despite differences in physical environment, social dynamics, and group composition.

The Leaders

Three children from each group could be characterised as the leaders of their groups in that they engaged in more directive, constituency, and receiving behaviour than any other group members. Selection of these six children is, in a sense, somewhat arbitrary, as many other children engaged in such behaviours at one time or another. These six children, however, did engage in leadership behaviours to a much greater degree than the others. Also, a significant number of children in their groups seemed to turn to them for leadership more often than they turned to any of the other members. The three children in Group I functioned quite independently of each other and had overlapping, but distinguishable, circles of followers. In Group II, the two males were part of the same subgroup (which was almost exclusively male). When both were present there was a tendency for J to act as M's "second-in-command". When M was absent, J frequently (though not invariably) assumed leadership of the male group.

Table 11.2 shows the frequency with which these six children engaged in directive, constituency, and receiving behaviour. A glance at the overall activity levels of the children clearly shows that there was much more uniformity in the gross number of behavioural instances across categories among children in Group I than in Group II. The reasons for this uniformity in Group I are not clear. The most plausible explanation is that each of the three children tended to move in their own sphere of influence and to avoid each other.

In contrast, the males from Group II were close friends. Also, in the judgment of the observer, A was capable of much more leadership behaviour than she ever attempted. Her activity level was low because of a lack of motivation; she simply preferred to be engaged in crafts or intellectual tasks (e.g., solving simple mathematical problems), and she ended up interacting with adults much of the time. This naturally limited her opportunities to act as a leader. Whenever she did choose to partici-

TABLE 11.2
Frequencies of Behaviours for Group Leaders

	Group I			*Group II*		
	S	*C*	*F*	*M*	*J*	*A*
Total Behaviours	*61*	*64*	*55*	*150*	*39*	*43*
Directive Behaviour	37(60%)	31(48%)	26(48%)	101(67%)	25(64%)	18(42%)
1. Regulation of activities	23(38%)	19(30%)	13(24%)	60(40%)	15(38%)	10(23%)
2. Inclusion or exclusion	6(10%)	2(3%)	13(13%)	20(20%)	5(5%)	3(3%)
3. Conflict resolution	0	4(6%)	0	12(8%)	3(8%)	0
4. Distribution of resources	6(10%)	4(6%)	0	2(1%)	1(3%)	2(5%)
5. Rule articulation	1(2%)	1(2%)	0	5(3%)	0	1(2%)
6. Rule interpretation	0	0	0	0	0	0
7. Rule enforcement	1(2%)	1(2%)	0	2(1%)	1(3%)	2(5%)
Constitutency Behaviour	7(12%)	12(19%)	9(16%)	17(11%)	5(13%)	15(35%)
1. Aid	5(8%)	7(11%)	4(7%)	5(3%)	5(13%)	9(20%)
2. Giving information	2(3%)	3(5%)	0	7(5%)	0	2(5%)
3. Protection	0	1(2%)	1(2%)	3(2%)	0	1(2%)
4. Retaliation	0	1(2%)	1(2%)	0	0	0
5. Giving attention	0	0	2(4%)	0	0	0
6. Showing affection	0	0	1(2%)	2(1%)	0	3(7%)
Receiving Behaviour	17(28%)	21(33%)	20(36%)	32(21%)	9(23%)	10(23%)
1. Proximity desired	9(15%)	12(19%)	16(29%)	13(9%)	7(18%)	5(12%)
2. Attention solicited	3(5%)	3(5%)	2(4%)	8(5%)	0	1(2%)
3. Imitation	4(7%)	6(9%)	2(4%)	10(7%)	2(5%)	4(9%)
4. Appeasement	0	0	0	1(1%)	0	0
5. Permission	0	0	0	0	0	0
6. Deference	1(2%)	0	0	0	0	0

pate in some peer group activity, she almost immediately became the centre of attention and the leader of the group.

The fact that "regulation of activities" is the model category for all but one of the six children is significant. "Regulation of activities" includes such behaviours as:

1. Choosing an activity for the self and at least one other child.
2. Changing the nature of an ongoing activity.
3. Monitoring another's activities (e.g., X supervises Y's work on a puzzle).
4. Persuading another to join in an activity.
5. Telling another child or children what to do during an interaction sequence (frequently fantasy play).
6. Getting another child or an adult out of the way of an activity.
7. Initiating repetition of an activity.
8. Assigning roles in fantasy play.
9. Directing another child's attention to something interesting (e.g., X induces Y to smell a perfumed, felt pen).
10. Preventing intended action(s) of another child (e.g., X dissuades Y from taking a toy out of his cubby).
11. Determining the physical location or companions of another child (e.g., X leads Y and Z to the swings).

One reason why these six children were leaders was that they were more likely than other members of the group to be consistently "where the action was". Not only were they usually involved in the most interesting available activity, but also they were commonly the focus of the group involved in that activity.

If these children shared one characteristic, it was the ability to think of interesting and attractive ways to spend time, combined with an ability to involve others in their activities. Fairly early in the life of both groups, the other members identified these children as leaders in this realm and, subsequently, tended to depend on them as sources of ideas and direction, especially during free play. Much of M's score in this category is attributable to daily games of "Batman", which the others seemed to anticipate with pleasure and which M orchestrated with skill and vigour. Every day he seemed to be able to think of some new adventure for Batman, and Batman's friends and enemies. Even being a "bad guy" seemed to be preferable to being excluded from the activity. Thus, M was frequently called upon to give direction to the on-going game of "Batman". In doing so, he effectively controlled the activities of his companions for long periods of time, and led them to act and to view themselves as a cohesive group.

This direction of play led quite naturally to a relatively high level of activity under the category "inclusion or exclusion". In other words, in his capacity as leader, M was frequently called upon to decide who should and should not take part in group activity. On the other hand, both C from Group I and A from Group II scored relatively low in this category. Unlike the other four, they had a strong preference for solitary or adult-centred activities. This limited their participation in those group activities where inclusion or exclusion of individuals became an issue. They were also the most intellectually inclined of the six. As mentioned earlier, A was strongly motivated to learn school-type skills, such as arithmetic. C seemed fascinated by nature and spend a good bit of time exploring and experimenting with the flora and fauna of the playground where the group was located. These two children scored the highest under the subcategory, "aid". Their aloofness and intellectuality were combined with a relatively greater maturity than was characteristic of the other leaders. Thus, when adult-type aid was needed, as when a child needed help in the manipulation and equipment or clothing, they seemed to be the preferred adult surrogates, and they also were usually able to perform the task that needed doing.

Conflict resolution is almost invariably by adults, even in situations of loose supervision, because children normally see this as a task that can only be performed by adults. Consequently, they will go to substantial lengths to seek out an available adult. M, however, was unusual in his ability to solve conflicts to the satisfaction of all parties involved. His technique was simply to talk to the parties until he had either mediated or distracted them from the source of the conflict. The other children (and, perhaps, M himself) became aware of this skill only gradually. As a result, most of his score in this area was accumulated late in the observation period. Thus, for a substantial subgroup of the children in Group II, he gradually became the preferred resolver of conflicts. He usually handled the situation at least as skilfully as the adults, and frequently arrived at a more satisfactory result, because he tended to have more information than did the adults about the origins and nature of the conflict, and about the personal idiosyncrasies of the parties involved.

Other variations in the frequencies in Table 11.2 might be commented upon. The ones already discussed, however, seem to be those that are either most striking, or most closely tied to the situation or the particular characteristics of the individuals involved. Also, as for Table 11.1, there is a considerable amount of uniformity in Table 11.2. This can be seen more clearly if one looks at the percentages. For example, in all three major categories of behaviour, there is a uniform tendency (with one exception) for the first subcategory listed to be the one with the highest level of activity. The activity level also tends to move downward from the first

TABLE 11.3
Correlations[1] Between the Behaviour Patterns of the Leaders

	S	*C*	*F*	*M*	*J*
S					
C	0.93				
F	0.75	0.73			
M	0.92	0.85	0.69		
J	0.94	0.93	0.82	0.92	
A	0.81	0.84	0.64	0.70	0.84

[1]Product moment correlation; $P < 0.001$.

subcategory listed to the last. The overall extent of the uniformity can be gauged by the information in Table 11.3

This Table contains the product moment correlations between the behaviour patterns of the six children taken as dyads. The coefficients are all high, indicating that despite the individual variations discussed earlier, there was considerable similarity in the overall behaviour patterns of the six children.

CONCLUSIONS

The sets of behaviours that have emerged during this research can be seen as being related to group cohesion in two of the meanings noted at the beginning of this chapter: (1), the level of co-ordination of activities in the group; and (2), the level of inter-personal attraction between and among group members. These, in turn, can be related to the two basic types of leadership orientation that have emerged in social psychological research on adult groups (e.g. Cartwright & Zander, 1960; Bowers & Seashore, 1966; Katz et al., 1950; McGregor, 1960). First, the co-ordination of activities within a group is usually in pursuit of some group goal and is performed by task-oriented leaders (Bass, 1981). Second, inter-personal attraction between and among group members is fostered and maintained by relational (socio-emotional) leaders (Bass, 1981).

In very small groups, the members may (depending on the circumstances) be able to co-ordinate their activities in an *ad hoc*, informal manner. As groups grow larger, however, the need arises for someone to perform the tasks of direction, control, and planning necessary if the group (or any of its subgroups) is to act effectively in relation to the environment or any internal tasks. In moderate-sized groups, such as those in this study, these governance tasks may be accomplished completely or in part by a tacit, informal, and somewhat fluid leadership structure. On the other hand, many such groups opt to have a more formal organisation and assignment of leadership roles. In large groups, formal arrangements are made more or less obligatory by the difficulty of co-ordinating the activities of large numbers of individuals. Both groups studied here were hybrids in the sense that the adult supervisors did constitute a formal leadership. The focus of the research, however, was on the peer group, so the behaviours studied were indicative of the less formal, tacit governance structure within the peer group.

Most directive behaviours, notably regulation of activities, were orientated toward structuring and co-ordinating the activities of the group or some sub-group. The task was to learn and enjoy through the medium of play, and the persons emerging as task-orientated leaders were those who were best able to think of the most interesting and enjoyable ways to pass the hours the children spent in the group setting. They were also the individuals who were able to keep group activities flowing as smoothly as possible. Thus, the group and its subgroups clustered around those myriad childhood tasks that are generally grouped under the vague designation, "play".

Cohesion based on taskes associated with play has a socio-emotional dimension. There are, however, other ways in which group membership needs to be emotionally rewarding in order for its satisfactions or rewards to balance the demands that the group often places on its members, and the limitations on personal autonomy that membership entails. Thus, task-orientated leadership is usually balanced by relations-orientated leadership. The various types of constituency behaviour foster inter-personal attraction, especially between group members and those in leadership roles at any given time. Receiving behaviour, especially proximity desired, tends to be an expression of inter-personal attraction, which is, at least potentially, gratifying to both parties to the interaction.

What this seems to suggest is that there are interesting parallels between the behaviour of children in natural settings and the behaviour of adults in the laboratories of social psychologists. Task-orientated and relations-orientated forms of leadership behaviour seem to emerge spontaneously in both settings. Because of the limited nature of this study, however, the relationship between the behaviour of children (and, perhaps, of other primates) in natural settings, and the concepts of task-orientated and relations-orientated leadership should be explored further.

My findings also point the way to other possible lines of future research. For example, is there a connection between the incidence of some of these behaviours and the level and characteristics of cohesion in similar groups? What is the difference (if any) between the behaviours emerging within involuntary groups, such as were studied here, and within groups that have formed as a result of voluntary choice on the part of the individuals involved? Finally, what is the relationship of such behaviours, or their absence, to the prevention or disruption of group cohesion?

ACKNOWLEDGEMENTS

Although they must remain anonymous to protect the privacy of the children, I wish to thank the personnel of the pre-school institutions in which this research was done. They were generous both with co-operation and with helpful suggestions. Most important, my thanks go to the children, who shared a bit of their lives with me, and to their parents for their friendly and interested co-operation. I would also like to thank Linda Rosenwein and the Editor of this volume for their comments and suggestions. Final responsibility for the content of this chapter remains, of course, with the author.

REFERENCES

Bass, B. M. (1981). *Stogdill's handbook of leadership: A survey of theory and research*. New York: The Free Press.

Berkowitz. (1975). *A survey of social psychology*. Hinsdale, Ill.: The Dryden Press.

Bowers, D. G. & Seashore, S. E. (1966). Predicting organizational effectiveness with a four-factor theory of leadership. *Administrative Science Quarterly*, *11*, 238–263.

Cartwright, D. & Zander, A. (1960). *Group dynamics—research and theory*. Evanston, Ill.: Row, Peterson.

Chance, M. R. A. & Larsen, R. R. (1976). *The social structure of attention*. New York: John Wiley & Sons Ltd.

Eckstein, H. & Gurr, T. R. (1975). *Patterns of authority: A structural basis for political inquiry*. New York: John Wiley & Sons Ltd.

Hallinan, M. T. (1978/79). The process of friendship formation. *Social Networks*, *1*, 193–210.

Hollander, E. P. & Julian, J. W. (1976). Contemporary trends in the analysis of leadership processes. In E. P. Hollander & R. G. Hunt (Eds.), *Current perspectives in social psychology* (4th edition) (Pp. 474–483). New York: Oxford University Press.

Katz, D., Maccoby, N., & Morse, N. C. (1950). *Productivity, supervision, and morale in an office situation*. Ann Arbor: University of Michigan, Institute for Social Research.

Larsen, R. R. (1976). Charisma: a reinterpretation. In M. R. A. Chance & R. R. Larsen (Eds.), *The social structure of attention*. New York: John Wiley & Sons Ltd.

Linton, R. (1936). *The study of man*. New York: Appleton Century.

McGregor, M. (1960). *The human side of enterprise*. New York: McGraw-Hill Book Co.

Nimmo, D. D. (1974). *Popular images of politics: A taxonomy*. Englewood Cliffs, N.J.: Prentice-Hall Inc.

Selltiz, C., Wrightsman, L. S., & Cook, S. W. (1976). *Research methods in social relations* (3rd edition). New York: Holt, Rinehart & Winston.

Shaw, M. E. (1981). *Group dynamics: The psychology of small group behaviour*. New York: McGraw-Hill Book Co.

Smith, M. (1945). Social situation, social behaviour, social group. *Psychological Review*, *52*, 224–229.

Verba, S. (1961). *Small groups and political behaviour: A study of leadership*. Princeton, N.J.: Princeton University Press.

Wright, H. F. (1960). Observational child study. In P. G. Mussen (Ed.), *Research methods in child development*. New York: John Wiley & Sons Ltd.

12 Nice Guys DON'T Finish Last: Aggressive and Appeasement Gestures in Media Images of Politicians

Roger D. Masters
Department of Government, Dartmouth College, Hanover, U.S.A

Biopolitics has begun to emerge as a new perspective in the study of politics (Somit, 1976; Watts, 1981; White, 1981; Wiegele, 1979). Building on the impressive and varied advances in the biological sciences over the last generation, political scientists have explored the political implications of psychophysiology (Tursky, Lodge & Cross, 1976; Wiegele, 1978; psychopharmacology (Somit, 1968; Stephens, 1970); genetic engineering (Blank, 1981), ethology (Barner-Barry, 1981; Corning, 1971; Schubert, 1981; Willhoite, 1971); and evolutionary theory (Corning, in press; Masters, 1975; in press; Pettman, 1981). A recent bibliography of works in "biopolitics" (Somit, Peterson, Richardson & Goldfischer, 1980) includes 288 items published since 1963. Indeed, it has been argued that biology can and should provide the unifying "paradigm" for the entire discipline of political science (Wahlke, 1979).

In earlier publications, I have outlined a research design utilising ethological concepts in the study of American Presidential elections (Masters, 1976; 1981; Note 1). Here, some of the findings in this project are reported. For brevity, hypothesis and methods will be only summarised on the assumption that those interested in details can consult earlier work. Results will be set out in three major categories: "attention" and photographic coverage; non-verbal cues of dominance and leadership; and evidence of editorial bias.

The research is based on componential analysis of photographs of American politicians in the printed media. Although easily extended to television coverage, analysis of the images of politicians—studying both overall frequency and non-verbal cues related to dominance and leadership—is facilitated by using newspaper and magazine photographs. Two election years were chosen for study: 1960 and 1972. As Richard M. Nixon lost one of these races and won the other, this choice was intended to permit control for such factors as individual, gestural repertoires and "personality". Pictures of American politicians were coded for 114 attributes—some being components of non-verbal gestures (lips open, lips closed, lips tucked in; arms above head, at head level, down, etc.), and some being aspects of the entire image or setting (politicians with other nationally known figures, with crowds, camera angle above or below head level, etc.) A more detailed description of the procedure has been published elsewhere (Masters, 1981).

All photographs of American politicians in the *New York Times* were coded for the entire 12 months of each election year. To control for differences between newspapers and other forms of journalism, photographs in weekly news magazines were also studied: *Newsweek* (for 1960) and *U.S. News and World Report* (for 1972). In addition, the *New Orleans Times-Picayune* was analysed (for 1972 only) to control for possible differences between nationally distributed media and a paper of primarily local and regional importance. A total of 4356 photographs were studied in the five data-sets.

"ATTENTION" AND POLITICAL SUCCESS

As M. R. A. Chance and others have shown (see Chance & Larsen, 1976), the structure of attention is a critical element in the social organisation of primate groups. As this approach has have very promising results in studies of leadership and dominance in groups of pre-school children (Barner-Barry, 1977; Montagner, 1978; Strayer, 1978; 1981), it is not unreasonable to assume that individuals capable of attracting and keeping public attention will also be more successful in the national, political arena. Hence it is possible to conceptualise an American Presidential campaign as a contest for attention, passing through phases (pre-Primary, Primaries, Convention, Election campaign, post-Election) that are defined in terms of the relevant audiences of principal concern to political rivals (Masters, Note 1)

In an introductory discussion of the relevance of ethology to the study of politics, (Masters, 1976, p. 211), it was hypothesised that, ". . . the more an individual is the source of communications which provide the focus of public attention, the more likely he is to establish or retain dominance". It follows that, ". . . candidates who cannot become the focus of public attention only win when mere partisan identification is sufficient to insure election" (ibid., p. 224). Hence I have suggested that there will be a high correlation between the candidate's proportion of photographs in the print media and their electoral success (as measured either by polling data or election results). This phenomenon should, moreover, be observed not only in the Presidential election itself (i.e., winners in November should have received more photographic coverage than losers), but in the pre-Primary and Primary process leading to the nominations of each major party.

Overall, the data confirm the correlation between successful leadership and public attention. The Presidential nominees received a large share of the photographic coverage in all periodicals studied—and in each year, despite the American doctrine of "equal time" in campaign coverage, the winner had a larger share of the coverage than the loser. Hence, of the 1960 pictures, 16.4% were of Kennedy and 13.5% of Nixon, whereas in the 1972 sample, 19.3% of the photos showed Nixon and 12.9% McGovern (Table 12.1).

The capacity to attract attention might also be measured by the content of photographs, as those that show a candidate with crowds or with other nationally known political figures imply that the subject is a focus for others in the political system. For each data-set studied, therefore, the percentage of pictures of the Presidential nominees was calculated for pictures showing a visible crowd or other nationally known personalities (see Tables 12.9, 12.11). In both cases, the candidates for President were much more likely than other politicians to be shown in these kinds of

TABLE 12.1
Total Photographic Coverage by Year—Presidential Candidates[1]

	Winner	*Loser*	
1960	*Kennedy*	*Nixon*	*Total Pictures*
N.Y. Times	270 (16.0%)	214 (12.7%)	1685
Newsweek	44 (18.7%)	45 (19.1%)	235
Year Total	314 (16.4%)	259 (13.5%)	1920
1972	*Nixon*	*McGovern*	
N.Y. Times	236 (20.5%)	186 (16.2%)	1149
N.O. Times–Picayune	108 (13.8%)	56 (7.1%)	785
U.S. News	127 (25.3%)	73 (14.5%)	502
Year Total	471 (19.37%)	315 (12.9%)	2436
Study Total	785 (18.07%)	574 (13.2%)	4356

[1]Total number of photographs (and percentages of all pictures of American politicians) for each year in each publication.

pictures. And except for Nixon in 1972 (who was shown with crowds less often than McGovern in two of the data-sets), the winner appeared more often than the loser in this type of picture. Perhaps more important than annual averages, however, is the breakdown of the data on attention by phases of the campaign.

Primaries and Nominations

During the period of each electoral year that culminates in the nominating conventions, politicians are competing with others in their *own* party for leadership. For the period from January 1st through to the Nominating Conventions, therefore, the number of pictures of each candidate was calculated as a percentage of the leading figures in his own party.

In all five data-sets (*New York Times* and *Newsweek* for 1960 as well as *New York Times*, *New Orleans Times–Picayune*, and *U.S. News and World Report* in 1972), the eventual nominee received more photographic coverage than his rivals (Tables 12.2, 12.3). To be sure, in 1960 Nixon received less coverage than Eisenhower—then an extraordinarily popular incumbent. But in 1972, when himself running for re-election, Nixon benefited from the similar advantage of incumbency. Hence the data show the

TABLE 12.2
Pre-Nomination Coverage of Leading Republicans[1]

1960	*Nixon*	*Eisenhower*	*Rockefeller*	*Other*	*Total*
N.Y. Times	21% (60)	62% (178)	15% (42)	3% (8)	288
Newsweek	37% (23)	40% (25)	16% (10)	9% (5)	63
Year average	23.6%	57.8%	14.8%	3.7%	351
1972	*Nixon*	*Rockefeller*	*Goldwater*	*Other*	*Total*
N.Y. Times	64% (156)	7% (16)	1% (2)	28% (69)	243
N.O. Times –Picayune	72% (68)	3% (3)	3% (3)	22% (21)	95
U.S. News	76% (82)	0% (0)	0% (0)	24% (26)	108
Year average	68.6%	4.3%	1.1%	26.0%	446

[1]Photographs of GOP leaders—January 1st through to Conventions. Photographs of nominee and selected rivals for attention expressed as percentages of pictures of leading Republicans during Pre-Primary, Primary, and Convention phases of year.

TABLE 12.3
Pre-Nomination Coverage of Leading Democrats[1]

1960	*Kennedy*	*Humphrey*	*Johnson*	*Stevenson*	*Symington*	*Truman*	*Total*
N.Y. Times	39% (88)	15% (33)	17% (39)	9% (21)	13% (28)	6% (14)	223
Newsweek	38% (21)	11% (6)	20% (11)	14% (8)	7% (4)	11% (6)	56
Year average	39.1%	14.0%	17.9%	10.4%	11.5%	7.2%	279

1972	*McGovern*	*Humphrey*	*Lindsay*	*Muskie*	*Wallace*	*Other*	*Total*
N.Y. Times	32% (118)	15% (56)	10% (38)	15% (55)	10% (37)	18% (68)	372
N.Y. Times –Picayuna	36% (34)	13% (12)	9% (9)	17% (16)	18% (17)	7% (7)	95
U.S. News	34% (46)	15% (20)	4% (5)	13% (17)	14% (19)	21% (28)	135
Year average	32.9%	14.6%	8.6%	14.6%	12.1%	17.1%	602

[1]Photographs of Democratic Party leaders—January 1st through to Conventions. Photographs of nominee and selected Rivals for attention expressed as percentage of pictures of leading Democrats during Pre-Primary, Primary, and Convention phases of year.

striking extent to which leadership is correlated with the frequency of photographic images during the Presidential nominating process.

Analysis of the relation between coverage and popular support as measured by public opinion polls is not yet complete. Preliminary data indicate, however, that changes in the proportion of attention focused on a politician correspond closely to changes in popularity (Masters, 1981). *It must be stressed that, even if confirmed, this would only indicate correlation, and is not intended to suggest causal relationships between media attention and success.* In effect, at some points the causal relation is clearly from leadership position to public attention (as is the case for incumbents), whereas at other times increases in media coverage may precede growing popular support.

The Presidential Campaign

During the campaign proper, the two major party candidates are in head-to-head competition. Hence it has seemed useful to consider each candidate's coverage compared only to his rival. For this purpose, the number of photographs of the eventual winner was calculated as a percentage of the total for the two parties' candidates by phase of the year (Table 12.4). At first sight, only two of the five data-sets seem to confirm the

TABLE 12.4
Winner's Share of Coverage[1]

1960	*Pre-Primary*	*Primaries*	*Conventions*	*Campaign*	*Post-Election*	*Total*
N.Y. Times	46%	56%	76%	44%	87%	56%
Newsweek	42%	48%	60%	39%	77%	49%
1972						
N.Y. Times	82%	50%	43%	49%	62%	56%
N.O. Times –Picayune	94%	87%	33%	67%	82%	70%
U.S. News	95%	64%	27%	56%	87%	63%

[1]Percentage of photographs of Presidential nominees by phase of campaign, derived from the number of photographs of winner (1960 = Kennedy; 1972 = Nixon) divided by number of photographs of winner plus loser for each phase.

hypothesis; from the last Nominating Convention to Election Day, the eventual loser is shown *more* often than the winner in the *New York Times* for both years studied, and in *Newsweek* for 1960. On closer inspection, however, the findings are not too hard to explain.

From July 28th to October 31st, 1960, the *New York Times* actually showed more photographs of Kennedy (91) than of Nixon (89). In the first days of November, 1960, however, the *Times* suddenly published 55 pictures of Nixon (compared to only 23 of Kennedy). Photographs of Nixon outnumbered those of Kennedy during the entire campaign (56% to 44%), so this difference was thus entirely due to the extraordinary flurry of Nixon pictures between November 1st and 8th. As is known from the polling data, during this same time Nixon's support in the polls rose sharply as he made a strong run in that "stretch" and almost overcame Kennedy's earlier lead (Masters, 1981).

In the *New York Times* coverage of the 1972 election, there is a similar phenomenon. From the last Nominating Convention to the end of October, there are more photographs of Nixon (51) than of McGovern (50). In the first two days of November, however, the *Times*—which had endorsed McGovern—ran three pictures of the Democratic nominee and none of his rival. One could conjecture that this last-minute series of pictures was a response to predictions of a Nixon victory. Whatever the explanation, it is evident that the precise timing of photographic coverage is of importance, as the hypothesis is confirmed in the *New York Times* for the campaigns of 1960 and 1972 up to the end of October, but not in the first days of November.

The remaining data-set seems rather different, as *Newsweek* was less likely than the *Times* to publish pictures of Kennedy for *all* phases of the

1960 election year. Why *Newsweek* had this tendency to give the Republican nominee more photographic coverage cannot be explained without further research. But, as will be shown below, such evidence of differences between publications provides a means of assessing the frequently expressed complaints that editorial bias influences ostensibly "objective" reporting of political events.

After the 1960 election, and for the year as a whole, the *New York Times* published more photographs of Kennedy than of Nixon (as predicted by the hypothesis). Indeed, except for *Newsweek* in 1960, which came close to giving equal photographic coverage to the two candidates, over the entire year the winning candidate was shown more often than the loser in all publications studied. Hence one could argue that the only data-set not confirming the basic prediction is *Newsweek* for 1960—and even there, Kennedy receives more coverage than his rival both during the Convention period and after his election.

Post-Election Period

In many respects, the period from Election Day to December 31st of each year is a critical test of the hypothesis. If photographic coverage is a measure of leadership, one would expect the winner to eclipse the loser as the focus of media attention as soon as the votes have been counted. This is exactly what the data show for all five data-sets. In the study as a whole, the winner's share of the Presidential nominees' photographs was over 88% during the post-election period—indicating a tendency to consign the loser to virtual oblivion (at least as compared to the limelight of the campaign). Even *Newsweek* in 1960, which up to Election Day had published 42 pictures of Nixon and only 34 of Kennedy, reversed this pattern during the last phase of the year.

This may seem like an obvious finding, not worthy of notice. But it provides strong evidence of the legitimacy of the American electoral process. The degree to which attention shifts to the winner does, after all, indicate a willingness to accept the election results as final and proper. Hence it is worth noting that while the *New York Times* published 24 pictures of Nixon and only 15 of McGovern in the post-election period of 1972, this proportion (62%) was the lowest in the sample—possibly reflecting different degrees of enthusiasm with the outcome. Similarly, after Nixon's defeat in 1960, he received more coverage in *Newsweek* than in the *Times*, thus matching *Newsweek's* tendency to give the Republican nominee more coverage throughout the year.

It would be interesting to use similar measures of attention to test the legitimisation of electoral outcome in other political systems. For example, it could be hypothesised that strongly ideological publications that deny the legitimacy of a regime would *not* show the tendency to shift photographic coverage to the winner observed in the United States. This is consistent with general impressions of the reporting in newspapers like the French Communist Party's organ, *L'Humanité*, but further research would, of course, be needed to determine whether attention in the media is indeed reliable as a measure of the legitimacy of free elections.

NON-VERBAL CUES OF DOMINANCE

Although the data will permit analysis of a wide variety of traits, it has seemed most useful to begin from the general presumption that leadership is somehow related to "dominance". Translated into ethological terms, this suggests the hypothesis that successful politicians will exhibit dominant non-verbal cues more frequently than losers. Or, to use an earlier form of this prediction (Masters, 1976, p. 225) "the frequency of dominance gestures will be correlated with a politician's success in attracting public attention and support, and will decline in the course of a manifestly unsuccessful electoral campaign".

Aggressive Gestures and Political Success

Among primates, Chance (1976) has identified two "modes" of social interaction related to group organisation and attention structure: "agonic" (mediated by gestures of threat and submission), and "hedonic" (characterised by display and facilitation rather than competitive behaviours). When dominance is discussed among humans, however, it is often assumed that the behaviours involved are principally those defined as aggressive or "agonic" (e.g., Lorenz, 1966; Maclay & Knipe, 1972). As the saying goes, "Nice Guys Finish Last".

In a study of the social behaviour of young children, Zivin (1977) demonstrated that such an approach permits a good prediction of the "winner" in competitive relationships. In her experiment, when two children competed for an object, the one exhibiting a threatening or "plus" face *before* the encounter won 66% of the time, whereas 51% of the losers had previously shown a submissive or "minus" face (Figs. 12.1, 12.2). Subsequent research has confirmed that the sight of a "plus" face does indeed seem to lower the rival's motivation for conflict (Zivin, Note 2).

It therefore seemed logical to test the extent to which the non-verbal images of Presidential candidates are characterised by similar dominance cues. As Zivin (1977, p. 717) defined these non-verbal facial gestures:

> the human 'plus' face . . . resembles various non-human primate faces that are commonly called 'threat' faces. The points to note about this human face are the raised brows, the eyes wide open in apparent direct eye contact, and

FIG. 12.1 The "plus face.

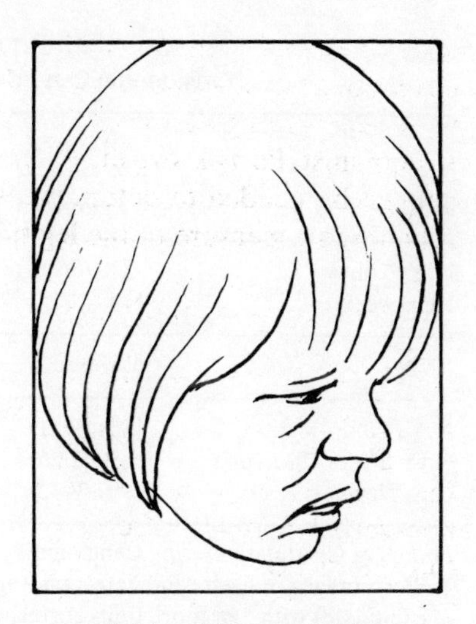

FIG. 12.2 The "minus" face.

> the raised chin. . . . the 'minus' face . . . resembles non-human primate 'submission' faces. Its features of interest are the gently furrowed brows, the eyes dropped in broken eye contact, and the lowered chin.

Hence for the five data-sets described in this study, the frequency of both "plus" and "minus" faces were caculated for both Presidential candidates.

Because Zivin's analysis used overall assessments of children's faces, whereas the coding protocol here is based on discrete components of the human face, her definition of the features of the "plus" face was coded as any picture having any five or more of the following traits: eyebrows and forehead raised, eyelids fully open or normal, eyeballs up or level, crow's feet present, teeth closed, lips closed or tucked in, mouth corners down, head up, tilted or turned. All pictures having five or more of these traits were considered "plus" faces, *unless* the picture also had more than two components of a "minus" face (in which case the image was considered to be a "blend" and discarded from the analysis).

Data for the frequency of "plus" faces in pictures of the Presidential nominees for all five data-sets are presented in Table 12.5. Neither the proportion of a candidate's pictures with a "plus" face nor the raw number (in parenthesis) confirms the hypothesis; in both 1960 and 1972, threat faces are exhibited in relatively small numbers and about equally often for both candidates. Indeed, in only one data-set (*Newsweek* for 1960) did the winning candidate have a higher proportion of his photos showing the "plus" face—and in that case the difference is due to a single picture in a small sample.

TABLE 12.5
Presidential Candidates and Zivin's "Plus" Face[1]

	Winner	Loser	
1960	*Kennedy*	*Nixon*	*All Others*
N.Y. Times	3% (3/114)	3% (5/144)	8% (24/290)
Newsweek	17% (2/12)	5% (1/19)	11% (2/18)
1972	*Nixon*	*McGovern*	*All Others*
N.Y. Times	18% (9/51)	21% (11/53)	13% (9/67)
N.O. Times–Picayune	8% (2/26)	15% (2/13)	5% (7/140)
U.S. News	9% (3/32)	8% (2/25)	11% (6/56)

Photos of Candidate During Campaign

[1]Percentages of each candidate's photographs (and, in parenthesis, number/total coverage of candidate) with 5 or more traits corresponding to "plus" or threat face (Zivin, 1977) during Campaign phase of year (end of Nominating Conventions to Election Day). Photos also having 3 or more traits of "minus" face ("blends") excluded. For definition of "plus" and "minus" face, see text.

In contrast, however, there does seem to be a difference in the frequency of "minus" faces (Table 12.6)—especially in the 1960 data. Kennedy was shown more often with this gesture than Nixon, both in absolute numbers and proportionally, in both the *New York Times* and *Newsweek*. Moreover, in 1972, Nixon as victor was shown with the "minus" face more often

TABLE 12.6
Presidential Candidates and Zivin's "Minus" Face[1]

	Winner	Loser	
1960	*Kennedy*	*Nixon*	*All Others*
N.Y. Times	30% (34/114)	15% (22/144)	13% (37/290)
Newsweek	17% (2/12)	5% (1/19)	0% (0/18)
1972	*Nixon*	*McGovern*	*All Others*
N.Y. Times	10% (5/51)	19% (10/53)	18% (12/67)
N.O. Times–Picayune	8% (2/26)	8% (1/13)	5% (7/140)
U.S. News	25% (8/32)	20% (5/25)	16% (9/56)

Photos of Candidate During Campaign

[1]Percentages of each candidate's photographs (and, in parenthesis, number/total coverage of candidate) with 5 or more traits corresponding to "minus" or submission face (Zivin, 1977) during Campaign phase of year (end of Nominating Conventions to Election Day). Photos also having 3 or more traits of "plus" face ("blends") were excluded. For definition of "plus" and "minus" face, see text.

than McGovern in two of the three data-sets; only in the *New York Times* (which had endorsed McGovern) did the loser appear more frequently than the winner showing this submissive facial cue. And in the *Times*, others were shown even more often with a "minus" face than was McGovern (Table 12.6). As the Democratic nominee was shown 21% of the time with a "plus" face and only 19% of the time with a "minus" face, the *Times* figure is probably due to the relative *absence* of appeasement or facilitating cues in photos of Nixon, rather than their *presence* in pictures of his rival.

Thus the data not only fail to confirm the correlation between aggressive cues and political success, but they point to the reverse: submissive or appeasement gestures seem particularly frequent among high-status politicians (Masters, 1981), and in the current study are generally correlated with winning rather than losing. This finding is not as paradoxical as it might appear. In affect, several studies of social behaviour in "natural" groups of pre school children have shown that the leader is *not* the most aggressive individual. Indeed, among primates as well as human children (Barner-Barry, 1977; Montagner, 1978; Strayer, 1978; 1981), it is typically the number 2 male in the dominance hierarchy that exhibits aggressive threat most frequently. The leader, in contrast, tends to be more self-assured—for example, capable of appeasing conflicts between others in the group as well as serving as a focus of imitation or observation (Maclay & Knipe, 1972; Montagner et al., 1978). As leaders may need to exhibit aggression on occasion, particularly when directly challenged, they are thus more frequently observed in appeasement or facilitating roles in human groups (and among some primates with comparable, face-to-face, social processes).

When reconsidered from this perspective, the data in Table 12.5 are consistent with ethological observations. In effect, of the unblended "plus" faces printed, a high proportion were exhibited by Presidential nominees. Combining both data-sets for 1960, for example, there were a total of 36 pictures coded as unblended "plus" faces; of these, 11 (31%) showed 1 of the Presidential nominees. Similarly, of the 51 unblended "plus" faces in the 3 1972 data-sets, 29 (57%) were of either Nixon or McGovern. Hence, one could say that an effective leader must be capable of showing aggression when threatened, but that such responses are not the typical or most frequent image of the successful candidate. Kissing babies and shaking hands are, in ethological terms, appeasement gestures and not threats.

Other traits analysed to date are consistent with this interpretation. It has been noted that pictures including crowds or nationally known figures are more frequently shown of Presidential nominees than of other politicians (Tables 12.9, 12.11). Such pictures presumably symbolise the capacity of an individual to draw other people together, not

TABLE 12.7
Percentage of Presidential Candidate's Photos showing Ekman's Happiness

	Winner	*Loser*	
1960	*Kennedy*	*Nixon*	*Others*
N.Y. Times	45% (121/270)	46% (98/214)	34% (413(1214)
Newsweek	14% (6/43)	20% (9/45)	19% (29/150)
1972	*Nixon*	*McGovern*	*Others*
N.Y. Times	51% (113/224)	54% (100/185)	32% (236/728)
N.O. Times –Picayune	42% (45/108)	33% (15/46)	32% (199/631)
U.S. News	26% (49/189)	45% (32/71)	25% (50/303)

to threaten or scare. During the campaign phase of each year (see Tables 12.10, 12.12), the frequency of these pictures is similar to the year-long averages, except that the winner is not consistently shown with crowds or nationally known people more often than the loser.

Note that the data also show individual differences. For the entire year and the campaign itself, Nixon was less likely to be shown drawing crowds, in both years studied. As with Zivin's "minus" face (Table 12.6) and pictures showing crowds during the campaign itself (see Table 12.10), when winning in 1972, Nixon seems to have been less effective than Kennedy in 1960 in projecting the image of a self-confident leader.

TABLE 12.8
Percentage of Presidential Candidate's Photos showing Ekman's Happiness (Campaign Period Only)

	Winner	*Loser*	
1960	*Kennedy*	*Nixon*	*Others*
N.Y. Times	42% (48/114)	48% (69/144)	40% (117/290)
Newsweek	8% (1/12)	26% (5/19)	11% (2/18)
1972	*Nixon*	*McGovern*	*Others*
N.Y. Times	55% (28/51)	59% (31/53)	46% (33/72)
N.O. Times –Picayune	31% (8/26)	8% (1/13)	23% (25/106)
U.S. News	56% (14/25)	56% (18/32)	32% (18/56)

The capacity of politicians to serve as the focus of attention does not seem to be primarily related to the overall characteristics usually associated with "good pictures" of an individual. For example, we tend to have a bias toward "smiling" images, perhaps because—as Paul Ekman has demonstrated—the smile is a universally understood non-verbal cue for happiness (Ekman, Friesen, & Ellsworth, 1976). Interestingly enough, however, pictures showing what Ekman defined as a "happy" facial image do *not* seem more frequent for winners, either in the year as a whole (Table 12.7) or during the campaign phase itself (Table 12.8). Hence, insofar as there are non-verbal cues that seem linked to winning, the "minus" face described by Zivin seems to be more reliable than most other traits studied to date.

EVIDENCE OF BIAS

Do editorial preferences influence images of candidates? The foregoing analysis suggests that potentially important bias could arise either from editor's choices in the *amount* of coverage, or in the *quality* of the non-verbal cues conveyed to the public. In both the allocation of photographic attention and in the images selected by the media, the current data provide a ready test of differences arising from partisanship or personal preferences.

Although all five data-sets are generally similar, several consistent differences are evident. In 1960, for example, the *New York Times* was *less* likely to print photographs of Nixon than was *Newsweek* (Table 12.1). That this was by no means a matter of *partisan* preference for Democrats is clear from Tables 12.2 and 12.3: during the pre-Primary and Primary phases of the 1960 campaign, the *Times* published 288 photographs of leading

TABLE 12.9
Percentage of Presidential Candidate's Photos with Crowds Visible

	Winner	*Loser*	
1960	*Kennedy*	*Nixon*	*Others*
N.Y. Times	22% (58/270)	20% (42/214)	10% (118/1214)
Newsweek	21% (9/43)	17% (10/58)	10% (10/105)
1972	*Nixon*	*McGovern*	*Others*
N.Y. Times	13% (31/236)	18% (32/179)	13% (96/731)
N.O. Times–Picayune	11% (12/108)	9% (4/46)	6% (37/631)
U.S. News	17% (22/127)	25% (18/73)	12% (37/301)

Republicans and only 223 of leading Democrats. But of Republicans shown during this period, the *Times* allocated only 21% to Nixon, compared to 37% of similar pictures in *Newsweek* (Table 12.2). As no similar contrast existed in the two publications' images of Democrats (Table 12.3), it is not impossible that editors differed in their responses to Nixon's personality.

Whereas *Newsweek* seems to have been less likely to publish pictures of Kennedy (the eventual winner) than of Nixon during all phases of 1960, in the 1972 sample *U.S. News* was more likely than publications in the other data-sets to show Nixon (Table 12.4). In both years, therefore, the news-weekly magazine studied had a tendency to give more photographic exposure to the Republican candidate. In 1972, the *New Orleans Times–Picayune*—perhaps not surprisingly—gave George Wallace more coverage during the period up to the Conventions than did the other two publications studied (Table 12.4). The data show other examples of such marginal differences in photographic coverage, presumably due either to the expected audience or to editorial decisions.

Similarly, it is not difficult to show differences in the non-verbal cues exhibited in politicians' pictures. As has been noted, in 1972 the *New York Times* was less likely to show Nixon with a "minus" face than the other publications. But if all instances of *either* a "plus" or a "minus" face are considered as a percentage of politicians' pictures during the campaign, the *New Orleans Times–Picayune* published fewer of these extreme gestures (11.7% of campaign photos) than either the *New York Times* (33.0%) or *U.S. News* (29.2%); even though unblended "plus" or "minus" faces were less frequent in the 1960 sample (*New York Times*, 22.8%; *Newsweek*, 16.3%), the one regional paper studied seems to have avoided a kind of picture that is particularly typical of highly dominant leaders.

TABLE 12.10
Percentage of Presidential Candidate's Photos with Crowds Visible (Campaign Period Only)

	Winner	*Loser*	
1960	*Kennedy*	*Nixon*	*Others*
N.Y. Times	30% (34/114)	24% (35/144)	12% (35/290)
Newsweek	58% (7/12)	21% (4/19)	6% (1/18)
1972	*Nixon*	*McGovern*	*Others*
N.Y. Times	16% (8/51)	28% (13/46)	19% (14/72)
N.O. Times–Picayune	12% (3/26)	8% (1/13)	10% (11/106)
U.S. News	38% (12/32)	60% (15/25)	21% (12/56)

TABLE 12.11
Percentage of Presidential Candidate's Photos with One or More Nationally Known Figure Visible

	Winner	*Loser*	
1960	*Kennedy*	*Nixon*	*Others*
N.Y. Times	33% (88/270)	29% (63/214)	22% (271/1214)
Newsweek	40% (19/48)	16% (7/43)	13% (19/144)
1972	*Nixon*	*McGovern*	*Others*
N.Y. Times	43% (102/237)	25% (46/186)	16% (109/676)
N.O. Times–Picayune	51% (55/108)	30% (14/46)	21% (122/591)
U.S. News	34% (43/127)	21% (15/73)	19% (56/302)

But do these differences reflect anti-Nixon feeling at the *New York Times*, or pro-Nixon sentiment at *Newsweek*, in 1960? Were differences in 1972 between the *New Orleans Times–Picayune* and the other publications due to editorial prejudice, market considerations, or other factors? Although more detailed evidence is needed to answer these questions, the findings reported here indicate that the method of componential analysis can discover both central tendencies in media coverage of politicians, and the bias introduced by editorial decisions. As a result, the extension of ethological concepts to the study of election campaigns should make it possible to improve our understanding of the extent to which non-verbal cues may influence the electorate.

TABLE 12.12
Percentage of Presidential Candidate's Photos with One or More National Known Figures Visible (Campaign Period Only)

	Winner	*Loser*	
1960	*Kennedy*	*Nixon*	*Others*
N.Y. Times	19% (22/114)	22% (33/144)	21% (62/290)
Newsweek	32% (6/19)	8% (1/12)	17% (3/18)
1972	*Nixon*	*McGovern*	*Others*
N.Y. Times	49% (26/53)	26% (14/53)	33% (24/72)
N.O. Times–Picayune	35% (9/26)	15% (2/13)	32% (34/106)
U.S. News	9% (3/32)	16% (4/25)	14% (8/56)

CONCLUSIONS

Although the conceptualisation for this study of candidate's images was derived from contemporary ethology, it provides a means of studying a phenomenon defined a generation ago by Harold Lasswell (1948, p. 91):

> The flow of activity between two or more interacting persons is guided by the presentation of cues at the focus of attention of the participants. Many situations are so highly specialised that the cue-giving function is concentrated in one person. . . . It is possible to "score" various situations according to the characteristic pattern of cue-giving and taking. This can be done by describing the focus of attention of each participant through the entire course of the interaction.

Whereas Lasswell gave the examples of an orchestra conductor and a legislative debate—i.e., situations in which all the participants are physically present—there is no reason why this perspective cannot apply equally well to electoral campaigns in which the process of focusing attention on "cues" of leadership is transmitted by the media.

It is particularly appropriate to consider facial gestures as leadership cues because of their immense importance in primate social interaction, as well as in human emotional behaviour (von Cranach, 1979). As stated by Maclay and Knipe, (1972, p. 76): "The face is the most obvious part of the body to use for symbolizing the dominance value of the intricate personalities we present to one another." Although important in a metaphorical sense ("facing up to situations"; "facing down an opponent"; "losing face"), it is far from simplistic to begin from the basic facts of how politicians' faces are *seen* by the electorate. In this regard, two major findings arise from the current research.

First, successful politicians translate their leadership positions into a large proportion of the visual images transmitted to the electorate. Americans have often commented negatively on the ever-present photographs of Stalin, Mao, or current Soviet or Chinese leaders as a symbol of totalitarian control. But even in our own society, where there is competition for power, a small number of politicians receive an awesome share of photographic coverage. Of the 4356 pictures included in the five data-sets studied here, 1359 (31.2%) were of the two major parties' Presidential candidates. In both years, incumbents received disproportionately large coverage early in the year, and the newly elected President was the subject of more coverage than the loser after Election Day. Like other primates and pre-school children, citizens in complex human societies follow individuals who serve as the focus of attention.

Second, the images of political leaders convey non-verbal cues of dominance that are *different* from those normally exhibited by other political actors. But contrary to the assumption that high-intensity gestures related to aggressiveness would be typical of successful leaders, pictures of winning candidates did not show these traits more often than losers—or other politicians. If anything, the cues shown more frequently by winners are those related to appeasement (Masters, 1981), and are similar to the submissive faces of children who *lose* in dyadic competition.

This is a far from trivial finding. On the contrary, it leads to a hypothesis that might explain why so many public opinion polls failed to predict Ronald Reagan's convincing victory in the 1980 election. Polls measure conscious, cognitive attitudes; non-verbal cues tap a deeper, "gut" reaction to leaders (Masters, 1976). Where voters are undecided—particularly if they have little information about issues—such images of leaders presumably play a disproportionate role in determining voting choices. In 1980, many voters were cross-pressured and made their decision late in the campaign. In particular, there seems to have been a swing toward Reagan just before Election Day. The current study raises the question of whether the non-verbal cues in images of Reagan and Carter—and especially their gestures during their TV debate—could have had a major influence on the outcome of these late decisions.

In effect, Reagan exhibited a high frequency of relaxed appeasement or submissive gestures during the debate: he looked down frequently, smiled frequently, and—in speech patterns—hesitated. It was Reagan who crossed the space between the candidates to initiate hand-shaking. In contrast, Carter stared directly at the camera, smiled rarely, and generally appeared tense and threatening. On the non-verbal level, therefore, Carter's *gestures* indicated that *he* was the threatening or aggressive candidate, not Reagan. For the cross-pressured voter, this non-verbal message contradicted Carter's verbal insistence that it was Reagan who was a "dangerous" man, hence negating the principal argument against the Republican challenger.

If this interpretation is correct, a single event at the focus of national attention could well determine popular images of candidates, and therewith have a disproportionate effect on electoral campaigns. Muskie's "sobbing" outside the *Manchester Union–Leader*, or Romney's admission that he was "brain-washed" would illustrate such events. Verbal and substantive statements of policy can have this influence, but it is sobering to realise that a non-verbal cue symbolising leadership or lack of self-confidence is equally likely to fix the attitudes of uncommitted or cross-pressured voters. Although it may wound our pride to admit it, our species is—to emphasise both words in Aristotle's phrase—the *political animal*.

REFERENCE NOTES

1. Masters, D. (1978) *Attention structure and political campaigns*. Paper presented to the Annual Meeting of the American Political Science Association, New York.

2. Zivin, G. (1977). *Stopped cold: Sight of two facial gestures differentially affects children's latencies during conflict*. Unpublished manuscript.

REFERENCES

Barner-Barry, C. (1977). An observational study of authority in a preschool peer group. *Political Methodology*, *4*, 415–449.

Barner-Barry, C. (1981). Longitudinal observational research and the study of basic forms of political socialization. In M. Watts (Ed.), *Biopolitics: Ethological and physiological approaches*: (*New Directions for methodology of social and behavioural science*, 7) (Pp. 51–60). San Francisco: Jossey-Bass.

Blank, R. H. (1981). *The political implications of human genetic technology*. Boulder, Col.: Westview.

Chance, M. R. A. (1976). The organization of attention in groups, in M. von Cranach (Ed.), *Methods of inference from animal to human behaviour* (Pp. 213–236). The Hague: Mouton.

Chance, M. R. A. & Larsen, R. R. (1976). *The social structure of attention*. New York: John Wiley & Sons Ltd.

Corning, P. A. (1971). The biological bases of behavior and some implications for political science, *World Politics*, *23*, 312–370.

Corning, P. A. (In press). *Politics and the evolutionary process*. New York: Harper & Row.

Ekman, P., Friesen, W. V., & Ellsworth, P. (1976). *Emotion in the human face*. New York: Pergamon.

Lasswell, H. (1948). *Power and personality*. New York: W. W. Norton & Co., Inc.

Lorenz, K. Z. (1966). *On aggression*. New York: Harcourt Brace Jovanovich, Inc.

Maclay, G. & Knipe, H. (1972). *The dominant man*. New York: Delacorte.

Masters, R. D. (1975). Politics as a biological phenomenon, *Social Science Information*. *14*, 7–63.

Masters, R. D. (1976). The impact of ethology on political science. In A. Somit (Ed.), *Biology and politics* (Pp. 197–233). The Hague: Mouton.

Masters, R. D. (1981). Linking ethology and political science: photographs, political attention, and presidential elections. In M. W. Watts (Ed.), *Biopolitics: Ethological and physiological approaches*: (*New directions for methodology of social and behavioural science, 7*) (Pp. 61–80). San Francisco: Jossey-Bass.

Masters, R. D. (In press). Is sociobiology reactionary? The political implications of inclusive fitness theory. *Quarterly Review of Biology*.

Montagner, H. (1978). *L'Enfant et la communication*. Paris: Stock.

Montagner, H., Henry, J. C., Lombardot, M., Restoin, A., Bolzoni, D., Durand, M., Humbert, Y., & Moyse, A. (1978). Behavioural profiles and corticosteroid excretion rhythms in young children, Parts 1–2. In V. Reynolds & N. G. Blurton-Jones (Eds.), *Human Behaviour and Adaptation*. London: Taylor & Francis.

Pettman, R. (1981). *Biopolitics and international values*. New York: Pergamon.

Schubert, G. (1981). The use of ethological methods in political analysis. In M. Watts (Ed.), *Biopolitics: Ethological and physiological approaches*: (New directions for methodology of social and behavioral science, 7) (Pp. 15–32). San Francisco: Jossey-Bass.

Somit, A. (1968). Toward a more biologically oriented political science: ethology and psychopharmacology. *Midwest Journal of Political Science*, *12*, 550–567.

Somit, A. (1976). *Biology and politics*. The Hague: Mouton.

Somit, A., Peterson, S. A., Richardson, W. D., & Goldfischer, D. S. (1980). *The literature of biopolitics*. DeKalb, Ill., Center for Biopolitical Research: Northern Illinois University.

Stephens, J. (1970). Some questions about a more biologically oriented political science. *Midwest Journal of Political Science*, *14*, 687–707.

Strayer, F. F. (1978). Social ecology of the preschool peer group. In W. A. Collins (Ed.), *The proceedings of the twelfth Minnesota symposium on child psychology*. Hillsdale, N.J.: Lawrence Erlbaum Associates Inc.

Strayer, F. F. (1981). The organization and coordination of asymmetrical relations among young children: a biological view of social power. In M. Watts (Ed.), *Biopolitics: Ethological and physiological approaches*: (*New directions for methodology of social and behavioral science, 7*) (Pp. 33–50). San Francisco: Jossey-Bass.

Tursky, B., Lodge, M., & Cross, D. (1976). A bio-behavioral framework for the analysis of political behavior. In A. Somit (Ed.), *Biology and politics*. (Pp. 59–96). The Hague: Mouton.

von Cranach, M. (1979). *Human ethology: Claims and limits of a new discipline*. Cambridge, England: Cambridge University Press.

Wahlke, J. (1979). Prebehavioralism in political science. *American Political Science Review*, *73*, 9–32.

Watts, M. (1981). *Biopolitics: Ethological and physiological approaches*: (*New directions for methodology of social and behavioral science, 7*). San Francisco: Jossey-Bass.

White, E. (1981). *Sociobiology and human politics*. Lexington, Conn.: Lexington Books.

Wiegele, T. C. (1978). The psychophysiology of elite stress in five international crises. *International Studies Quarterly*, *22*, 467–511.

Wiegele, T. C. (1979). *Biopolitics*. Boulder, Col.: Westview.

Willhoite, F. (1971). Ethology and the tradition of political thought. *Journal of Politics*, *33*, 615–641.

Zivin, G. (1977). Facial gestures predict preschoolers' encounter outcomes. *Social Science Information*, *16*, 715–729.

13 The Two Dimensions of Sociality

Theodore D. Kemper
Department of Sociology, St. John's University, New York, U.S.A

Every generalising science requires a valid and reliable set of categories for its descriptive and analytical functions. Without these, it is not possible to cumulate knowledge. Investigators cannot know whether their observations correspond with those of others, and theory lags behind because there can be no certainty that propositions in different theories are about the same phenomena.

The problem of categories has been particularly difficult in the behavioural sciences, especially when these sciences are broadly defined to include not only humans but also other primates. Differences between species seem especially insuperable in this regard but, even within the human domain, common features are often concealed under the blanket of specific cultures. Cultural differences are so great that it is attractive to think that although there may be some pan-human qualities these will not easily be revealed.

A contrary proposition is that regardless of the manifest differences between cultures (and between species) there are certain fundamental categories upon which behavioural science theories can be built. This chapter is devoted to discussion of a set of categories of this type, which are useful in characterising social relations and social structures across diverse cultural and species domains.

First, I will set out some of the materials that lead us to conclude that there is currently a set of descriptive and analytic categories suitable for theoretical purposes. Then I will give two illustrations of the use of these categories in the human domain, one at the level of individuals and the other at the level of social structure. The solution then arrived at entails two dimensions that are able to characterise both social relations and social structure. They are called here power and status-accord, (or status, in brief). The meaning of these dimensions will emerge in the discussion.

EVIDENCE FOR THE POWER AND STATUS DIMENSIONS

The fundamental question is how, usefully, to characterise the patterns of activities that two actors may engage in with reference to each other. This requires observation, but although informal observational schemes are of long-standing, systematic observations of human and non-human behaviour have only been undertaken in the present century. Ordinarily, what is required is a coding scheme that allows an observer to record a meaningful portion of the on-going behaviour.

In the era immediately after World War II investigators, armed with such coding protocols, observed numerous small groups of humans, in search of the fundamental dimensions of social behaviour. They subjected the results of their observations to factor analysis, and this joining of systematic observation with such analysis proved extremely fruitful. Carter (1954) provided the first major review of results from the wedding of these two methods. He concluded that three dimensions were found to describe the behaviour of members of small groups; he named these dimensions Group Goal Facilitation, Individual Prominence and Achievement, and Group Sociability. In order to understand these, we must look at the items that load the different factors, keeping in mind that factors are named according to the disposition and interests of the particular investigator.

Group Goal Facilitation was derived from items (aggregated across different studies) that included efficiency, co-operation, adaptability, "work with" skills, behaviour directed at group solution, and the like. This dimension reflects the task the actors have gathered to do.

Individual Prominence and Achievement was based on such items as aggressive, authoritarian, leadership, bold, forceful, not timid, physical ability, quick to take the lead. This set of behavioural characteristics appears to reflect a controlling and dominating stance toward others. We call it the power dimension.

The final factor, Group Sociability, emerged on the basis of such items as sociability, behaviour that is socially agreeable to group members, behaviour directed toward group acceptance, genial, cordial, and the like. We call this the status factor.

Before explicating these factors further, it is important to understand that, since Carter's presentation, these three dimensions have been found repeatedly in several dozen studies of social interaction, and in analyses of social dimensions of human personality (for a review, see Fromme &

O'Brien, 1982; Kemper, 1978). Hence I am dealing here with dimensions that have been repeatedly confirmed by diverse investigators. Let us now look at the three dimensions more closely.

We refer to Group Goal Facilitation as the task factor, reasoning that all social aggregations occur because the individual actors, by genetic design or through intentional assessment of events, conclude in some manner that they are not able to attain certain goals or undertake efficiently certain tasks alone. Among humans (and other primates), this is evidence of fundamental interdependence. We cannot survive at birth without care from others. Nor could we have been conceived and brought to birth without the co-operative sexual engagement of a male and female. From making love, to making war, to making automobiles, humans must engage in co-operation in regard to tasks.

Hence we may judge that humans in interaction are always engaged in some task. The task factor uncovered by Carter reveals in the behaviour of the actors how their interactions are orientated toward the task they are together to do. We may judge further that without the need to accomplish a task involving interdependent actors, there would be little reason for such actors to aggregate in the first place.

But all is not task interaction in the limited sense so far discussed. When we consider the two additional factors (Individual Prominence/ Achievement and Sociability) found by Carter, we see that they are of a different order. Foremost is the fact that the behaviours in these two factors are orientated not to task completion, but to the other actors. To be forceful or authoritarian, or sociable or cordial, is to orientate one's conduct towards others. Hence the power and status factors are fundamentally descriptive of relational conduct.

Indeed, when actors are together for a task they not only work on that task but relate to each other. One actor wants another to do something either faster or slower, or to stop doing something, or to start it. One actor wants to take precedence in the consumption of the rewards that accrue from the task activity. Other actors may not agree so there is some need to resolve this matter. If it cannot be settled by status-accord (essentially voluntary compliance, as will be discussed, there is strong likelihood it will be settled by forced (involuntary) compliance, or power.

In many cases it will be seen that the task for which the actors have come together is the relationship itself. A parent feeding a child is engaged in a task as well as a loving relationship; the physical and verbal interplay of sexual interaction is both task activity as well as expressive of relationship. Action in terms of the two relational factors, power and status, can be precisely the task the actors are together to engage in. We must now consider these two modes of relation, the power mode and the status mode, more formally.

Power is defined as the ability of one actor to realise his/her interests against the opposition of another actor. This follows closely the important formulation of the sociologist, Max Weber (1946). It signifies that in interaction one actor obtains involuntary compliance from one or more others. The others do not wish to comply, but they do so because to them the costs of non-compliance at that moment outweigh the costs of submitting to the more powerful actor. This interaction and its costs may involve the threat of physical punishment, verbal abuse, deprivation of benefits, loss of prized possessions, and so forth. The resources of power are widespread and involve both noxious and deprivational elements.

Once a number of decisive power exchanges have occurred, i.e., where one actor has compelled another to comply involuntarily, we may expect that the same outcome will recur on the next occasion of interaction. The more powerful actor will once again compel the less powerful to conform to his/her will. At this point there is justification in referring to a social structure of power. The notion of structure indicates pattern and predictability. Power relations are among the more stable patterns that can be observed in societies of both human and non-human actors.

But it is important to see that power structures are founded on power behaviours, which may be regarded as the processes of power relations. Actors may compete with each other for dominance in structural power. This is what we ordinarily mean by conflict. In such competition, all the resources of the power processes may be brought to bear, their use depending on conditions such as the simple availability of weapons, the fear of deploying them, conscience over their use, and so forth. The purpose and consequence of the use of power is precisely to coerce the other, who is unwilling to comply. After one or more successful applications of coercion, if the less powerful actor cannot leave the field, the more powerful one usually need only to indicate his/her desire in order to elicit compliance. Should compliance not be forthcoming, then coercion will ensue.

Given that the actors have arrived at the state of power structure in their relationship it will not be necessary to engage in enforcement activity through the frequent use of power processes. Relations between actors may then become ritualised. The mere presence of the dominant one will evoke gestures of submission, which indicate that compliance will be forthcoming without need to resort to further process of power. In this case only symbolic gestures of power may be used, for example, a particular tone of voice, or a stare, or a pause in behaviour to allow the compliance to begin. Once in place, power structures are relatively stable and predictable.

If power relations were all that characterised human interaction, we could say justly (after Hobbes) that life is merely "nasty, brutish, and [probably] short". For it would mean that compliance occurs only involuntarily, that actors never value the purposes of other actors more highly than their own, or that, in the borderline case, other actors' purposes are only worthy of endorsement if they are the same as one's own. This does not appear to be the case.

Human actors and, as we shall see, non-human as well, are capable of voluntary compliance with the wishes, desires, and interests of other actors, even at what may appear to be some sacrifice to themselves. Carter's factor of Group Sociability, as described earlier, presents only tepid examples of this, but stands for the whole range of behaviours in which actors willingly accord each other benefits and compliance.

Just as power relations tend to become stable and predictable power structures, so do status relations become status structures, equally stable and predictable. Relations of voluntary compliance between actors are designated friendships, alliances, and, in the case of ultimate status conferral, love. Again, as with power structures, the structures of status are based on processes, but in this case, those of status.

Status as process entails the whole range of behaviours that are intended to gratify and enhance the other, and provide experiences that evoke good feeling in them. In authentic status conferral, the intention is only to gratify. The goal is not to deceive the other about one's intentions; this would be to manipulate and exploit the other, which is a power tactic. Rather, certain qualities of the other evoke the voluntary compliance.

These qualities are culturally variable. They include wealth, intellect, physical strength, beauty, moral excellence, aesthetic skill, technical competence—in fact, any quality endorsed by the culture. These qualities form the basis for the status structures that characterise a given society. In addition to these, voluntary compliance is also evoked when the other accords status first. This is the pleasant state in which another has gratified us voluntarily, has freely accorded us benefits and compliances without coercion on our part. Although there is a social norm of reciprocity (Gouldner, 1960), it is possible that the positive response we give to others who provide us with status is a result of early socialisation. Care-taking responses elicit similar responses, which become integrated into the behaviour pattern. Status-accord is a general notion that also includes disinterested care-taking, as in childhood.

If we conceive of the ultimate in status-accord, that is, an essentially open-ended amount of status, then we touch on what is ordinarily understood as love in Western societies. In fact, a love relationship can be usefully defined as one in which at least one actor is according unlimited amounts of status to the other. (The ramifications of this definition for the number and types of love relationships, and for their dynamics of growth and devolution, are found in Kemper, 1978.)

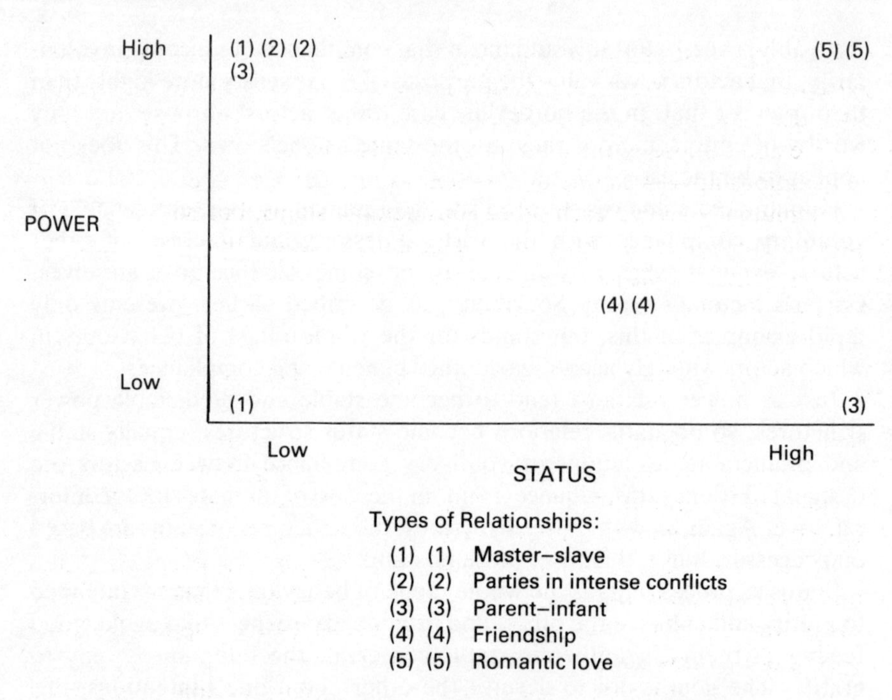

Fig. 13.1 Some relationships depicted in the power–status space.

If power and status are the two dimensions of social structure and relationship according to the repeated findings of the large volume of factor analytical studies, then it follows heuristically that all social relationships can be represented by the standing of the actors in relation to each other on these two dimensions. Because of the orthogonal rotations in the factor results that produced the power and status dimensions, construction of a Cartesian two-dimensional space is possible, of which the axes are power (ordinate) and status (abscissa). Any social relationship can be located in the space so formed (Fig. 13.1).

We see in Fig. 13.1 some relationships that are discursively defined. The 1–1 relationship may be understood under the heading of master–slave, or victor–vanquished. Neither accords the other voluntary compliance or benefit. But one of the actors has a very great deal of power over the other, whereas the other is entirely powerless.

The 2–2 relationship can stand for any parties engaged in intense conflict. As in 1–1, neither gives any benefit to the other voluntarily, but each is engaged in efforts to coerce the other.

The 3–3 relationship can be seen as representing parent and infant. The infant, in the lower right corner of the Figure, receives all the status (love), i.e., all the benefit and necessary compliance to maintain health and life.

But the infant has no power to coerce the parent. On the other hand, the parent, who receives no status from the infant, has enormous power over it, indeed the power of life and death. This is to say that the parent can coerce and dispose of the infant how he or she wills.

Relationship 4–4 can be understood as one between two good friends. They mutually accord each other considerable status, but both are low in power. They rarely try to coerce each other. It should be noticed that this relationship approaches, but does not reach, the end of the status scale. Hence it differs from love, as indeed most, though not all, friendships differ from love.

Finally, relationship 5–5 is the full-blown love relationship, because it is in accord with the definition provided earlier, namely at least one actor giving extreme amounts of status to the other. Here both are mutually according the extreme amount of status, which is delightful enough. But it is noteworthy that in this particular form of love relationship both actors have enormous amounts of power as well. It is a type of love relationship in which there is full recognition of one's dependency on the other, hence that one is vulnerable to the power of the other, should the other choose to use the power. Few love relationships avoid this stage (see Kemper, 1978, for a description of how such relationships progress).

It should be apparent that there is dynamic co-variation between the power and status dimensions. There is strong evidence (Kemper, 1978) that power behaviour is ordinarily met with counterpower behaviour. Thus aggressive action, once begun, tends to develop an autonomy that leads to an escalating round of power acts until one actor is decisively defeated, and thus acquiesces to the power of the other; or until one actor desists through fear or from conscience, as suggested earlier; or until there is an intervention from outside the system.

In general, actors in relationships that do not provide them sufficient status are prone to leave those relationships, as long as there are no constraints of power to prevent this. (See Hirschman (1970) concerning the dimension of "exit", which is the escape mode when human relations turn sour or become psychologically unsupportable.) Similarly, whether or not there is sufficient status, actors are prone to leave relationships in which they are excessively subjected to the power of the other.

In these two statements we encompass a large amount of what can be understood in relational terms about divorce, about runaway children, about job-changing, about the vagaries of friendships that bloom and decline, and so forth. The essential point is that if power and status are useful dimensions for descriptive and analytic purposes, they should also be useful in accounting for social processes in which relationships are the focus. A societal-level analysis using these dimensions will be presented later.

ADDITIONAL SOURCES OF THE POWER AND STATUS MODEL

Up to this point it may appear that I have relied over much on the outcome of factor analyses of behaviour in small groups of humans in a limited population (namely the U.S.A.). Are the dimensions perhaps less general than has been claimed so far? The answer must be "no" for, as will be seen, the power and status dimensions have cross-cultural as well as cross-species relevance. They are also quite ancient in origin, and so cannot be dismissed as merely the product of a contemporary consciousness produced by specific social formations, e.g., Capitalism.

Empedocles, the pre-Socratic Greek philosopher, was the first to propose the two dimensions. He saw all matter as comprising the familiar tetrad of earth, air, fire, and water. But he required some principles by which the constant changes of state in and between these elements could be explained. He proposed two forces: love and strife. Love creates unities, binding elements together, whereas strife creates divisions, tearing apart what was previously solid. Mutatis mutandis, love and strife, are the status and power dimensions, respectively.

Freud saw quite clearly the correspondence between love and strife and his two instinctual principles, Eros and Thanatos, and for this reason he (Freud, 1937, Pp. 349–50) referred to Empedocles as "my great predecessor".

Others in the philosophical tradition, who have used the same ideas, are Hegel, with his notions of noble (status) and base (power) grounds for action; and the Utilitarians, with their hedonic calculus of pains (power) and pleasures (status).

Three other domains of enquiry provide evidence for the two relational modes. The first is cross-cultural, the second is from the cognitive domain of meaning, and the third is from studies of non-human primates.

A common method of establishing the pan-human validity of a concept is to demonstrate its cross-cultural universality. For example, in recent years, researchers have domonstrated the universality of certain facial expressions of emotion. In a review of studies of interaction in various cultures, Triandis (1972, p. 270) concluded that two relational factors—"superordination–subordination" [power] and "intimacy" [status]—are ". . . the fundamental dimensions of human social behaviour [that] are obtained with different methods of human investigation". Subsequently, Wish, Deutsch, and Kaplan (1976) have found the same two dimensions in data from Greek studies, and White (1980) has repeated the finding with Melanesian and Indian data, using descriptors of personality that reflect social interaction. By employing multidimensional scaling, Lutz (1982) found that emotion words among the Ifaluk, a people of the South Pacific, fall along two scales that reflect the power and status dimensions.

The second domain of supporting evidence for the two relational modes is that of semantic meaning. Osgood, Suci, and Tannenbaum (1957) evaluated this by means of a device called the Semantic Differential. Essentially, this requires the rater to evaluate stimulus words or concepts according to polar scales, e.g., warm–cold, active–passive, happy–sad, and the like. The results of the ratings are then factor analysed. Three factors repeatedly emerge named: activity, potency, and evaluation. Clearly, the activity factor is equivalent to the task dimension discussed earlier, the potency factor is equivalent to power, and the evaluation factor to status. This is a striking confirmation, from an entirely different domain of thought, of the broad reach of the two relational modes, and of the task dimension. These results have been frequently replicated; also the two modes are found to constitute the fundamental dimensions of meaning cross-culturally (Osgood, May, & Miron, 1975).

It has sometimes been argued (D'Andrade, 1965) that results such as these are inevitable, in as much as we are conceptually imprisoned in a certain "meaning space". That is, we "see" what is already in our heads (as conceptual apparatus expressed in language), and so no other results are possible. Therefore we view the phenomenal world in the terms of our conceptual structure; *a priori* meaning and interaction must thus be identical.

This argument is fundamentally self-defeating. Either it is wrong, in which case language and conceptual structure are not prisons that constrain our observations of the world; or it is correct, but with no effect, as, if it is correct, there is then no way for us to get at the "real" world. Any "real" world that is different from our conceptual world could never be apprehended. So, we are left with the correspondence that has been found in semantics, in human interaction, cross-culturally, and, as we shall now see, in the interaction patterns of non-human primates.

It is of the utmost importance for the utility and significance of the power–status formulation, as prefigured in early social thought, and manifested clearly in the findings of dozens of empirical investigations of human conduct and cognitive meaning, that the same dimensions have also emerged in careful observations of non-human primate societies.

The two relational factors—designated "the two modes" by Chance (1976; 1980; see Introduction and Chapter 1)—are named "agonic" and "hedonic" for power and status, respectively. Derived from observations of a number of non-human primate societies, these two forms appear to characterise the pattern of social cohesion in those societies. In the agonic mode, attention is focused on a central, dominating animal or small set of such animals. All behaviour is undertaken with reference to the possibility of aggression from the dominant(s). Ordinarily, relatively low amounts of actual aggression take place. The group has achieved a power structure, hence the active display of power as process is obviated, though it may emerge at any time.

By contrast, in hedonic groups, social organisation is a flexible mix of attention to more dominant animals combined with indifference to them. The group may disperse widely for a time, hence dominance itself is only occasional. Even when dominance is manifested, it is not by strictly power-based threat behaviour but by "displays", which the less dominant animals reflect with displays of their own. In hedonic groups, there is also a good deal of grooming and hugging, analogues of status-conferral at this species level.

Chance (1984) has also extended to human studies the implications of the two-mode attentional structure and has found in Montagner's observations (see Chapter 10) of children's leadership–followership patterns in play, close variants of the agonic and hedonic styles apparent in non-human primate social attention structure. Chance has further advanced our understanding by noting the compatibility of the two modes with the personality development models posited by Pearce and Newton (1969), and by Vaillant (1977).

Chance (1980; 1984) has also recognised the important parallels between the two modes and the neurophysiological and autonomic theories of Gellhorn (1967; 1968). From an entirely different body of data, and with an aim toward a different level of theoretical construction, Gellhorn proposed that the infrastructures of emotion reside in two major neurophysiological systems, called the "ergotropic" and the "trophotropic". The first of these, essentially linked to the sympathetic nervous system, is involved in the kind of flight-or-fight arousal underlying fear and anger. The second system, essentially connected to the parasympathetic branch of the autonomic, is responsible for the calm, quiet, and relaxed emotions of security and satisfaction (see Kemper, 1978, for details).

These several discoveries of dimensions, or of articulated structures or processes, that are isomorphic or compatible with the power–status model encourage the prospect of a set of universal constructs that can capture not only theoretically relevant materials within analytic domains, but also allow for serious comparisons and translations between domains. It is equivalent to having a yardstick that is not limited to a single modality, e.g., extension, but can measure a number of modalities.

So far I have been content to describe and define the power and status dimensions, and to construct a case for their reliability and validity in a number of analytic domains. It remains now to indicate some uses of the power–status model. First, I will present an application of the model to the domain of the emotions in a more direct, socially relevant sense than did Gellhorn, but one which is importantly integrated with Gellhorn's work. Then I will show how the model applies to the macrostructures of society.

POWER, STATUS, AND EMOTIONS

Most human emotions result from outcomes of social interaction. This suggests that we would gain significant understanding of emotions if we knew how to describe such outcomes. The discovery of the power–status modes of relationship simplify this effort considerably. If the model is valid, it means that all interactional outcomes can be characterised in the power and status modes. The description of these outcomes can follow a relatively simple analytic assumption: each actor's power and/or status may increase, decrease, or remain the same, as a result of an interaction. This leads to a paradigm of 12 possible outcomes (2 actors × 2 relational modes × 3 possible outcomes), only 4 of which will actually occur. That is to say, each episode of interaction will reflect on one's own power and status in some way, and on the power and status of the other. What are the emotional outcomes of interaction likely to be?

In broad outline (details are in Kemper, 1978), the following will ordinarily result. Elevation of one's own power will lead to a greater sense of ease and security, as will decline in the power of the other. Elevation of the other's power will lead to fear/anxiety, as will a decline in one's own power. The effect of no change in power for either self or other is complicated by the state of anticipatory emotions (Kemper, 1978). If one anticipated an increase in one's own power or a decrease in the other's and it did not occur, this should instigate fear. On the other hand, if one anticipated that the other's power would increase or that one's own would decrease, and it did not, this should enhance ease and security.

In the status dimension, increase in status obtained from the other will lead to satisfaction, happiness, or contentment. This naturally follows from the definition of status, namely voluntary compliance received from others. The one normal exception here would be where the amount of increase was less than expected or deemed to be merited. The emotional results of such an interaction are tantamount to what may be expected from a decrease of status.

Any decrease in the amount of status that was expected, deserved, or usually received from another follows a more complicated course than in status increase. The fundamental difference involves the importance of agency, namely the locus of responsibility for the interactional outcome. Although agency has some importance in the outcomes of status increase, its more striking effects are found with decrease. If agency for such a decrease is assigned to the other party in the interaction—they acted in an arbitrary or unjustified manner in not providing the status desired—the resulting emotion is most likely to be anger. On the other hand, if agency is assigned to self—one feels oneself flawed, responsible, or disabled, and essentially unlikely to be able to recoup the status lost—the most probable emotion is depression.

Clearly, anger and depression can both result from an episode of status loss. It may be difficult to assign agency, hence one may alternately contemplate self or other as responsible for the loss. Or responsibility may be realistically apportionable to both self and other, hence there is double agency that leads to the experience of both anger and depression. This clearly fits well with the usual, psychoanalytical theories of depression.

Price (Chapter 7) reasons that depression is an evolutionary mechanism that helps to accommodate those who lose rank—or status in the terminology of my chapter—to their lower status. This gives yet another entrée into the understanding of this emotion, which is otherwise perplexing because its syndrome may disable an organism. By so doing, depression causes withdrawal from interaction that, in the particular circumstances either may be, or will imminently become, conflictful. Hence depression protects from competition or conflict over status when one may not have the resources to compete effectively, or not be likely to emerge a victor. The disabling effect of depression prevents engagement in a struggle that one would more than likely lose. In the withdrawal stage, one can reintegrate the self at the new status rank, either permanently accommodating to it, or marshalling one's resources for another effort to gain higher status.

Turning now to the status of the other, interaction that results in an increase in that status—one is giving the status to the other—leads to satisfaction for the self. This is normally so when we voluntarily confer benefits on others, that is, when we want to give, and we do. When the status of the other declines, that is, we do not confer status when we ought or desire to, the outcome for self is guilt or shame.

When the status of either actor does not change, it depends again on the anticipatory emotions. Ordinarily, anticipation of a status increase that is then confirmed will lead to somewhat lower intensity of satisfaction than for a status increase that was unanticipated. There appears to be nothing like the joy of unexpected rewards. When status decreases in the face of anticipation of the opposite outcome, the anger or depression will ordinarily be more severe than when such decrease was anticipated. The unexpected reversals often take the worst toll (see Kemper, 1978).

This brief summary does not fully address the complexity of the issues involved in relating power and status outcomes to emotions, but gives some notion of the manner in which this can be done. It is important that power–status oucomes, emotions, and neurophysiological and autonomic systems are firmly bound together. The work of Gellhorn, as discussed earlier, and of Funkenstein (1955), Funkenstein, King, and Drolette (1957), and others (see Kemper, 1978), supports the theoretical model of a single arc that connects power and status relations at the social level, via emotions at the psychological level, with neural and endocrine systems at the physiological level. This over-arching framework, which has its sources in observations of human and non-human species, in interaction and in personality, within cultures and cross-culturally, in behaviour and in the symbol systems of language, and in the psychological and the physiological, offers important possibilities for the integration of many elements of behavioural systems.

POWER AND STATUS DIFFERENTIATION OF SOCIETY

A final example of the utility of the two-dimensional model comes from the domain of the macrosocial, namely society itself. Evidence is accumulating for an efficient model of societal structure that is rooted in the power and status dimensions. Virtually all known human societies differentiate their members along both vertical and horizontal dimensions (Collins, 1975; Lenski, 1966; Weber, 1946). The vertical dimension is the differentiation both of individuals and groups by such distinctions as age, gender, ethnic origin, and the like, into higher and lower strata with respect to two criteria: power and benefits. Power is precisely the "power mode" that I have discussed. It allows some individuals or groups to control other individuals or groups. To greater or lesser degree this power is regarded as legitimate (see Weber, 1946) and thus is, within strictly defined limits, acceptable even if it entails unwilling compliance. When power is accorded legitimacy, we commonly refer to it as authority. Societal power structures are complex mixes of legitimate and non-legitimate power.

Benefits comprise whatever rewards the society has to distribute, whether they are material, symbolic, or psychic; social relations of power centrally determine the distribution of these on a societal basis. Indeed, this is fundamentally why power is sought. Groups with sufficient resources of power—sometimes these are bayonets, sometimes access to productive opportunities in work—can assure themselves of more of the benefits than competing groups.

But power is not the only possible determinant of the way benefits in society are distributed. Status, as discussed earlier, entails voluntary compliance. Indeed, individuals in society are ordinarily willing to "pay more", that is, give more status to some persons and groups than to others. This has something to do with the value system of the society, and also with the way in which power operates on the value system. An example will indicate the degree to which these can be combined.

In most Western societies, health is highly valued and physicians are highly valued as a result. This often leads to an acceptance of high fees for medical services. People are willing to pay for health. In relational terms, this means that patients give voluntary compliance, i.e., status defined in monetary terms, to physicians. On the other hand, it is well known that physicians, much in the manner of all professionals (Freidson, 1970), try to control their clients' attitudes and values in regard to the profession. This involves control of information about health and illness, and secrecy about such matters as rates of success and failure of individual practitioners, as well as about other aspects of practice that might affect the willingness of clients to engage in voluntary compliance. Hence, even in areas where members of society may be choosing "freely", so to speak, power may enter to determine the limits of the choice.

The vertical dimension of society is understandable as a mix of power and status relations in which the resulting stratification distributes both control and opportunities for benefits. At the horizontal level of societal differentiation are the groups whose members share a similar power and status level in comparison to other groups with which they may be engaged in more or less intense struggles for power and benefit (Weber, 1946). However, within these groups, by virtue of common membership, relations are relatively status-oriented. That is, there is enhanced voluntary compliance among members. This is a necessary consequence of the fact that all members of these groups depend on fellow members for support and sustenance in the face of actual or potential conflict with outsiders. Within-group solidarity is heightened through the promotion of a maximum of voluntary and freely-given relations between members. Morale is always a desideratum in groups that are engaged in conflict. Status-exchange among members promotes the emotional bonding that enables the group to attain solidarity and to compete best (Collins, 1975; 1981).

The horizontal axis of societal differentiation, with its reliance on status relations, also accounts, in the main, for the phenomenon called "social support" (Thoits, 1984). Social support is seen as a key factor in mental health and in the ability to ward off the deleterious consequences of a variety of negative stresses, from loss of loved ones to the disappointments that are endemic in the pursuit of most occupations and careers in modern societies.

CONCLUSION

It can be seen that developments in a number of separate fields and in the work, both empirical and theoretical, of many independent investigators, have revealed a set of dimensions that can be used to characterise social behaviour. Furthermore, these dimensions articulate in important ways with systems at different social and biological levels. This is an emergent, comprehensive, biosocial perspective, but a cautionary note must be sounded, especially in the light of the extensive claims made on behalf of sociobiology (Wilson, 1975). Although there appears to be phylogenetic continuity in the power and status modes between human and non-human forms, and despite the fact that the modes of relationship are articulated with neural and physiological support structures—as of course they must be, otherwise no behaviour would be possible—it must not be assumed that these elements are under genetic control. It is easy to see that all species face problems of adaptation that are similar in important respects. Hence, certain gestures and emotions are adaptive: for example, offering a threat when one is about to be attacked or deprived of some benefit in one's possession; or responding with care to the cries of one's offspring. It is entirely possible for learning via modelling and association to explain a good deal of what we observe in actual behaviour. Studies of imprinting (Harlow, 1962) have shown that simple genetic control does not exist even in so fundamental a species requirement as reproduction.

REFERENCES

Carter, L. F. (1954). Evaluating the performance of individuals as members of small groups. *Personnel Psychology*, 7, 477–484.

Chance, M. R. A. (1976). Social attention: society and mentality. In M. R. A. Chance & R. R. Larsen (Eds.), *The structure of social attention* (Pp. 315–33). London: John Wiley & Sons Ltd.

Chance, M. R. A. (1980). An ethological assessment of emotion. In R. Plutchik & H. Kellerman (Eds.), *Emotion: Theory, research and experience* (vol. 1). New York: Academic Press Inc.

Chance, M. R. A. (1984). A biological systems synthesis of mentality revealing an underlying functional bimodality (hedonic and agonic). *Man–Environment Systems, 14(4)*, 143–157.

Collins, R. (1975). *Conflict sociology*. New York: Academic Press Inc.

Collins, R. (1981). On the microfoundations of macrosociology. *American Journal of Sociology*, *86*, 984–1014.

D'Andrade, R. G. (1965). Trait psychology and componential analysis. *American Anthropologist, Part II*, *67*, 215–228.

Freidson, E. (1970). *Professional dominance*. New York: Atherton.

Freud, S. (1937). Analysis terminable and interminable. In J. Riviere (Translator), *Sigmund Freud, Collected Papers*. New York: Basic Books Inc.

Fromme, D. K. & O'Brien, C. S. (1982). A dimensional approach to the circular ordering of emotions. *Motivation and Emotion*, *6*, 337–363.

Funkenstein, D. (1955). The physiology of fear and anger. *Scientific American*, *192*, 74–80.

Funkenstein, D., King, S. H., & Drolette, M. E. (1957). *Mastery of stress*. Cambridge, Mass.: Harvard University Press.

Gellhorn, E. (1967). *Principles of autonomic–somatic integrations*. Minneapolis: University of Minnesota.

Gellhorn, E. *Biological foundations of emotion*. Glenview, Ill.: Scott Foresman.

Gouldner, A. (1960). The norm of reciprocity. *American Sociological Review*, *25*, 161–178.

Harlow, H. F. (1962). Development of affection in primates. In E. L. Bliss (Ed.), *The roots of behavior*. New York: Harper & Row.

Hirschman, A. O. (1970). *Exit, voice, and loyalty*. Cambridge, Mass.: Harvard University Press.

Kemper, T. D. (1978). *A social interactional theory of emotions*. New York: John Wiley & Sons.

Lenski, G. (1966). *Power and privilege*. New York: McGraw-Hill Book Co.

Lutz, C. (1982). The domain of emotion words on Ifaluk. *American Ethnologist*, *9*, 113–128.

Osgood, C., Suci, G. J. & Tannenbaum, P. (1957). *The measurement of meaning*. Urbana, Ill.: University of Illinois Press.

Osgood, C., May, W. H., & Miron, M. S. (1975). *Cross-cultural universals of affective meaning*. Urbana, Ill.: University of Illinois Press.

Pearce, J. & Newton, S. (1969). *The conditions of human growth*. New York: Citadel.

Thoits, P. (1984). Coping, social support, and psychological outcomes: the central role of emotion. In P. Shaver (Ed.), *Review of personality and social psychology* (vol. 5, Pp. 219–238). Beverley Hills, Calif.: Sage.

Triandis, H. (1972). *The analysis of subjective culture*. New York: John Wiley & Sons.

Vaillant, G. (1977). *Adaptation to life*. Boston, Mass.: Little, Brown & Co.

Weber, M. (1946). In H. H. Gerth & C. W. Mills (Eds.), *From Max Weber: Essays in sociology*. New York: Oxford University Press.

White, G. (1980). Conceptual universals in interpersonal language. *American Anthropologist*, *82*, 759–781.

Wilson, E. C. (1975). *Sociobiology: The new synthesis*. Cambridge, Mass.: Harvard University Press.

Wish, M., Deutsch, M., & Kaplan, S. J. (1976). Perceived dimensions of interpersonal relations. *Journal of Personality and Social Psychology*, *33*, 409–420.

14 Organisational Culture and the Agonic/Hedonic Bimodality

Felix Wedgwood-Oppenheim
Institute of Local Government Studies, University of Birmingham, U.K

The importance attached to organisational values and culture has increased considerably following the publication of *In search of excellence* by Peters and Waterman (1982) and the subsequent rash of books and articles stressing, among other things, the central importance of culture to the success of an organisation. An important characteristic of "excellent" firms is a set of values and a climate that encourage independent action, personal autonomy, exploration, and risk-taking without the threat of censure within the organisation if the risks do not pay off. Formal hierarchy is played down in favour of a very rich pattern of communication. A high value is placed on individual contributions wherever in the formal hierarchy the individuals are. The organisation is to a large extent steered and controlled by creating a strong loyalty to its values rather than by the exercise of personal or positional power. It seems that, to use the language derived from the study of non-human primate behaviour, (see Introduction), organisations with hedonic cultures are more successful than those with agonic cultures. This appears to be the case at least in environments and times of considerable change.

This chapter gives a brief exploration of the relation between different organisational cultures and the modes of achieving group cohesion that have been described in this book. In doing this I take up ideas developed primarily in relation to non-human primates, living in relatively small troops, and I look at the parallels with theories developed about modern, often large, work organisations. In as much as theories about primate behaviour can be seen also to apply to large organisations, they may be fundamental to the nature of human social behaviour, and should be understood when considering the working and design of human institutions.

"How to describe your organisation" (Harrison, 1972) has been widely used in thinking about management, particularly since being given a wider readership by Handy (1976) in *Understanding organisations*. Roger Harrison differentiates four organisational "ideologies", which Charles Handy renames "cultures" in order to convey their pervasiveness throughout an organisation. They are the power, role, task, and person cultures.

Wherever there is a classification involving more than two categories it may be useful to explore whether the categories can be better explained by placing them on more than one dimension. In this case, the four cultures can be classified along two dimensions: that of the agonic–hedonic social control mode, and that of the organisation versus individual orientation.

THE FOUR CULTURES

The *power culture* is the culture of the King and his barons, of politics, of the industrial-revolution entrepreneurs, or of some of the current marauding financial or property companies. The key figure is the central power source and everyone else's power derives from that source. What matters is what The Boss wants. He (the masculine pronoun is used intentionally in this context) determines the direction that the organisation takes, and it is incumbent upon subordinates to know what direction he is taking. Succession to the top is a major issue and determines the continued success of the organisation. Members of the organisation play organisational politics and jockey for power. It is indisputably an agonic culture as it dispenses power from the centre of a hierarchical attention-binding structure.

The *role culture* is the culture of the classic bureaucracy. In this culture, personal power is replaced by rules and procedures as a determinant of behaviour. People do what their job description and the rules require them to do. It is the role that counts rather than the individual who fills it. The hierarchy is important and everyone is quite clear where they are in that hierarchy.

In a role culture the insecurity of jockeying for power does not exist, or is at least depersonalised, but it can nevertheless be seen as an agonic culture. The hierarchy is important in determining a whole range of privileges. Deference is payed by those lower down the hierarchy to those higher up. All employees have constantly to be concerned to work within the rules. Individual initiative and exploration are penalised, either directly or by diminishing the prospect of climbing up the ladder. In a sense, it is not the dominant person that has to be pleased, but the organisation itself.

The *task culture* is the culture to which a large majority of managers say they would like to belong. It is the dominant culture of large firms working in dynamic technological environments, such as in electronics, and it is the culture to which many organisations, both public and private, aspire as they attempt to cope with pressures and uncertainties in a time of dramatic economic and political shifts.

In a task culture, people are grouped to perform tasks or carry out projects. The group comprises people appropriate to the task. Recognition is accorded on the basis of what the individual can contribute rather than on their formal position in the hierarchy. Personal status derives from having appropriate knowledge. The task group lasts as long as is required by the task, whether this be weeks or years. There is a constant process of regrouping to tackle new issues, and the membership of groups changes as different skills are required. Individuals are less bound by having to observe rules than in a role culture. Neither do they have to watch the vagaries of their superiors as they would do in a power culture. They are freed and encouraged to devote their efforts to pursuing particular tasks. They are able to respond to new situations and to innovate much more successfully than in organisations with a role culture. In fact, their mode of operating has many characteristics of the hedonic mode.

The *person culture* is rarely found as the dominant culture in an organisation. It is of interest, however, because of the number of people whose values are appropriate to such a culture, even though they are working in organisations in which one of the other cultures dominates.

This culture is one in which individuals pursue their own interests: earning, providing goods or services, developing their capacities, "doing their own thing". The role of the organisation is to assist the individuals in this pursuit, to provide common services and shared resources. It is the culture of archetypal partnerships in the traditional professions, of co-operatives of skilled people, of the commune, and of the "modern" family.

Social cohesion in an organisation with a personal culture is certainly not maintained by agonic behaviour. The question remains whether it is truly hedonic. In its consequences it is, in that individuals and small subgroups are free to explore and carry out independent activities without constant reference to a dominant person or to the rules imposed by the organisation. Organisations that maintain their person culture without evolving, as most do, into one of the other cultures, probably do so by engaging in a large amount of relationship-maintaining communication, placing them firmly in the hedonic category.

Actual organisations, of course, although they have one dominant culture may well also have elements of the other three cultures. In particular, certain units may have a different dominant culture from that of the organisation as a whole. Thus a large organisation that has primarily a role culture might have a planning department that has a task culture, and a research unit where most of the researchers try to adhere to a person culture. There may also be individual senior managers who dominate a departmental empire, which they have embued with a power culture. A conglomerate like the General Electric Company for instance, may include individual operating divisions with role or task cultures, whereas at the corporate, inter-division level there is a power culture. These conflicting cultures bring about tensions within the organisation, with different assumptions being made about the reasons for task allocation, promotion, rewards, time-keeping, etc.

INDIVIDUAL VERSUS ORGANISATION ORIENTATION

Thus the four cultures can be shown to be divided between the two modes of social cohesion. The power and role cultures are agonic; the task and person cultures are hedonic. Burns and Stalker (1966) distinguish two main types of organisational structure—organismic and mechanistic—that parallel the task and role cultures. Indeed most current large organisations can be broadly allocated into one of these two categories. Again, the organismic forms are characterised by more hedonic modes of social interaction, whereas the mechanistic form is associated with the agonic mode.

However, the cultures can also be arranged along another dimension (Fig. 14.1). The power and person cultures are centred on the individual. To know what the goals of these organisations are you have to know what the goals of the individuals (the key individuals in power organisations) are. To talk about the organisation's activities is to talk about the activities of (key) individuals. By contrast, organisations with a role or task culture are much more organisation centred. They have organisational goals that are independent of the goals of individuals, and the organisation has an identity that is to a large extent independent of that of its members—although this is less so in a task culture. Thus any particular organisation can be located on a two-dimensional plane depending on the extent to which its cohesion is based on agonic or hedonic modes of behaviour, and the extent to which the focus is on individuals or the organisation.

FOCUS	AGONIC	HEDONIC
ORGANISATION	ROLE CULTURE	TASK CULTURE
INDIVIDUAL	POWER CULTURE	PERSON CULTURE

FIG 14.1 The four cultures

HEDONIC AND AGONISTIC BEHAVIOUR RELATED TO ORGANISATION CULTURE

Some characteristics of the hedonic behaviour of chimpanzees have interesting parallels in organisations with task cultures, in particular those successful, "excellent" firms described by Peters and Waterman (1982). Chimpanzees maintain a high level of communication between dispersed groups, and provide large amounts of reassurance when they meet. Similarly, excellent organisations have a high level of information flow around the organisation, which goes far beyond the limited requirements of information needed for the purpose in hand. The role of communication is not just the passing on of information but maintaining social cohesion, giving reassurance that all is well. These firms are also characterised by MBWA—management by walking about—which can be seen as an effective means of reassurance that individuals and subgroups are not isolated, and of maintaining the identity and coherence of the organisation.

Chance (1973) points out that chimpanzee groups are brought together, and their coherence maintained, through "mutual display binding the individuals' attention from time to time", and he cites the chimpanzee "carnivals" described by Reynolds and Reynolds (1965), and by Margaret Power in Chapter 4. The excellent firms parallel this in frequent celebrations, Friday afternoon "beer busts", prize givings, sporting activities, etc., which bring large numbers of employees together in an enjoyable atmosphere.

Cohesion in hedonic primate groups and organisations is maintained through communication and celebration but, by contrast, in agonic cultures it is maintained through the exercise of power. The exercise of power in a power culture is clearly parallel to that in baboon troops. In role cultures, it is more subtle as it has been institutionalised in a set of rules and role relationships. Nevertheless, the power of these forms maintains the organisational cohesion.

People who work in organisations that have task or individual cultures (hedonic) need to be capable and skilled while at the same time they are given the opportunity to learn and to develop themselves. Like in chimpanzee groups, the hedonic mode allows the development of knowledge and skill. As a result the organisation itself, at least if a task organisation, can be regarded as developing its own capacities and learning. In contrast, agonic organisations, certainly those with a role culture, are bad at responding to a changing environment because they do not provide a framework within which individuals can develop and learn, except within the strictly defined confines of their roles. In a power culture it is more important to know what the leader wants than to learn independently about the environment. The one context in which agonic organisations are good at coping with an uncertain environment is when power organisations are responding to threats and dangers as in take-over battles.

In a role culture, not only is the development of individual abilities not facilitated but, in much of the organisation, less ability is needed than would be the case in a task organisation, because of the routine nature of the jobs. To quote the old saw: routine allows simpletons to do what it took a genius to invent.

MOVEMENT BETWEEN CULTURES

Organisations can be subject to a number of pressures that will tend to change their culture. It seems that in organisations, as elsewhere, hedonic cultures are particularly vulnerable to pressures toward more agonic modes of relating. Agonic cultures offer security. Power organisations offer the security of being protected by a powerful leadership, even if that leadership is itself threatening to transgressors within the organisation. Role organisations offer the security of knowing exactly what is expected of one and that one will not be exposed to unfamiliar situations. Although many employees value the freedom of hedonic cultures, many also look for the security offered by agonic cultures. This tendency is so strong that considerable leadership is needed to maintain hedonic cultures.

Person cultures rarely survive the departure of the original enthusiasts, whose constant efforts to maintain multi-lateral communications sustain the organisation in its culture. They can become dominated by one individual and move towards a power culture. Alternatively, the house rules can begin to take over and develop an importance of their own, rather than merely serving the needs of the individuals, and thus developing into a role culture. The top management of the "excellent" task organisations devote a high proportion of their time and energies to ensuring that everyone knows what the organisational values and goals are, to ensuring the quality and training of the staff, and to sustaining the culture itself. Without this continuous effort the likelihood is that the organisation will move towards an agonic, role culture.

Power (1986; also Chapter 4) shows how a chimpanzee troop, in which there is not normally direct competition for food, can evolve from its hedonic form of social control to an agonic form when valued foods are scarce. There can be a similar movement towards more centralisation of detailed control in organisations when resources become scarce and there is a need to conserve them or even cut back on activity. This was well-illustrated by changes in British local authorities in the late 70s and early 80s when severe financial cuts imposed by central government usually led to central control, and much tighter rules and procedures.

More recently, however, many local authorities in the U.K., both Conservative and Labour controlled, have seen the appropriate reaction to resource shortage not as stronger control but as a greater need to innovate. They have thus reduced the number of detailed controls, decentralised their services, and encouraged a greater task orientation among their staff. So, there are not only conditions that tend to move organisations in a more agonic direction, there are also those that will encourage them to move in the opposite direction, or penalise them if they do not. In particular, as the pace of change in the environment increases, so the rigidities of a role culture become an inhibition to innovation and adaptation, and it becomes necessary to adopt a more open, free-ranging style.

Organisations move not only along the agonic—hedonic dimension but also along the individual—organisation dimension. As individuals in a person culture tackle more complex tasks, so differentiation of skills and team working become necessary, and the organisation may move toward a more structured form with a task culture. Better established is the move from a power to a role culture. As the organisation increases in size so the leader of a power organisation can no longer maintain control without greater "routinisation" and depersonalisation of the managerial processes. There is a well-established historical tendency in which new, entrepreneurial firms with a power culture develop into mature, routinised and efficient firms with a role culture, which may later transform themselves into adaptive organisations with a task culture.

Movement in the opposite direction—from organisation centred to individual centred cultures—has been less well analysed. However, as Naisbitt (1984) describes in *Megatrends*, there are now certain trends in the direction of person cultures. Whole organisations may not move from task cultures to person cultures, but units and increasing numbers of individuals are doing so. The ideals of personal growth and self-actualisation are having their impact on organisational life. There is a movement towards the setting up of co-operatives and consortia of skilled or professional workers. Some large organisations, both private and governmental, are experimenting with an arrangement whereby individuals or whole units are no longer employed by the organisation but instead work freelance with certain guarantees of continuing contracts. Very well known is F International, a computer software firm, that does all its contract work through women specialists working freelance from home.

CONCLUSIONS

The parallel between the classification of modes of social interaction in primates and of organisational cultures is not total. It is nevertheless sufficiently strong to suggest that the basic organisational culture types are not just products of industrial society, but may be based on fundamental social forces that human beings share with other primates.

It is, of course, risky to draw conclusions about human nature from animal studies. Culture itself is seen as one of the distinguishing characteristics of our species. Culture determines our behaviour as well as do any inherent characteristics. It is interesting, at the very least, therefore to realise from analyses by Power (1986) that chimpanzees may be seen as having different "cultures", even if these are determined by environmental factors. However, there are widespread beliefs about the "true nature" of human beings, and therefore about the necessity for certain forms of organisation, for certain relationships of the individual to the state, and for the nature of international relationships. These beliefs (about our innate aggression, territoriality, hierarchical nature, etc.) are more than likely influenced by our understanding of animal, and particularly primate, behaviour. It is important, therefore, to know that non-human primates have more than one mode of behaviour, that these have parallels with modes of behaviour in human organisations, and that, at least under some conditions, a co-operative, non-aggressive, free-ranging, non-hierarchical, non-territorial mode is successful. If this is possible in some primate societies, in foraging societies (see Chapter 4), and in innovative organisations, who knows, it may be possible in society at large, and even at an international level.

REFERENCES

Burns, T. & Stalker, G. H. (1966). *The management of innovation*, London: Tavistock.

Chance, M. R. A. (1973). The dimensions of social behaviour. In J. Beenthall, (Ed.), *The limits of human nature*. London: Allen Lane.

Handy, C. B. (1976). *Understanding organisations*. Harmondsworth, England: Penguin.

Harrison, R. (1972). How to describe your organisation. *Harvard Business Review*, *50*, 119–128.

Naisbitt, J. (1984). *Megatrends*. London: Macdonald.

Peters, T. J. & Waterman, R. H. (1982). *In search of excellence*. New York: Harper & Row.

Power, M. (1986). Foraging adaptation of chimpanzees, and the recent behaviours of the provisioned apes in Gombe and Mahale National Parks, Tanzania. *Human Evolution*, *1*, 251–266.

Reynolds, V. & Reynolds, F. (1965). Chimpanzees of the Budongo forest. In I. DeVore, (Ed.), *Primate behaviour*, New York:Holt, Rinehart & Winston.

Glossary

Absolute RHP (see also RHP). An estimate an individual makes of his or her own RHP (fighting capacity) in relation to other individuals in general. A concept related to self-confidence, ego-strength, and structural power.

Advertence. The deferential attention shown towards a dominant by a subordinate individual.

Affect. A person's emotional-feeling disposition and its outward manifestations. Affect and emotion are commonly used interchangeably.

Agonic Mode. An affect state consisting of an inhibited form of agonism (q.v.) so that overt conflict is prevented. Two forms of inhibition exist; one in which the muscles are tensed and another in which the brain motor arousal is braked so that arousal is restricted to the ideational and emotional system.

Agonism. The term applied to the behaviour repertoire of conflict, which includes flight, withdrawal, freeze and submission, as well as elements of aggression.

Agonistic Behaviour. Inter-individual conflict based on agonism (q.v.)

Allo-mothering behaviour. Same as adoptive or fostering behaviour, also called aunting behaviour, where females other than the mother care for an infant.

Alpha psalic. A communicational propensity state (see Psalic) in which the individual is dominant in a social rank hierarchy and/or defends a territory or aggression field. Pathological human variations include manic episodes (see Mania), early stages of alcohol intoxication, and sociopathy.

Alpha-reciprocal psalic. A communicational propensity state (see Psalic) in which the individual is responsive in behaviour to the signals of others in the same in-group. Other influencing individuals may be exhibiting alpha psalic (q.v.) or, like the individual in question, they may be exhibiting alpha-reciprocal psalic. Pathological human variations include loss of values as in a mob or in conversion disorder.

Anaclitic depression. An acute and striking impairment of an infant's physical, social, and intellectual development that sometimes occurs following a sudden separation from the mothering person.

Anaemia, sickle cell. An anaemic condition confined to blacks whose ancestors originated from Africa. Red blood cells assume a sickle-like or crescentic shape.

Analogous evolution. Evolutionary process whereby two species have separately evolved similar features.

Anathesis. The act or process of sending anathetic (q.v.) signals, delivering anathetic messages or carrying out anathetic behaviour towards another individual.

Anathetic. A type of signal, message or behaviour that has the effect of raising the RHP (q.v.) of the recipient.

Antidepressants. Drugs used in psychiatric medicine that reverse depressive illnesses, especially major depression (q.v.).

Arashiyama troop. One of the well-studied troops of the Japanese monkeys (q.v.) where research was started in 1954. Arashiyama is a noted place of Kyoto and is located in the western suburbs of the city.

Archers. Mbuti pygmies who depend for their subsistence on bow-and-arrow hunting. They live in northern and eastern parts of the Ituri Forest, Eastern Zaire.

Asymmetrical relationship. A relationship in which one member consistently gets his or her own way.

Asymmetrical signal. A signal which is customarily passed only one way in relation to rank.

Attachment. The relation found between the infant and the mother (or other care-giver), first analysed by John Bowlby.

Attention structure. The organisation and distribution of social (q.v.) and environmental attention among individuals and within social units.

Band. A residential group of the nomadic hunter–gatherers.

Basic rank and dependent rank. Kawai recognised two kinds of social rank in the society of Japanese monkeys (q.v.). The *basic rank* is observable between two monkeys without the influence of the other individuals. If there is any interference by a third high-ranking individual, the basic rank of the two changes due to the dependence of a subordinate on the third individual. Kawai called the latter *dependent rank*.

Begging. The behaviour of asking or requesting food items from other individuals. Begging and sharing behaviours have been observed in chimpanzees (q.v.) and pygmy chimpanzees.

Bilateral social unit. The social unit (q.v.) that allows both sexes to emigrate and immigrate. The composition of the group is multi-male and multi-female.

Bipolar mixed state. An episode of affect (q.v.) illness in a person with bipolar or manic-depressive (q.v.) disorder characterised by simultaneously present manic and depressive symptoms and signs.

Bushmen. Hunter–gatherers of the Kalahari Desert; they call themselves "San".

Catathesis. The act or process of sending catathetic (q.v.) signals, delivering catathetic messages, or carrying out catethetic behaviour towards another individual.

Catathetic. A type of signal, message or behaviour that has the effect of lowering the RHP (q.v.) of the recipient. The onset of catathetic behaviour is a signal of favourable relative RHP on the part of the sender, and it takes the form of conventional fighting such as the pecking of birds (in man it may take the form of a statement of the sender's superiority over the recipient).

Chimpanzee. *Pan troglodytes*: one of the African great apes (q.v.); classified into three sub-species. Distribution is from West Africa through Central Africa to Western Uganda and Tanzania.

Circumscription. A situation in which an individual motivated to leave a group cannot do so because there is no satisfactory place to go.

Complementary relationship. A form of asymmetrical relationship (q.v.) between A and B in which A responds to B's catathesis (q.v.) with catathesis, whereas B responds to A's catathesis with a reduction of catathesis or with anathesis.

Conspecific. A member of the same species.

Cortex (cerebral). An enormous sheet of highly differentiated cells covering the brain; in the human, it has a volume of about 300 cubic centimetres.

Corticosteroids. Steroid hormones secreted by the adrenal cortex that have profound systemic and metabolic effects.

Delusions. A firm, fixed idea not amenable to rational explanation. Maintained against logical argument despite objective contradictory evidence. Types include grandiose, jealous, persecutory, and/or somatic.

Dementia praecox. An early term for a psychiatric disorder now known as schizophrenia (q.v.).

Dependent personality. Describes a person who exhibits a pervasive pattern of dependent and submissive behaviour.

Depressed mood. Feelings of sadness, tearfulness, misery, helplessness and hopelessness; reduced interest in and appetite for life, sometimes to the extent of a wish for escape or death.

Depression, major. The major depressive syndrome is defined as depressed mood or loss of interest, of at least two weeks' duration, accompanied by several associated symptoms, such as sleep difficulty, appetite disturbance, fatigue, guilt, and difficulty concentrating.

Depression, melancholic. This is a subcategory of the major depressive syndrome with several associated specific symptoms that include loss of pleasure in all, or almost all, activities, lack of reactivity to stimuli usually pleasurable, worsened mood in the morning, early morning awakening, and psychomotor retardation or agitation.

Depressive episode. A period lasting weeks or months (sometimes years) in which there is diffuse discomfort and malfunction of mind and body. There is subjective (and to a lesser extent objective) impairment of perceptual, cognitive, and executive activity, alteration of vegetative function such as sleep, appetite and libido; a negative appraisal of the self, the world, the past, and the future; and usually, but not always, depressed mood.

Descriptive psychiatry. A system of psychiatry based upon the study of readily observable external factors.

Dopamine. A catecholamine known to have neurotransmitter (q.v.) functions especially in the central nervous system.

D.S.M.-III. Diagnostic and Statistical Manual of Mental Disorders, Third Edition. This landmark manual was first published in 1980 by the American Psychiatric Association and featured, for the first time, operational definitions of psychiatric disorders as official nomenclature. In 1987, it became replaced by DSM-IIIR (R = revised) with a number of minor modifications.

Dysphoria. Disorder of mood.

Ego dystonic. Aspects of a person's behaviour, thoughts and attitudes that they view as repugnant or inconsistent with their total personality.

Ego syntonic. Aspects of a person's behaviour, thoughts, and attitudes that they view as acceptable and consistent with their total personality.

Elemental society. Itani called the society of solitary species, or a society without social units (q.v.), the *elemental society*.

Environmental attention. The portion of an individual's attention directed to significant aspects of its environment.

Environmental attention structure. The organisation and distribution of environmental attention among individuals and within social units (q.v.).

Equality and inequality. Rousseau assumed that the transformation from man's natural state to his civil state entailed that from the society of equality to the society of civil (or political/moral) inequality. The author assumed that the equality and inequality originated in the stage of non-human primates, and set up four stages from original equality, fundamental inequality, conditional equality, and lastly to social (or civil) inequality).

Ethogram. Part of ethological method. Any diagrammatic representation of the proposed relationship between observed elements. Now replaced by sequence analysis of observed elements, e.g., acts, postures, gestures, etc.

Ethology. A branch of biology featuring study of the behaviour of animals in their natural surroundings by observation including human behaviour.

Folie à deux: A condition in which two persons in a close relationship, often in the same family, share the same delusion(s).

Fornix. A neuro-anatomical structure consisting of fibres connecting the mammillary bodies in the hypothalamus to the fimbria of the medial temporal lobe of the cerebrum. It has an arch-like shape, lies under the corpus callosum, and branches of it connect the cerebral hemispheres.

Genito-genital contact. Japanese primatologists who studied pygmy chimpanzees at Wamba, Central Zaire, observed various types of this interaction among both sexes, which is a kind of appeasement behaviour and not sexual behaviour.

Gorilla—Gorilla gorilla: one of the African great apes (q.v.); it is classified in two subspecies, lowland gorilla and mountain gorilla.

Graded psalics. A modification of the concept of psalic (q.v.) that allows variations in the extent to which the propensity state influences the behaviour of the individual.

Great apes. Anthropoids (Pongidae) are classified into the great apes (Ponginae) and lesser apes (Hylobatinae). The former includes the gorilla, chimpanzee, pygmy chimpanzee, and orang-utan (-utang) (q.v.).

Hadza. A hunting and gathering tribe of North-central Tanzania. Their territory is east and north of Lake Eyasi. The population of the Hadza is less than 500. They speak a kind of click language.

Hallucinations. False sensory perceptions in the absence of actual external stimuli. These are characteristic of psychosis (q.v.); may be induced by emotional and other factors such as drugs, alcohol, and stress; and may occur in any sensory modality.

Hamadryas baboon. Papio hamadryas: a species of savannah baboon, distributed in the Ethiopian high plateau and in limited areas of Arabia; males have a well-developed mane.

Hedonic Mode. An affect (q.v.) state of the mind and, at one and the same time, a set of affiliative social relations. Arousal in the individual is low, and with attention, fluctuates as part of the operational activity of the moment. This mode facilitates the exploratory, integrative and systems-forming aspects of the intelligence.

Homology. A phenotypical feature that individuals of two species have inherited from the same feature of the common ancestor.

Imaginary play. In a paper on play in juvenile chimpanzees Hayaki stated: "Juveniles and adolescents often engage in solitary play. They used various objects in their environment during solitary play. One played with the branch as if he had been playing with a friend."

Infanticide. Sugiyama first recorded infanticide by non-human primates in the Hunuman langur (*Presbytis entellus*) of India. Recently many observations have been reported especially in the societies that have one male/multi-female unit groups.

In-group. Members of a species that are tolerant of each other. The members are usually acutely aware of other conspecifics that do not belong within the group.

In-group omega psalic. A communicational propensity state (see Psalic) in which the individual exhibits an exaggerated, self-humbling submissive display. This is the communicational display of the lowest ranker in an in-group. A pathological, human example includes major depression (q.v.).

Inhibition. Kummer used the term "social inhibition" as the opposite of social facilitation. He induced this concept from the observation of male hamadryas baboons. He stated, as follows: "The usually peaceful life within the troop is the effect of specific male inhibitions which prevent them from touching each other's females." But, in this book, authors have used this term more generally for the intra-group social relationships.

Inter-specific communication. Communication that occurs between members of different species. Examples include confusing signals of potential prey exhibited to predators, as well as human–pet interactions.

Intra-specific communication. Any kind of information sent and registered among or between conspecifics.

Jie. A pastoral people of Eastern Uganda. Their territory is neighbouring to Turkana-land (q.v.). Both repeatedly raid livestock from the other.

Japanese monkey. *Macaca fuscata*: one of the macaque species and the only endemic monkey species in Japan. They are short-tailed and red-faced monkeys.

Lithium carbonate. A drug used to treat mania (q.v.) directly and as a prophylaxis against future occurrences of mania and depression (q.v.) in bipolar disorder.

Locus coeruleus. A nucleus with norepinephrine-containing neurones located in the dorsal-pons part of the brain stem. Neurones originating in this nucleus send axons to many parts of the rest of the central nervous system.

Loss model of depression. This postulates depression (q.v.) to be based on behaviour patterns that occur with loss of a significant other conspecific(s), similar to the anaclitic depression of the immature organism.

Mania. A mood disorder characterised by excessive elation, hyperactivity, agitation, and accelerated thinking and speaking, sometimes manifested as "flight of ideas". A manic episode will also contain an inflated self-esteem, or grandiosity, decreased need for sleep, pressure of speech, distractibility, increases in goal-directed activity, and excessive involvement in pleasurable activities that have a high potential for painful consequences.

Manic-depressive disorder. Also known as bipolar disorder, this major affective disorder is characterised by recurrent mood swings that are disabling, as well as a tendency to remission and recurrence. Bipolar refers to the patient having experienced manic as well as depressive episodes (unipolar depression consists of recurrent depressions without manic episodes).

Matrilineal social unit. Males emigrate from the natal group and immigrate to other groups, but females do not leave their natal group. Almost all of the species belonging to the Cercopithecidae have this type of social structure (see also Patritineal social unit).

Mbuti pygmies. A hunting and gathering people in Africa living in the Ituri Forest of the Eastern Zaire. The smallest people in the world, ranging from 150cm in height. Their population may be 40–50,000.

Minoo-B troop. One of the well-studied troops of Japanese monkeys (q.v.) in the early period of the research. Minoo is located north-west of Osaka City.

Monoamines. Dopamine and norepinephrine are two catecholamines and serotonin is an idoleamine that have biogenic actions in the body and in the central nervous system. They are widely thought to be involved with major illnesses (dopamine with schizophrenia (q.v.); norepinephrine and serotonin with affective disorders) and their metabolism affected by medications that ameliorate these illnesses.

Mood. A pervasive and sustained emotion that, in the extreme, markedly colours the person's perception of the world. Common examples of mood include depression, elation, anger, and anxiety.

Motor theory of communication. Neural structures responsible for motor apparatus primarily determine communicational expression, e.g., in humans, for articulation of speech phonemes and in birds, of song syllables.

Multi-layered social structure. There are a few non-human primate species that have multi-layered social structures. Gelada baboons and hamadryas baboons are examples.

Natural selection. The commonly agreed (Darwinian) theory of the mechanism of evolutionary process. Favourable variations of organisms that fit most effectively into their eco-niche survive and perpetuate their genes whereas less favourable variations do not.

Negative referent. A monkey that another monkey always avoids.

Negative symptoms (schizophrenia). These include affective blunting or flattening, poverty of speech, apathy, anhedonia and attentional impairment, and are thought to be less responsive to antipsychotic (dopamine-blocking) medications than are positive symptoms.

Net hunters. Mbuti pygmies who depend for their subsidence on net-hunting. They live in the southern and western part of the Ituri Forest of Eastern Zaire.

Neurotransmitters. Chemicals released in minute amounts from the terminals of nerve cells in response to the arrival of nervous excitation. These then diffuse across the synaptic gap between the nerve terminal and the target cell, which may be another nerve, muscle, or glandular cell. There are about 40 transmitters, including the monoamines (q.v.), some amino acids and a number of small peptides.

Neurotransmitter disregulation. A relatively new theory of the cellular–molecular mechanisms of mental illness whereby pacing and rhythm effects of neurotransmitters (q.v.) produce psychiatric disorders including their timing.

Niche. A particular, often localised, set of environmental conditions to which a species is well-adapted.

Nocturnal prosimian. More than half of the prosimians are nocturnal animals. They are the tarsiers, lorises, galagos, mouse lemurs and aye-aye. Almost all of the nocturnal prosimians are solitary animals and forest dwellers in the tropics. They are distributed in South-east Asia, India, Madagascar and Africa.

Norepinephrine. (Also noradrenaline): A member of the family of catecholamines (others include dopamine and epinepherine), this primary amine is a neurotransmitter (q.v.) and hormone with many effects and wide distribution in the brain and peripheral nervous system; as a hormone, it is secreted into the blood by the adrenal medulla. Norepinephrine is implicated in the organism's reaction to stressful stimuli.

Nurturance psalic. A communicational propensity state (see Psalics) in which the individual assumes caretaking functions for another conspecific. The prototype is feeding of offspring and parenting but the state is not limited to inter-generational communication.

Nurturance–recipience psalic. A communicational propensity state (see Psalic) in which the individual elicits caretaking functions from another conspecific. This typically represents communications from a younger member of a species, as directed towards a parent but, as for nurturance psalic, is not limited to inter-generational communication. An abnormal human example is dependent personality disorder.

Ontogenetic (ontogenic) learning. Learning within the *developmental* life-span of the individual organism.

Orang-utang. Pongo pygmaeus: one of the great apes (q.v.); they occur in Borneo and Sumatra.

Organic brain disorders. This designates a particular organic mental syndrome (see below) in which the aetiology is known or presumed. There are a heterogeneous group and no single description characterises them all.

Organic brain syndrome. A constellation of abnormal psychological or behavioural signs and symptoms associated with transient or permanent dysfunction of the brain. The most common such syndromes are delirium, dementia, intoxication, and withdrawal. Common symptoms include disturbance of attention, disorganised thinking, memory impairment, and disorientation.

Out-group. Conspecifics that are ejected from a territory or aggression field, but remain related to the home groups (see also In-group).

Out-group omega psalic. A communicational propensity state (see Psalic) in which the individual is very cautious, intently aware but wary of conspecifics and expectant of persecution. "Normal" human instances include members of a persecuted group who are indeed persecuted; abnormal instances include paranoid individuals whose persecution is delusional (see Paranoia).

Paedophilia. A sexual deviation involving sexual activity by adults with children as objects. It may involve any form of heterosexual or homosexual intercourse.

Pair-type society. A society that has social units (q.v.) formed by a particular male and a particular female and their offspring.

Panic states (also panic attacks). Discrete periods of sudden onset of intense apprehension, fearfulness, or terror, often associated with feelings of impending doom.

Paranoia. A psychiatric syndrome in which a person feels him or herself to be persecuted, such as being conspired against, cheated, spied upon, poisoned or drugged, or harassed. Small actual inter-personal slights are often exaggerated.

Paranoid psychosis (in DSM-IIIR (q.v.), delusional disorder). Paranoia (q.v.) experienced to a delusional extent.

Para-play. Play bouts are initiated by invitation, approach, or direct contact; however, some bouts terminate without the play interactions, and Hayaki called this type of bout para-play.

Paternal care. Itani found special relationships between particular dominant males and particular one-year-old infants in a Japanese monkey (q.v.) troop during the birth season. The male takes care of the infants like their mother.

Pathogenesis. Mechanisms for the development of a disease.

Patrilineal social unit. The mirror image of the matrilineal social unit (q.v.). Only females transfer between groups and males do not leave their natal group. Chimpanzees (q.v.) and pygmy chimpanzees have this type of social structure.

Peck order. Term for a linear dominance hierarchy stemming from the original description in domestic chickens.

Peculiar–proximate relationship (PPR). In a study of Japanese monkeys (q.v.) in Arashiyama, Takahata found particular pairs that kept affinitive relations during the non-mating season, but in the mating season they avoided mating relations.

Pheromones. Substances secreted to the outside of an individual. When experienced by a conspecific, a particular reaction is released, e.g., a definite behaviour or developmental process.

Play bout. A bout begins when one directs any behaviour to its partner and ends when the participants stop their activities or one of them moves away. Hayaki called this a play bout (see also Para-play). A bout includes various kinds of behaviour, such as wrestling, chasing, etc.

Play face. A special expression for inviting a partner to play.

Play session. Play and para-play (q.v.) bouts often occur in series; these bouts can be clustered together in time. The clusters of bouts are called a *play session*.

Pongidae. A family in the Anthropoidea. The Pongidae are divided into two subfamilies, Ponginae and Hylobatinae. The former includes the gorillas, chimpanzees, pygmy chimpanzees and orang-utangs (q.v.) and the latter includes eight species of gibbons.

Pons. A part of the brain that connects the cerebrum, cerebellum and medulla-oblongata part of the brain stem.

Popperian approach. In Karl Popper's philosophical view of science, hypotheses are not to be proved; rather, clear-cut alternative hypotheses should be ruled out, leaving less and less likelihood for how the phenomenon in question can be explained. The process of science does not so much demonstrate truth as converge on it.

Positive referent. In monkey societies, the individual of the same of higher rank with which an individual habitually associates and from which support is solicited and often given in an agonistic encounter with another (see Negative referent). In humans, a habitual and supportive friend.

Positive symptoms (psychosis). Psychotic symptoms that are obvious, such as delusions, hallucinations, or being convinced that thoughts are being broadcast from the person. These seem to be more amenable to treatment with antipsychotic drugs than are the negative symptoms (q.v.).

Pre-band hypothesis. Itani hypothesised that the unit group of chimpanzees (q.v.) and the band of hunter–gatherers is sociologically homologous, and he called the former a "pre-band". Therefore, the human family is a lower unit of the human basic social unit (q.v.).

Presentation. Standing with hindquarters close to or directed towards another. It is a sign either of sexual solicitation or of submission.

Primate. A member of the class of mammals to which we, as an ape, and also monkeys belong.

Prosopagnosia. An inability to recognise famous or familiar faces, usually secondary to a brain lesion.

Protean. Unpredictable change of shape or sequence of behaviour.

Protected threat. Threatening another monkey close to and in front of a positive referent (q.v.).

Prototherians. Mammals that are examples of the subclass Prototheria. They are the most conserved or primitive mammals and include monotremata, e.g., platypus and echidna, and marsupials, e.g., opossum and kangaroo. The other subclass is Theria to which all modern mammals belong.

Psalic. A primitive communicational state mediated by deeply homologous neural structures which, when stimulated and activated, causes the organism to demonstrate unusual readiness to assume distinctive roles relating to functional activities that involve one or more of its conspecifics. Psalics can overlap (i.e., two or more psalics can be simultaneously active in the same individual at the same time) and can override, to varying extents, other factors influencing the organism's behaviour and perceptions.

Psychiatric disorder (also mental disorder). A clinically significant behavioural or psychological syndrome or pattern that occurs in a person and that is associated with present distress (a painful symptom), or disability (impairment in one or more important areas of functioning), or with a significantly increased risk of suffering death, pain, disability or an important loss of freedom. Deviant behaviour, conflict, or expected responses to stressors are not automatically considered psychiatric disorders.

Psychoanalysis. A method of psychological treatment in which unconscious material is made conscious by means of free association, dream interpretation and similar techniques. Also the theory of mental function derived from the use of this method.

Psychopharmacology. This field of study concerns the effects of drugs on animal and human behaviour, encompassing any chemical that influences behaviour by a direct or indirect effect on the central nervous system.

Psychosis. A psychosis is a human behavioural state characterised by delusions, hallucinations, incoherent communication, loosening of association between ideas, marked poverty of content of thought, markedly illogical thinking, or behaviour that is grossly disorganised or catatonic.

Psychotropic drugs. Drugs used in the treatment of psychiatric disorders. Examples are antidepressants and antipsychotic drugs (q.v.).

Pygmy chimpanzee. *Pan paniscus*: one of the African great apes (q.v.);

living in the tropical rain forest of the left bank of the Zaire River; smaller and more slender than common chimpanzees.

Raphe nuclei. A specific set of midline, brain-stem, serotonin-containing neuronal cell bodies that project widely throughout the central nervous system.

Recursive. Literally "turned back upon"; a comprehensive term referring in organic systems to all processes that feed back onto their origins.

Reified concepts. Abstractions that are treated as concrete material things.

Relative RHP. The estimate an individual makes of the difference between their own RHP and the RHP they ascribe to an adversary in a ritual agonistic encounter (q.v.). In the simplest case the estimate may be either favourable (own RHP equal to or superior to adversary's) or unfavourable (own RHP inferior to adversary's).

Resource Holding Power (RHP). An intervening variable estimating fighting capacity defined on the input side by size, strength, skill, weapons, allies, and other resources; and on the output side by probability of escalating a ritual agonistic encounter (q.v.) (rather than withdrawing or submitting).

Reverted escape. The movement back towards a dominant monkey after the actor has withdrawn from a threat from the dominant. Originally termed, incorrectly, reflected escape.

Reward sites (also Olds' self-stimulating areas). Brain areas in which animals, if chronically implanted with stimulation electrodes and provided with stimulation opportunities, will repetitively stimulate themselves.

Ritual agonistic behaviour. A process of signalling between two individuals that converts a symmetrical relationship (q.v.) into an asymmetrical (q.v.) complementary relationship.

Ritual agonistic encounter. An episode involving two (or more) individuals in which ritual agonistic behaviour (q.v.) takes place. A stage of mutual assessment may be followed by a stage of engagement.

Ritual behaviour. Behaviour that is symbolic or conventional or stands for something else, implying the capacity (either genetic or cultural) to make a similar interpretation of its meaning on the part of both actor and observer. Ritualised behaviour describes a signal evolved from behaviour that lacked signalling functions.

Savanna woodland. Typical vegetation of Southern Africa. Trees are taller and stands are thicker than in savanna.

Schizophrenia. A major mental illness with characteristic psychotic symptoms during active phases of illness, and function below the highest level previously achieved or anticipated. Subtypes include catatonic, disorganised (hebephrenic), paranoid, undifferentiated and residual.

Self-handicapping behaviour. According to Fagen, the term "self handicapping" denotes an animal's behaviour that reduces the probability of its achieving a tactical objective in play and thereby prolongs the play interaction. However, Hayaki said that this includes not only activity reduction by the strong side but also activity facilitation by the weak side.

Sexual perversions (also paraphilias). Disorders characterised by sexual arousal in response to objects or situations that are not part of normative arousal–activity patterns. Examples include paedophilia (q.v.) voyeurism, exhibitionism, sexual masochism, sexual sadism, and fetishism.

Sexual psalic. A communicational propensity state (see Psalic) in which the individual is sexually excited about or with another conspecific. Abnormal human examples are the sexual perversions (q.v.) or paraphilias.

Social attention. That portion of an individual's attention directed to socially significant individuals (see also Environmental attention).

Social attention structure. The organisation and distribution of social attention (q.v.) among individuals and within social units (q.v.) in animal societies.

Social dynamics. The continuously changing and/or potentially changing aspects of social interactions.

Social geometry. The pattern of the totality of interactions between vectoral and spatial relations among individuals and between social units (q.v.).

Social unit or basic social unit (BSU). Stable social groups consisting of both sexes can be seen in several prosimian societies and all simian societies except the orang-utan (q.v.). The composition and structure of a BSU is species-specific. Itani recognised six types of BSU in non-human primates. These are the monogamous, bilateral, polyandrous, polygynous, matrilineal (q.v.) and patrilineal (q.v.) types.

Sociobiology. The particular way of studying behaviour based on the assumption that the survival of the gene ultimately determines the form of behaviour.

Spacing-avoidant psalic. A communicational propensity state (see Psalic) in which the individual specifically avoids direct contact with other conspecifics.

Spatial relations. Metrical and configurational aspects of social geometry (g.v.).

Specia. Imanishi's term (presented in 1950) referring to a species society.

Substantia nigra. A bilateral, crescent-shaped, dark-coloured nucleus in the midbrain portion of the brain-stem containing neurones that project to the basal ganglia of the cerebrum and contain dopamine. Damage to this nucleus causes Parkinsonism.

Symmetrical relationship. A relationship in which the two members cannot be differentiated according to their agonistic behaviour (q.v.). The opposite of a complementary relationship (q.v.), in that each member responds to the other's catathesis (q.v.) with catathesis. The opposite of an asymmetrical relationship (q.v.), in that one member does not consistently get his or her own way.

Telencephalic. Referring to the telencephalon, embryologically the furthest forward part of the central nervous system. It includes the cerebral cortex and basal ganglia.

Territoriality. A defended area by an individual animal or group of conspecifics. A method of social spacing.

Thought collective. A community of persons mutually exchanging ideas or maintaining intellectual interaction. This exchange provides the special "carrier" for the historical development of any field of thought, as well as for the given stock of knowledge and level of culture.

Turkana. A pastoral people of North-western Kenya. Turkana-land is near to the Kenya–Uganda border, about 80,000km^2. They are nomadic with five kinds of livestock in the arid semi-desert. Population about 150,000.

Two Modes. Agonic and Hedonic modes (q.v.) are at one and the same time a state of mind (brain state) and of the corresponding social relations.

Vectoral relations. Orientational and directional aspects of social geometry (q.v.).

Yakushima. An island south of Kyushu. Yaku monkeys (*Macaca fuscata yakui*) occur here in subtropical forest, at the southernmost limit of the Japanese monkeys (q.v.).

Yakushima-M troop: One of the well-studied troops of the Japanese monkeys (q.v.).

Youngest ascendancy principle. Kawamura found that in the dominant/subordinate relationship between sisters of the same mother, the younger always becomes dominant over the older; first identified in the Minoo-B troops (q.v.) of Japanese monkeys.

Author Index

Subject Index